T0318531

COMPREHENSIVE GUIDE TO HETEROGENEOUS NETWORKS

COMPREHENSIVE GUIDE TO HETEROGENEOUS NETWORKS

Edited by

KIRAN AHUJA
Electronics and Communication Engineering Department,
DAV Institute of Engineering and Technology, Jalandhar,
Punjab, India

ANAND NAYYAR
Assistant Professor, Scientist, Vice-Chairman (Research), Director (IoT and Intelligent Systems Lab), School of Computer Science, Duy Tan University, Da Nang, Vietnam

KAVITA SHARMA
Computer Science and Engineering Department, Galgotias College of Engineering & Technology, Greater Noida, Uttar Pradesh, India

Academic Press is an imprint of Elsevier
125 London Wall, London EC2Y 5AS, United Kingdom
525 B Street, Suite 1650, San Diego, CA 92101, United States
50 Hampshire Street, 5th Floor, Cambridge, MA 02139, United States
The Boulevard, Langford Lane, Kidlington, Oxford OX5 1GB, United Kingdom

Notices
Knowledge and best practice in this field are constantly changing. As new research and experience broaden our understanding, changes in research methods, professional practices, or medical treatment may become necessary.

Practitioners and researchers must always rely on their own experience and knowledge in evaluating and using any information, methods, compounds, or experiments described herein. In using such information or methods they should be mindful of their own safety and the safety of others, including parties for whom they have a professional responsibility.

To the fullest extent of the law, neither the Publisher nor the authors, contributors, or editors, assume any liability for any injury and/or damage to persons or property as a matter of products liability, negligence or otherwise, or from any use or operation of any methods, products, instructions, or ideas contained in the material herein.

ISBN 978-0-323-90527-5

For information on all Academic Press publications
visit our website at https://www.elsevier.com/books-and-journals

Publisher: Mara Conner
Acquisitions Editor: Carrie Bolger
Editorial Project Manager: Emily Thomson
Production Project Manager: Prasanna Kalyanaraman
Cover Designer: Miles Hitchen

Typeset by STRAIVE, India

Contents

Contributors

Kiran Ahuja
DAV Institute of Engineering and Technology, Jalandhar, Punjab, India

Komal Arora
CGC Group of Colleges, Mohali, India

Indu Bala
Lovely Professional University, Phagwara, Punjab, India

Oussama Bazzi
Lebanese University, Beirut, Lebanon

Pradeep Bedi
Department of Computer Science and Engineering, Galgotias University, Greater Noida, Uttar Pradesh, India

Fayssal Bendaoud
LabRI-SBA Lab, Ecole supérieure en Informatique, Sidi Bel Abbes, Algeria

Elif Bozkaya
Department of Computer Engineering, National Defence University, Turkish Naval Academy, Istanbul, Turkey

Matthieu Crussière
INSA-Rennes, Rennes, France

Levent Eriskin
Department of Industrial Engineering, National Defence University, Turkish Naval Academy, Istanbul, Turkey

S.B. Goyal
Faculty of Information Technology City University, Petaling Jaya, Malaysia

Jean-Francois Hélard
INSA-Rennes, Rennes, France

H. Hannah Inbarani
Department of Computer Science, Periyar University, Salem, India

Tanu Kaistha
I.K.G Punjab Technical University, Jalandhar, India

Mumtaz Karatas
Department of Industrial Engineering, National Defence University, Turkish Naval Academy, Istanbul, Turkey

Okuthe P. Kogeda
Department of Computer Science and Informatics, Faculty of Natural and Agricultural Sciences, University of the Free State, Bloemfontein, South Africa

Jugnesh Kumar
St. Andrews Institute of Technology and Management, Gurgaon, India

Shailesh Kumar
BlueCrest College, Freetown, Sierra Leone, West Africa

Danvir Mandal
Lovely Professional University, Phagwara, Punjab, India

Topside E. Mathonsi
Department of Information Technology, Faculty of Information and Communication Technology, Tshwane University of Technology, Pretoria, South Africa

Rashmi Mishra
Krishna Engineering College, Ghaziabad, UP; Delhi Technological University, Delhi, India

Youssef Nasser
Huawei, Paris, France

Thomas O. Olwal
Department of Electrical Engineering/F'SATI, Faculty of Engineering and the Built Environment, Tshwane University of Technology, Pretoria, South Africa

T. Poongodi
School of Computing Science and Engineering, Galgotias University, Greater Noida, Uttar Pradesh, India

Kavita Sharma
Galgotias College of Engineering & Technology, Greater Noida, India

Ahmad Shokair
INSA-Rennes, Rennes, France

Kiran Singh
School of Computing Science and Engineering, Galgotias University, Greater Noida, Uttar Pradesh, India

T. Sudhakar
School of Computer Application, Lovely Professional University, Phagwara, India

Rajesh K. Yadav
Delhi Technological University, Delhi, India

Suman Avdhesh Yadav
Amity School of Engineering & Technology, Amity University, Greater Noida, Uttar Pradesh, India

Preface

Heterogeneous networks have gained momentum in industry and the research community, attracting the attention of standardization bodies such as 3GPP LTE and IEEE 802.16j, whose primary objective is to improve the quality of service and channel capacity. Heterogeneous networks are an evolved network topology that improves spectral efficiency in a geographical area using a combination of macro-cells, pico-cells, femto-cells, and relay nodes. It means carriers get greater network efficiencies and end users get the benefit of high-performing mobile devices. The next-generation wireless systems [also sometimes referred to as fourth-generation (4G) systems] are being devised with the vision of heterogeneity in which a mobile user/device will be able to connect to multiple wireless networks (e.g., WLAN, cellular, WMAN) simultaneously. For example, IP-based wireless broadband technology such as IEEE 802.16/WiMAX (i.e., 802.16a, 802.16d, 802.16e, and 802.16g) and 802.20/MobileFi will be integrated with 3G mobile networks, 802.11-based WLANs, 802.15-based WPANs, and wireline networks to provide seamless broadband connectivity to mobile users in a transparent fashion. Heterogeneous wireless systems will achieve efficient wireless resource utilization, seamless handoff, global mobility with QoS support through load balancing, and tight integration with services and applications in the higher layers. After all, in such a heterogeneous wireless access network, a mobile user should be able to connect to the Internet in a seamless manner. The wireless resources need to be managed efficiently from the service providers' point of view for maximum capacity and improved return on investment. Load balancing and network selection, resource allocation and admission control, fast and efficient vertical handoff mechanisms, and provisioning of QoS on an end-to-end basis are some of the major research issues related to the development of heterogeneous networks.

There is an urgent need in both industry and academia to better understand the technical details and performance insights that are made possible by heterogeneous networks. To address that need, this edited book covers comprehensive research topics in heterogeneous networks. This book also focuses on recent advancements, trends, and progressions in heterogeneous networks. This book can serve as a useful reference for researchers, engineers, and students to understand the concept of heterogeneous networks in order to design, build, and deploy highly efficient wireless networks.

The scope of topics covered in this book is timely and will grow in future. The book contains 10 referred chapters from researchers working in this area around the world.

Chapter 1, authored by Bozkaya et al., titled "Heterogeneous wireless sensor networks: deployment strategies and coverage models" provides a detailed overview of heterogeneous wireless sensor networks (HWSNs) with a focus on deployment strategies and coverage types with respect to different application areas. A detailed discussion is provided on HWSN application areas and approaches for implementation. All the existing algorithmic solutions with feasibility are also examined.

Chapter 2, authored by Singh et al., titled "Efficient multitasking in heterogeneous wireless sensor networks" highlights a review of utilization in the area of HWSNs with unique concentration on signal processing aspects and the primary difficulties that must be handled for the plan of HWSNs.

Chapter 3, authored by Bendaoud, titled "Network selection in a heterogeneous wireless environment based on path prediction and user mobility" discusses the network selection decision problem and provides in-depth discussion of supporting standards and distinct approaches. In addition, a new paradigm for network selection is proposed with a focus on user path prediction and mobility. In addition, experiments are done to compare the proposed work with existing approaches on the basis of throughput, latency, packet loss, and jitter.

Chapter 4, authored by Sudhakar and Inbarani, titled "Reducing control packets using covering rough set for route selection in mobile ad hoc networks" highlights a novel covering rough set (CRS) approach for route selection in wireless ad hoc networks. And based on the experimental results, the proposed approach is found better in PDR, throughput, fewer RREQ packets as compared to conventional rough set theory, DSR-based RST, and AODV-based RST algorithms.

Chapter 5 authored by Shokair titled "Optimization of hybrid broadcast/broadband networks for the delivery of linear services using stochastic geometry" focuses on user-sharing hybridization, in which each subnetwork is responsible for serving a section of the users and is defined on the basis of a BC deployment scheme, a BB operation mode, and the sharing criteria between the networks. A thorough analysis is conducted to obtain equations that can help in estimating performance metrics such as the probability of coverage and power efficiency. Experimental results conclude that an optimal setting in between a complete BC system and a complete BB system can be obtained.

Chapter 6, authored by Bala et al., titled "A comprehensive survey on heterogeneous cognitive radio networks" highlights some of the major challenges regarding the coexistence of heterogeneous cognitive radio networks in TV white space that need immediate attention for various standardization activities. Various resource allocation schemes and MAC protocols have been discussed with various security aspects and algorithms. In addition, the chapter elaborates various standardization activities around the globe for the unlicensed use of the TV white space to support dynamic spectrum sharing among heterogeneous networks in TV white space.

Chapter 7, authored by Mishra, titled "Evaluation and analysis of clustering algorithms for heterogeneous wireless sensor networks" surveys various clustering techniques based on remaining energy. A comparison of protocols, i.e., HEED, HCA, and EEUC, is done based on parameters such as energy efficiency, number of dead nodes, number of alive nodes, number of packets sent to the base station, and stability period. Experimental results show that HEED, HCA, and EEUC are better protocols in bonds of numeral of alive nodes as compared to the other protocols by approximately 30% and that DWEHC is improved in terms of residual energy approximately for the heterogeneous environment. MRPUC improves the node dies as per the rounds. It achieves 251.7% improvement over HEED and 34.4% improvement over MRPEC.

Chapter 8, authored by Bedi et al., titled "Analysis of energy-efficient cluster-based routing protocols for heterogeneous WSNs" reviews the existing energy-efficient routing protocols for homogeneous as well as heterogeneous WSNs. In addition, the chapter proposes the application of machine learning, swarm optimization, and an evolutionary approach to compensate for these limitations and to make routing decisions more cost-effective in terms of energy.

Chapter 9, authored by Kaistha, titled "Imperative load-balancing techniques in heterogeneous wireless networks" surveys various load-balancing algorithms and strategies with different parameters.

Chapter 10, authored by Mathonsi et al., titled "Intelligent intersystem handover delay reduction algorithm for heterogeneous wireless networks" proposes an intelligent intersystem handover (IIH) algorithm by integrating gray prediction theory (GPT), multiple-attribute decision-making (MADM), fuzzy analytic hierarchy process (FAHP), and multiobjective optimization ratio analysis (MOORA). The proposed algorithm is tested using an NS-2 simulator to evaluate the performance relative to the fuzzy logic-based vertical handover (FLBVH) and adaptive neuro-fuzzy inference system (ANFIS) algorithms. The proposed IIH algorithm has shown an average of 1.9 s handover delay, 4.9% packet loss, 4.6% probability of the ping-pong effect, and 95.1% better throughput performance compared to the FLBVH and ANFIS algorithms at 100 s time intervals.

This book has been made possible by the exceptional efforts and contributions of many people. First, we thank all the contributors for their excellent chapter contributions. Second, we thank all the reviewers for dedicating their time to review the book, and for their valuable comments and suggestions to improve the quality of this book. Finally, we appreciate the advice and support of the Elsevier Editorial Project Managers for putting this book together.

Kiran Ahuja[a], Anand Nayyar[b], and Kavita Sharma[c]
[a]DAVIET, Jalandhar, Punjab, India
[b]Duy Tan University, Da Nang, Vietnam
[c]Galgotias College of Engineering and Technology, Greater Noida, Uttar Pradesh, India

CHAPTER 1

Heterogeneous wireless sensor networks: Deployment strategies and coverage models

Elif Bozkaya[a], Mumtaz Karatas[b], and Levent Eriskin[b]
[a]Department of Computer Engineering, National Defence University, Turkish Naval Academy, Istanbul, Turkey
[b]Department of Industrial Engineering, National Defence University, Turkish Naval Academy, Istanbul, Turkey

1. Introduction

With the continuous development of sensor technology, wireless sensor networks (WSNs) have gained considerable importance due to their flexibility and improved performance in terms of various metrics compared to the traditional sensor technologies. Considering that these sensors are employed in several challenging missions, including disaster management, battleground surveillance, border security, and health care, they should be extra reliable and flexible to perform these tasks. Thus, heterogeneous wireless sensor networks (HWSNs) are introduced to improve the overall network performance in terms of different metrics, such as lifetime, stability, and coverage [1]. With the inclusion of different types of sensors, the advantages posed by WSNs are further enhanced.

HWSNs comprising different types of sensor nodes each having a distinct capability and functionality, different power supply or energy consumption, and different software design and hardware architecture have diverse implementation areas. Being widely used by planners and decision makers in various real-world scenarios, HWSN technology offers several benefits that fall into categories such as coverage, detection, tracking, deployment, lifetime, energy management, topology, and network architecture. In particular, this technology is altering our vision of monitoring, measuring, distributing, and analyzing sensor data to use applications and services such as Internet of things (IoT), robotics, nanotechnology, and security. In addition, the integration of low-power wireless networking technologies with inexpensive hardware in HWSN applications is further improving the network performance. Due to their various sensing and coverage capabilities as well as communication ranges and other features, HWSNs are becoming more effective and flexible. Therefore, HWSNs are preferred to the homogeneous WSNs since the former is more fitted for most practical applications.

These benefits come at the cost of increased system complexity and additional challenges in terms of network management, performance evaluation, and planning of

Comprehensive Guide to Heterogeneous Networks
https://doi.org/10.1016/B978-0-323-90527-5.00009-5

system architecture. Although the heterogeneity of the network can potentially increase the overall system performance and flexibility, requirements such as task sharing, collaborative coverage, and fusion of data collected from different sensor types raise the issue of building reliable and stable networks. Due to the limitations on battery lifetime and increased complexity of computations, meeting the quality of service (QoS) and quantifying the coverage performance against various target types also become more challenging. Therefore, the success of HWSN technology heavily depends on (i) effective sensor deployment, (ii) use of appropriate sensor types in line with coverage objectives, (iii) establishment of a reliable and flexible communication network among nodes, and (iv) coordinated operations management.

Sensors are responsible for sensing and monitoring the environment, and then collecting and transmitting the data. In addition to these operations, they also perform data filtering and data fusion operations to prevent data duplication, cluster data, reduce noise, and increase reliability. The deployment of heterogeneous sensors is considered as a challenging task, especially when performing real-sensitive missions or applications. The challenge becomes more complex when the density of heterogeneous sensors increases. In addition to the aforementioned issues, timeliness and reliability are other essential components of HWSNs. In other words, a reliable and timely information flow from sensors to access points (APs) and application domain is of utmost importance for the effective use of HWSNs. While wired sensor networks are generally immune to these problems thanks to their wired topology, the connectivity of HWSNs is not guaranteed [2].

In light of the discussion earlier, the success of an HWSN strongly depends on the implemented deployment strategy as well as the desired coverage type and objective of the network. Hence, in this chapter, our main ambition is to provide readers with an overview of the main characteristics of HWSNs with a focus on the deployment strategies and coverage objectives.

In Fig. 1, we summarize the main issues that characterize the deployment of HSWNs and pertaining application areas. In the following sections, we will elaborate these issues

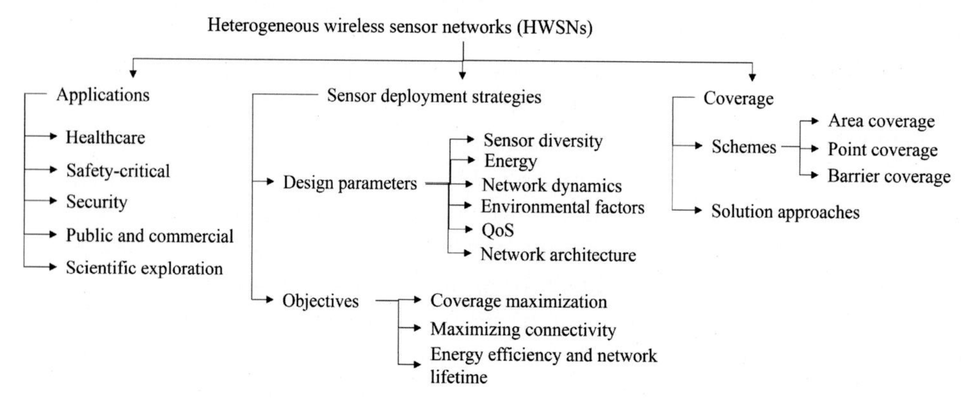

Fig. 1 Main issues that characterize the HWSN deployment and pertaining application areas.

thoroughly. The remainder of this chapter is organized as follows: In Section 2, we discuss the impact of employing HWSNs in improving the network quality with respect to different application areas such as health care, safety, security, commercial and public, and scientific exploration. Next, in Section 3, we address commonly used sensor deployment strategies and investigate their characteristics in terms of coverage, energy efficiency, network lifetime, and connectivity. In Section 4, we provide the basic characteristics of coverage types and problems as well as common solution approaches implemented in the literature. In particular, we focus on the three main coverage types, namely area, point, and barrier, and discuss on the advantages and disadvantages of each type with the recent examples from the literature. We finally conclude with a few remarks in Section 5.

2. HWSN applications

Sensors are functionally simple devices that convert the physical properties of the environment into electrical signals. With the use of different types of sensors in HWSNs, their use has increased in many application areas. Thanks to their low cost, small size, and low energy consumption features, new generation sensors (and integrated sensing techniques) play a crucial role in improving the effectiveness of decision-making processes.

Recently, IoT is a technology based on the communication of smart devices, which have many application areas from simple home appliances to smart vehicles, from agriculture to smart cities. IoT concept has become a trend with the transfer of data collected through heterogeneous and smart sensors over the Internet. Depending on the area of use, IoT devices allow data sharing with communication equipment, collect/store data from the environment through the sensors, and make analysis on the data. More accurate and reliable data have been monitored and collected by including different types of sensors, that is, more powerful processor and higher battery life. While the number of "things" connected to the Internet has increased significantly, their application areas in many industries are also expanding. In this section, we will review some examples of HWSN application areas.

2.1 Health-care applications

IoT has changed the way patients' health status and vital signs are monitored. Rapidly growing IoT technologies have enabled building health-care ecosystems consisting of HWSNs, most of which are wearable smart sensors, and cloud services. These ecosystems are usually accompanied with big data analytics for collecting, analyzing, classifying, and storing the data obtained from HWSNs. Within these ecosystems, patient-centric health-care services have been made possible to enhance the quality of patient life.

As part of the health-care ecosystems, HWSNs collect real-time and precise data, which improve the accuracy of predictions made regarding the health status of individuals. In this regard, these devices can be considered as preventive health-care tools for reducing the risks of potential illnesses. Moreover, historical data obtained with HWSNs enable medical staff to make accurate diagnosis and apply personalized treatment for illnesses.

HWSNs have also been deployed in workplaces and homes for monitoring the health status of workers, elderly, and patients. Humidity, temperature, and carbon dioxide levels are among many factors that affect the medical health status and comfort of people of interest. Therefore, in these applications, the main aim is to keep these factors at their nominal levels for a productive and healthy environment.

Deployment of HWSNs in health-care ecosystems has also contributed to the goal of building efficient *e-health* programs for the public health. The World Health Organization describes *e-health* as "the use of information and communication technologies" for the health care. E-health programs have paved the way for faster and easier access to services for patients and health-care providers.

There exists a large body of literature regarding the deployment of HWSNs in health care. For instance, Trinugroho and Baptista [3] proposed an IoT infrastructure in an effort to build a patient-centric health-care service structure, where numerous portable and wearable HWSNs as well as cloud-based services are deployed. ElSaadany et al. [4] developed an early prediction system of cardiac arrest deploying HWSNs and IoT. The main motivation for the work is that sudden cardiac arrest out-of-hospital has a low survival rate, hence, they proposed collecting relevant data such as heart rate and body temperature to trigger an alert on the eve of an attack. Massaro et al. [5] developed a decision support system comprising of wearable HWSNs and big data analytics to monitor health status of individuals. Utilizing support vector machines and long short-term memory algorithms, the developed framework provides decision support to decision makers for generating multidimensional risk map of patients. Deploying technologies such as radio frequency identification, smart mobile devices, and HWSNs, Catarinucci et al. [6] considered building a smart e-health system within a hospital, where real-time environmental and patient data are collected and transmitted to a control center. Then, gathered data are made accessible within the hospital for medical staff.

Among applications focusing health care in workplaces and homes, Mattsson et al. [7] developed a multidimensional human performance measurement system for enhancing the health and safety of workers in a workplace. They demonstrated how novel technologies such as HWSNs, cloud computing, and IoT can be utilized for combining physiological data with other work environment data of workers. Wu et al. [8] proposed a framework, where environmental data (i.e., temperature, humidity) and vital signs of individuals (i.e., hearth rate, body temperature) are collected by HWSNs and transmitted to a gateway. The primary task of gateway is to trigger alerts based on the gathered data. They particularly aimed at reducing the health risks in the construction industry in an effort to improve the safety of the working environment.

2.2 Safety-critical applications

Heterogeneous sensors have huge potential in safety-critical roles, such as detection of forest wildfires and volcanic explosions, earthquakes, landslides, floods, and hazardous gas emissions. The monitoring and surveillance of large areas (e.g., forests, volcanoes, maritime zones) usually require hundreds or thousands of sensors [9]. Hence, their low cost, low power, and small size make it feasible to embed these sensors into various environment monitoring tags to be used in safety-critical missions. In general, these sensors are densely deployed for monitoring, detecting, and communicating those events at a smaller granularity. Detection of unexpected critical events is crucial for establishing situational awareness and enabling rapid reaction. In environmental monitoring, any node failure or disruption in the network may result in the interruption of normal activities and loss of service. In addition, it is difficult to replace or recharge sensor batteries in harsh environments. Hence, reliability and connectivity will be the most critical parameters in these environments. Since these applications require endurance and long-term deployments, it is important to utilize relatively high-capacity, reliable, and energy-efficient sensors.

One other way of improving the performance of sensor networks is to add the option of mobility. Although mobility brings additional challenges in routing, data fusion, and coverage assessment, with this approach, some of the traditional problems associated with static sensor networks can be mitigated. In their study [10], Erman et al. discuss the impact of mobility in the context of disaster response and propose the integration of WSNs with unmanned aerial vehicles (UAVs) and actuators for fire detection scenarios. In a similar study, Tunca et al. [11] attempt to evaluate the performance of HWSNs for monitoring forest fires. Considering various types of densely deployed sensors used for measuring the temperature and humidity of a particular forest, the authors carry out experiments on both simulated data and realistic fire propagations and present results concerning the impact of environmental conditions, sensor quality, and number to the overall performance.

As another example, sensors can be located into the concrete structure of bridges, buildings, dams, etc., with the purpose of measuring the stability of those structures during earthquakes. In [12], the authors propose the use of an HWSN comprising of vibration and strain sensing devices and a conventional P2P for inspecting and evaluating bridges. In a more recent work, Wang and Hong [13] propose the use of HWSNs for monitoring earth buildings in the rainy villages of China. Huang and Rodriguez [14] propose a software framework capable of collaboratively operating with a set of heterogeneous environmental sensors and microcontrollers by fusing real-time environmental data (e.g., temperature, humidity, CO_2) in buildings.

2.3 Security applications

HWSNs are widely used in the security domain and deployed in both hostile and friendly environments. In this regard, the environment where HWSNs are deployed dictates the security protocols and topology of the network. In security applications, a combination of different types of sensors are deployed in an effort to obtain various types of data such as electromagnetic waves, pressure, noise, seismic, light, and image. Hence, these data are fused and processed for a reliable, timely, and accurate target evaluation. Having these features, HWSNs have found many application areas in border protection, force protection, critical infrastructure protection, and reconnaissance-surveillance tasks. Other remarkable application examples include battle damage assessment, and nuclear, biological, and chemical (NBC) attack detection [15].

One of the main applications of HWSNs in the security domain is the barrier coverage. Barrier coverage problems deal with providing adequate surveillance against possible illegal intruders. These barriers can be formed to protect critical infrastructures or an area of interest. Hence, barrier coverage applications have received significant attention of researchers. Among the large body of barrier coverage literature, Bhansali et al. [16] proposed a new concept for ultrasmall, ultracompact, and unattended multiphenomenological HWSN for surveillance of area of interest. The network comprises of both acoustic and seismic sensors. The concept also employs air assets to provide a reliable and timely communication within the network. Abhilash et al. [17] proposed two protocols in random deployment of HWSNs in a barrier coverage to enhance the lifetime. Benkoczi et al. [18] considered covering a barrier represented by a line segment with a mobile HWSN. Their study aimed at minimizing the total distance traveled by the sensors. Karatas and Onggo [19] focused on detecting intruders within a barrier. They proposed a mathematical model that maximizes the probability of detecting an intruder. Karatas and Onggo [20] developed two mathematical formulations considering multiple target types, unreliable sensors, and budget constraints to assess the performance and optimize the locations of HWSNs in a barrier. Karatas [21] presented a multiobjective bilevel HWSN location problem for an integrated area, point, and barrier coverage problem. They developed a multiobjective mixed integer nonlinear program (MINLP) and an equivalent mixed integer linear program to solve the model.

Ensuring security of the HWSNs is also an important issue in the security domain. Possible hostile attacks in HWSNs will result in unreliable and low-quality data transmission. Moreover, some severe attacks may even make these networks totally inactive. Therefore, implementing new secure routing mechanisms for these networks has also been a popular research area among researchers. As an example, Brown and Du [22] focused on providing security of HWSNs deployed in the hostile environments. In these environments, the attacker can interfere the network and launch a selective forward attack. To hedge against these attacks, the authors proposed a scheme for detecting such interference. Kumar et al. [23] particularly considered deployment of HWSNs in a military environment for border protection and enemy object tracking. Emphasizing that network security is of utmost importance in these environments, they proposed a new routing protocol (NetScreen redundancy protocol—NSRP) for secure data transmission.

2.4 Public and commercial applications

Advances in wireless networking technologies and sensor technology have widened the scope of HWSN applications. This has enabled to evolve into a massive IoT environment in our daily life and industry ranging from smartphones, wearable devices, and tablets to smart city infrastructure management, smart vehicles, and intelligent transportation systems. Implementing these coveted applications of IoT requires reliable delivery of data, at high data rates and ultralow latency, and efficient wireless connectivity between multiple IoT devices. IoT combines multiple heterogeneous networks, including HWSNs, vehicular networks, mobile networks, and wireless mesh networks [24]. IoT environment can be implemented under certain restrictions, such as reliability, connectivity, and latency, throughput depending on the application areas. Thus, HWSNs would have different requirements in terms of sensor mobility (stationary or mobile nodes), sensor deployment strategy, frequent sensor addition and removal, power restrictions, intermittent or continuous connectivity, and desired QoS.

In [24], a four-layer network architecture consisting of sensing layer, networking layer, cloud computing layer, and application layer is proposed for heterogeneous IoT. In sensing layer, different types of sensors (i.e., color sensor, sound sensor, camera, motion sensor, gas sensor, temperature sensor) are deployed in the monitoring area, the topology can be constructed in the form of "self-organizing." The collected data are transferred to the AP and used in the decision-making process at the cloud servers.

Most public and commercial applications (i.e., home automation, transportation, agriculture, environmental monitoring) today can be performed more efficiently with heterogeneous sensors. Heterogeneous sensors have been of particular interest for these applications due to easy and fast deployment, capabilities, and low maintenance costs.

For example, consider a home security system. Once an event occurs (i.e., illegal home entry attempt), sensor nodes detect the event and report it to the destination. Then, the report is transmitted to homeowner, for example, via smartphone interface. The objects are constantly monitored and controlled by the central server [24] and intelligently analyzed in real time. Here, AP manages all users and sensor nodes [25, 26].

Agriculture is another exemplary area of applications that benefits from HWSNs [24]. With environmental parameters, such as temperature, carbon dioxide concentration, light, and humidity, the meaningful sensing data are instantly or continuously obtained for intelligent agriculture management. Ideal environmental conditions for crop growth or improvement of agricultural areas are analyzed. Real-time monitoring especially of regions far from settlements is possible.

2.5 Scientific exploration

As another important application area of HWSNs, scientific exploration missions, has attracted attention from several scholars. Deploying HSWNs on the surface of a planet or under water for scientific research aims at quantitative measurement and analysis of changes in the environment.

Undersea exploration investigates underwater creatures, deep-sea life, physical and chemical properties of sea water, and geological properties of the Earth's crust. The obtained information allows scientists to make predictions on, for example, long-term weather, climatic changes, global warming, and the consumption of resources in the Earth. They are also used for monitoring the pollution level in an environment by regularly reporting the conditions in a particular field.

Similarly, space exploration investigates resources beyond the Earth's atmosphere for the benefit of humanity and increases knowledge of the cosmos through a crewed and uncrewed spacecraft. In [27], the authors investigate the challenges and sensor deployment solutions in Mars planet. Different from the traditional sensor deployment approaches, environmental parameters, such as air resistance, prediction for the wind gusts, and the gravitational field, are also included in the process for the extraction of aerial delivery sensor dispersion profiles. The sensors are located to employ self-organizing. The sensing data are transferred to the sink nodes hop-by-hop. Sink nodes are more powerful devices with faster processor and higher storage memory. Then, the collected data can be relayed back to Earth, through satellite or Mars Science Laboratory.

Consequently, sensors are used at every stage of a space or underwater mission. The more complex the task, the more advanced the sensors need to be. Designing more advanced heterogeneous sensors will lead to the easier management of HWSNs.

3. HWSN deployment strategies

In the scenario of deploying homogeneous sensors, it is easier to present effective solutions since the sensors have the similar characteristics (e.g., power, communication range, etc.). On the other hand, deployment algorithms offered with homogeneous sensors may not be effective with heterogeneous sensors due to their different characteristics. In HWSNs, one of the fundamental research questions is how to deploy and organize the heterogeneous sensor nodes within the given some constraints, such as energy, connectivity, and transmission range.

In this section, we first discuss topology and architectural issues to emphasize the effects on the network performance of HWSNs.

- *Sensor diversity*: The data collected from heterogeneous sensors have different characteristics. For each application area, the transmission frequency, priority, or data types will be different from each other. For example, in Internet of health things (IoHT), heart rate and location information cannot be evaluated in the same category. Therefore, the collected data characteristics should be evaluated according to the data types, interdependence, and independence from each other.

- *Sensor deployment*: Due to constraints and differences in power, size, features, and cost of heterogeneous sensor nodes, the deployment algorithms must satisfy the requirements and challenges in HWSNs.
- *Energy*: Sensors are devices with limited battery capacity. In many application areas, the change of sensors' positions or batteries is not possible due to the large scale of such networks, which contradict the advantages of wireless networks. Therefore, efficient management of energy consumption is one of the main research areas of WSNs. Intelligent management mechanisms are inevitable when heterogeneous sensors are included in the problem. Energy consumption will be the most important design consideration affecting the network lifetime.
- *Network dynamics*: In a dynamic network environment, a reliable and adaptable network configuration is needed. When the battery level is low due to the energy constraints of the sensor nodes, the node is out of service or goes into sleeping/charging mode. Nodes may be either static or dynamic. It is important how the nodes will be managed in such a dynamic network environment.
- *Environmental factors*: Depending on the environment conditions, for example, indoors or outdoors, the deployment of heterogeneous sensors or the network architecture differs.
- *QoS*: In some application areas, time-critical and high-priority data delivery are required, while in others, periodic data transfer is sufficient. Therefore, data delivery and reliability control mechanisms support different service quality needs and packet delivery rate, which in turn affects the energy consumption.
- *Network architecture*: Different architectures can be proposed in HWSNs according to the application areas, QoS requirements, and constraints. We will summarize two network architectures for HWSNs here:
  - *Hierarchical architecture*: In hierarchical architecture, sensors are divided into clusters and each cluster has a coordinator, named as cluster heads (CHs) so that they manage the clusters for aggregating, fusing, and filtering or delivering the data to the sink node or other CHs. Sink node can be base station (BS) or AP. Data transmission can be either one hop or multihop.
  - *Layered architecture*: In layered architecture, the data collected from heterogeneous sensors are designed in a layered structure. The architecture mainly consist of (i) physical layer, (ii) data link layer, (iii) network layer, (iv) transport layer, and (v) application layer. The physical layer provides the bit stream over the physical channel. In this layer, different types of data are transferred from heterogeneous sensors. Data link layer is responsible for medium access, error control, frame detection, data streams, reliability of point-to-point connection, or point-to-multipoint connection. Network layer supports the routing mechanism. Especially, energy-efficient routing is one of the main objectives in HWSNs. Transport layer is responsible for congestion avoidance and reliability since different application areas need

different requirements. Application layer provides the management-level functionality based on the different types of applications.

For example, in [28], IoHT architecture includes five layers: (i) perception layer, (ii) mist layer, (iii) fog layer, (iv) cloud layer, and (v) application layer. With the combination of HWSNs and IoHT, the data collected from different sensors in the perception layer (e.g., step counter, heart rate and position information from the smart watch, heartbeat rhythm from electrocardiogram [ECG] sensor, etc.) are transmitted to the mist layer. In the mist layer, data preprocessing and fusion processes are applied. Then, data filtering and analysis processes are applied in the fog layer.

In HWSN, deployment problem is one of the critical issues. In this section, we will discuss deployment strategies in terms of coverage maximization, energy efficiency and network lifetime, and maximizing connectivity in HWSNs.

3.1 Coverage maximization

The use of heterogeneous nodes in HWSNs provides a reliable and stable network architecture. The major design challenge that needs to be addressed is the coverage maximization such that multiple nodes operate with each other, aim to minimize overlapping regions, and provide maximum coverage in a given target area. One of the challenges is that when an event is detected, a large number of sensors will begin data transfer to report. In this case, while increasing data repetition occurs, it also leads to the consumption of the limited energy [29]. In addition, considering real-world problem, the environment conditions, such as trees, buildings, obstacles, and objects, can block the activities of sensors and QoS can decrease. The wireless signal is attenuated along its propagation path. Because of the obstacles, the signal can be scattered, reflected, or absorbed. The direct path loss (PL) can be simplified as follows [30]:

$$PL = 10\eta \log_{10} d + \sum_{i=1}^{M} \chi(i) + \beta(i)^{i-1} \tag{1}$$

where η is the PL coefficient, d is the distance between transmitter and receiver, M is the number of obstacles, $\chi(i)$ is the exponent of attenuation for ith obstacle, and $\beta(i)$ is the penetration rate of ith obstacle.

Deployment strategies in terms of coverage problem and the proposed algorithmic solutions will be detailed in Section 4.

3.2 Maximizing connectivity

In HWSNs, heterogeneous sensor nodes are located to send the collected data to the outside world through APs [31]. Continuous connectivity is one of the challenging issues since the sensors communicate with each other through the wireless channel. If the collected data are not transmitted to the AP, then it will be unusable. It is difficult to establish a continuous connectivity with the numerous sensors, limited energy capacity, and bandwidth. Energy problems, frequent failures, and lack of infrastructure need an efficient configuration.

HWSN can be represented by an undirected graph $G = (V, E)$, where V and E represent the set of sensor nodes and set of edges, respectively. If two nodes (u, v) are within the transmission range of each other, then there is connection between them. A weight is defined between edges, which represent cost. Cost can determine as consumed energy, transmission range, connectivity, etc. In practice, there is an upper bound on the possible flow along any edge. Thus, the problem can be also modeled as a connectivity problem, where maximum flow from a source node to the sink node is constructed. Hence, for each edge (u, v) $0 \leq f(u, v) \leq c(u, v)$ is valid, where $c(u, v)$ represents the bandwidth capacity. Accordingly, we can focus on finding a flow of maximum value in any given HWSN.

Theorem. (Max-flow min-cut) *For a given network, the maximum possible value of the flow is equal to the minimum capacity of all cuts.*

$$\max F(G) = \min \left(K(T, \bar{T})\right) \tag{2}$$

where $K(T, \bar{T})$ *is the capacity of a cut. A cut of* $G = (V, E)$ *is denoted as* $(T, \bar{T})$, *where source node is included to* T, *sink node is included to* $\bar{T}$ *and* $T \cap \bar{T} = \varnothing$ *and* $T \cup \bar{T} = V$.

Connectivity of a HWSN is a measure of its connectedness [32]. If two sensors can reach each other mutually, then this is a "strongly connected" graph. If HWSN is connected "loosely," this means that failure of a sensor destroys the connectedness of the network. In this section, we will address the solutions of deploying the sensor nodes to achieve higher network connectivity.

3.2.1 Random deployment strategies

The use of random sensor deployment algorithms is possible when the application environment is not known exactly, the number of sensors is too large, it is a hostile region or inaccessible to human. Among these, uniform random deployment is one of the most used approaches. Consider a HWSN scenario that sensors are distributed uniformly and traffic is transmitted to the AP via sink nodes. In this case, each sensor will provide traffic flow to the nearest sink node. According to the distribution, among the sink nodes, those closest to AP will always consume the most energy. Because if these nodes are not connected to the AP by one hop, the closest sink nodes will also be responsible for forwarding the incoming traffic to the AP.

Xu et al. [33] considers three types of devices: sensor nodes, relay nodes, and a BS. Sensors and relay nodes are randomly deployed according to the uniform distribution. Relay nodes are CHs responsible for aggregating the traffic and then transmitting to the BS by adjusting its transmission power so that energy wastage is also prevented. In the paper, the connectivity is defined as how much of the generated data have reached to the BS. The probability of reaching relay node, p_c, with at least one hop for any sensor node over a sensing area, A is given in Eq. (3) [33]:

$$P = 1 - \left(1 - \frac{\pi r^2 p_c}{A}\right)^N \tag{3}$$

where p_c represents the probability of connecting a sensor node to a relay node and N is the number of relay nodes.

In a weighted random deployment strategy, consider two regions. Let one of these regions be close to the BS and the other far away. In this case, the relay node in the area farther from the BS will run out of energy faster. Therefore, more relay nodes are deployed to balance energy consumption in the area farther from the BS. Another random deployment approach is the hybrid deployment that maximizes the network lifetime, while satisfying the connectivity requirement [33].

3.2.2 Lloyd algorithm-based approach

Lloyd algorithm, also known as k-means algorithm, is an iterative method and used to cluster points into groups represented by a centroid. Every data point is assigned to the closest center. The procedure includes two processes. (1) When a set of centroids is defined, the clusters are updated to include the points closest to each center of the centroid. (2) Given a set of clusters, the centroids are calculated as the mean of all points belonging to a cluster. This two-step procedure continues until the assignments of clusters and centroids do not change. In order to cover a given target area, the Lloyd algorithm is used in the sensor deployment problem.

Guo and Jafarkhani [31] address the sensor deployment problem for homogeneous and heterogeneous mobile WSNs using Lloyd-like algorithms to optimize sensing quality under the constraint of network lifetime and communication range. In the given scenario, a centralized BS collects all data from the sensors and determines the sensor locations. Accordingly, sensors move to the defined locations and the consumed energy is calculated in terms of communication, sensing, and computation. The authors propose two Lloyd-like algorithms: centralized constrained movement Lloyd (CCML) algorithm and backward-stepwise centralized constrained movement Lloyd (BCCML) algorithm to optimize the sensor deployment. CCML algorithm consists of two steps for partition optimization and location optimization. In this manner, each sensor moves to the closest point to its centroid. BCCML algorithm focuses on low battery energy sensors, which have limited movement. Therefore, nodes neighbor to this sensor ensure that it remains within the communication range of this sensor, even if it has high battery power.

3.2.3 Deterministic annealing algorithm-based approach

Simulated annealing (SA) is a metaheuristic to approximate the global optimum of a given objective function (energy function) and a parameter, called temperature. In each iteration, search space moves from the current state to a neighbor state. It probabilistically accepts the solution or state. While better solutions are always accepted, worse solutions are accepted with a probability of acceptance criteria [34]. However, Guo and Jafarkhani [2] defend that SA ignores the characteristics of the objective function and requires additional calculations. Thus, a deterministic annealing (DA) algorithm is designed in Guo and Jafarkhani [2] that generates two candidate sensor positions deterministically, but unlike SA, it randomly selects one of these two options. The authors analyze the homogeneous and heterogeneous sensor deployment problem by simulating Lloyd algorithm, virtual force algorithm (VFA), DA algorithm, and restrained Lloyd algorithm. Every sensor calculates its own approximate desired region and moves one by one after a new position has been determined with maximum local performance. Here, when a sensor moves to the desired region, the aim is to maintain the connectivity of the backbone network after movement.

3.3 Energy efficiency and network lifetime

The energy efficiency of sensors depends on a variety of factors, such as sensing, transmitting, communicating, processing, and computation. An energy consumption model is given in Heinzelman et al. [35]. Accordingly, the consumed energy is mainly divided into four parts: transmitting energy, $E_{Tx}(k, d)$; receiving energy, $E_{Rx}(k)$; amplifier energy, E_{amp}; and communication energy, E_{com}.

The consumed transmitting energy for k-bit packets is calculated depending on the Euclidean distance, d as follows [35, 36]:

$$E_{Tx}(k,d) = \begin{cases} kE_{Tx-elec} + kE_{Tx-amp}d^2, & d < d_0 \\ kE_{Tx-elec} + kE_{Tx-amp}d^4, & d \geq d_0 \end{cases} \tag{4}$$

If the distance is less than a threshold value, d_0, then the free-space channel model is used, otherwise, multipath channel model is used. $E_{Tx\text{-}elec}$ represents the energy dissipation at which the transmitter dissipates the energy to run the radio electronics and E_{amp} is the transmission amplifier. The consumed receiving energy for k-bit packets is calculated as follows [35, 36]:

$$E_{Rx}(k) = kE_{Rx-elec} \tag{5}$$

where the receiver dissipates energy to run the radio electronics. While electronic energy depends on the digital coding of the signal, modulation, and filtering, the amplifier energy depends on the distance between the receiver and the transmitter and the bit error rate [35]. In many recent studies, the heterogeneous sensor deployment problem has been addressed as an optimization problem that minimizes the total energy consumption [37]. In this section, we will describe the energy-efficient sensor deployment strategies that prolong the network lifetime in HWSNs.

3.3.1 VFA-based approach

Sensor nodes can be located in a given target area to gather information [38]. Finding the appropriate location of the sensors is the first problem to be solved to obtain efficient data. For the deployment problem, random or deterministic deployment can be considered. Random deployment of the sensors may not be an effective solution to cover the entire area. On the other hand, deterministic deployment consists of (1) defining the area of interest, (2) designing an algorithm for appropriate sensor locations, and (3) placing the sensors according to step (2) [38]. The problem is more complicated when environmental obstacles are involved such as buildings and trees.

VFA is presented in Zou and Chakrabarty [39] to enhance the coverage area. Initially, sensors are randomly deployed and a sequence of virtual motions is defined. When the effective position is determined, the sensor moves to this position. The consumed energy is also included in the algorithm. Virtual force is defined based on the distance of the sensors to each other. Sensors that are close to each other (proximity is determined by a threshold value) exert a negative force on each other. It means that clustering leads to poor coverage. Sensors that are far from each other exert positive force on each other. This indicates that a global sensor deployment has been achieved. Kuawattanaphan et al. [38] proposes a deployment algorithm using Delaunay triangulation method and extended virtual force algorithm (eVForce) with the adaptive parameter (AP) tuning mechanism in HWSNs.

3.3.2 Game theory-based approach

Liu et al. [40] investigate the energy efficiency problem to increase network lifetime in HWSNs. Packet transmission among nodes is modeled using game theory. The authors consider three types of sensors: CH, sensor node, and BS. Sensor node and CH have different energy capacity. BS is in the central location in a given target area and nodes are distributed randomly and uniformly. The behaviors of sensors are modeled as a game. A sensor node as a player can send the data to the CH with a probability between [0–1]. For instance, CH receives the data from the sensor nodes with a defined probability or there is no data transmission from the sensor nodes. Each sensor determines a strategy for the data transfer process aimed at maximizing the utility function to extend the network lifetime.

3.3.3 Particle swarm optimization algorithm-based approach

Particle swarm optimization (PSO) is metaheuristic and population-based optimization technique motivated by behavior birds or fish [41]. PSO is started with a population of random solutions and the search for the optimal solution is carried out by updating the generations. The potential solutions move in the search space by following the current optimum potential solutions, called as particles. In this algorithm, initialization and updating the particles are performed randomly. Thus, Wang and Wang [42] proposed a virtual force directed coevolutionary particle swarm optimization (VF-CPSO) algorithm for homogeneous and heterogeneous WSNs, which consists of static and mobile nodes. Initially, sensors are placed in a random manner. After the proposed deployment algorithm is executed, mobile sensors move to the defined locations to maximize the coverage while minimizing energy. CPSO is performed by the collaboration of n swarms, where each swarm is responsible for the location of one sensor. CPSO is used for global searching and virtual force (VF) is implemented to update the particles. In the paper, the lifetime of WSN is defined as the shortest lifetime of all sensor nodes and the uniform distribution is chosen to improve QoS and decrease interference. Thus, uniformity is defined for the network lifetime as follows [42, 43]:

$$
v = \frac{1}{Kl} \sum_{i=1}^{K} \left(\sqrt{\frac{1}{\tau_i} \left(\sum_{j=1}^{\tau_i} (D_{i,j} - M_i)^2 \right)} \right) \tag{6}
$$

where K is the number of sensors, l is the length of the sensor region, and τ_i is the number of sensor within the communication range of sensor i. $D_{i,\ j}$ represents the distance between sensor i and neighbor sensor j and M_i is the average distance between sensor i and neighbors' nodes.

4. Coverage concept, coverage schemes, and solution approaches

In this section, we will provide preliminaries for the coverage concept, coverage schemes, and various solution algorithms proposed in the literature to solve coverage problems, respectively.

4.1 Preliminaries

The *coverage* concept was first introduced by Church and ReVelle [44] in the location set-covering context. In location modeling, the coverage objective aims to ensure that each demand node is "served" or "satisfied" by a set of facilities that are within a predefined coverage radius of the demand node. Ever since proposed by Church and ReVelle [44], the set-covering concept has been applied to many areas such as the location of emergency services, retail facilities, hospitals, signal transmission facilities, water purification facilities, as well as many others [45].

Coverage is one of the most important metrics used for assessing the performance of WSNs. In the WSN context, the coverage can be interpreted as how well the physical conditions of an area of interest or the environment are monitored by wireless sensors [46]. It is also defined as the surveillance quality provided by the network, while maintaining communication and connectivity requirements [47]. Since each type of sensor in a HWSN has a unique coverage and detection performance against different types of targets, the coverage assessment problem for HWSNs becomes a challenging topic for both researchers and practitioners. In all settings, coverage-related problem are generally centered on the fundamental question: "How well an area is monitored by the sensors?" [48]. In the previous works, it has been recognized that the coverage is binary, in other words, a point is fully covered provided that there exists a sensor node within a predefined distance R and not covered if the closest sensor node is beyond that distance. This coverage phenomenon is represented with *disk* or *cookie-cutter* sensing models. These sensing models, sometimes referred to as "all or nothing" or "deterministic" models, may not reflect the nature of sensing event, where the coverage actually decays with distance [49]. For instance, the signal quality of a cell-phone tower decays with distance. The closer the tower, the better the communication quality. As another example, the probability of detection of an acoustic underwater sensor (i.e., sonobuoy) diminishes with distance. These shortcomings of deterministic sensing models led to an extension called "gradual" or "probabilistic" models, where the coverage level depends on the distance between the sensor and the point of interest. In this modeling approach, a point is fully covered when both sensor and point are located at the same position and is partially covered otherwise. The partial coverage level is quantified with a nonincreasing decay function. Sometimes two distance parameters are defined where a point is assumed to be covered within a distance r and not covered beyond distance R. Between these two distances, on the other hand, the point is partially covered.

As remarked by some authors [45, 50], a conceivable way to interpret the probabilistic coverage is the probability that a full coverage will occur. This particular interpretation of the probabilistic coverage gives rise to the question of how to aggregate the gradual cover of multiple sensors on a point. In the earlier studies, considering multiple sensors assumed that the closest sensor provides coverage, while the remaining sensors do not contribute to the coverage of a point. As Berman et al. [45] remarked, this assumption is particularly problematic if the coverage is interpreted as a probability. For instance, fixed radar stations are established on shore for sea surveillance and in order to increase the probability of detection of a target, the coverage of multiple radars usually overlaps with particular areas where radar signal decays. The probability of detection is a nonincreasing function of distance, therefore, the target will most probably be detected by the closest radar. However, detection can be made by another radar due to atmospheric conditions, sea state, or terrain. Hence, the closest sensor assumption may not be applicable to situations, where multiple sensors are present. For a more realistic modeling, the collective contributions of individual sensors should be combined to designate if a point is covered or not. To address this issue, which is prevalent in cases, where sensors emit/receive signals such as light or sound, the cooperative coverage concept is proposed. According to this modeling framework, a point is determined to be covered if the aggregated signal levels received exceed a certain threshold. Inherently, this modeling framework corresponds to a binary coverage framework even though probabilistic sensing models are assumed. Berman et al. [45] refer to models assuming binary coverage as *cooperative models* and models assuming gradual or partial coverage as *joint models* [49].

Various decay functions have been proposed in the literature to represent the partial coverage of a sensor. Among those, Fermi-type sensing model is commonly used in the literature [21, 51–53]. Let I be the set of points to be covered, J be the set of sensor nodes, and S be the set of sensor types, then the Fermi probability of covering of a point located at node $i \in I$ by a sensor of type $s \in S$ located at node $j \in J$ is represented as

$$p_{ijs} = \frac{1}{1 + 10^{\left(\frac{d_{ij}}{\rho_s} - 1\right)/b}}, \quad \forall\ i \in I,\ j \in J,\ s \in S \tag{7}$$

In this formula, ρ_s corresponds to the distance from the sensor to a point, where the coverage probability is 0.5 and b denotes the diffusivity parameter that governs the shape of the Fermi curve. Other commonly used decay functions are the exponential [20], Gaussian [54], and hybrid [55].

Assuming that the covering events of sensors are independent, the joint probabilistic coverage level of a point i provided by multiple heterogenous sensors can be computed as

$$JPC_i = 1 - \prod_{j \in J} \prod_{s \in S} (1 - p_{ijs}) \tag{8}$$

As can be seen from Eq. (7), the diffusivity parameter governs the shape of the Fermi curve. For illustration, Fermi-type sensing curves for different diffusivity parameters and joint coverage of two sensors having Fermi-type sensing model are given in Fig. 2.

Aforementioned binary coverage framework implies a 1-coverage problem, where a point or region is determined to be covered provided that at least one sensor provides coverage on that point or region. In some environmental monitoring applications such

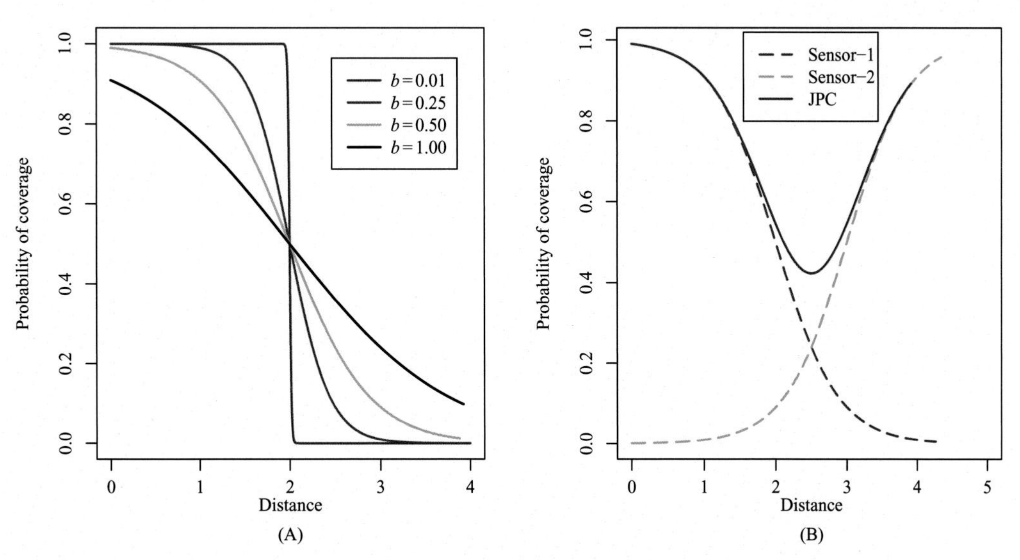

Fig. 2 (A) Fermi-type sensing curves for different diffusivity parameters and (B) joint coverage of two sensors.

as fire, gas leakage, and explosion, it is usually not enough to ensure that each point or region is covered for once because some sensors may fail or get blocked due to physical damage. In these situations, redundant coverage is desired to reduce the risk of failure. In order to address this particular problem, k-coverage concept is developed. k-coverage concept requires that each point or region is covered by at least k sensors simultaneously, where k is larger than one. As an extension to the classical 1-coverage problem, homogeneous k-coverage problems have been studied well in the literature, while k-coverage problems considering heterogeneous sensors are relatively few [56–60].

Several previous works tackled the coverage-related problems, presented novel problem definitions, and developed innovative modeling and solution approaches. While some studies targeted the problem from the coverage perspective directly with a focus on performance assessment and/or improvement, some others approached the problem indirectly by devoting to the connectivity and/or reliability of the network. Also, a number of niche applications have been proposed by some researchers although the main idea is still related to the coverage phenomenon. In this context, various covering schemes have been developed in the literature depending on the application and objective of the problem.

4.2 Coverage schemes

Coverage phenomenon can be categorized into three schemes with respect to the network design objective as: (1) *area* coverage, (2) *point* coverage, and (3) *barrier* coverage. In the following sections, these coverage schemes will be explained and prominent studies in these fields will be summarized.

4.2.1 Area coverage

The area coverage problem is the most common coverage problem type in HWSNs, and is widely studied by several scholars for decades. It basically considers the objective of achieving the highest possible sensing level across relatively large areas, sometimes referred as "fields." It can also be regarded as the coverage type, which seeks to physically cover every accessible point of a field by at least one sensor in the network. The coverage fields can be forests, cities, maritime zones, specific habitats, etc. [61].

A *coverage hole* refers to an area in the surveillance field that is not covered by any sensor node. Holes may occur due to various reasons such as: (1) resource-limited settings where there exists an insufficient number and quality of sensors to accomplish a full coverage, (2) sensor-dense fields that are formed by random aerial deployment creating void, (3) the existence of manmade or natural obstructions, and (4) sensor failures [62]. A full coverage implies that every point in the field is at least covered by a sensor node such that there is no coverage hole in the field. Fig. 3A depicts an example rectangular-shaped field with randomly deployed sensors of different types (and sensing radii). Fig. 4, on the other hand, displays (A) a coverage hole and (B) a fully covered region with a coverage degree of $k = 1$.

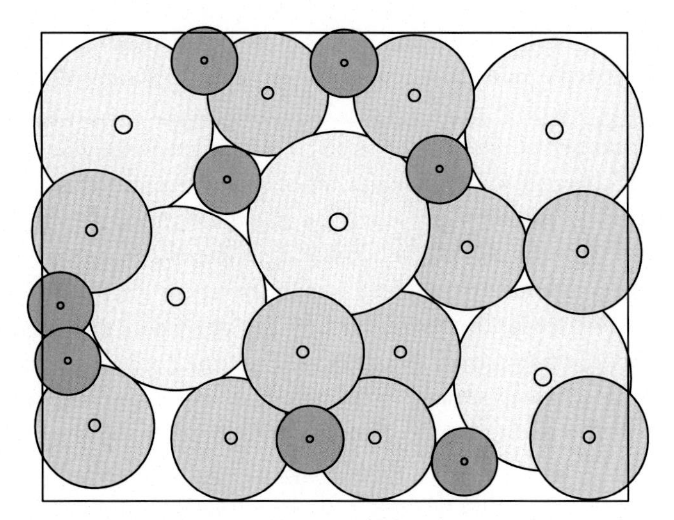

Fig. 3 Randomly deployed deterministic sensors used for area coverage.

There are several ways of formulating an area coverage problem and metrics to assess the coverage performance. A general formulation in continuous space can be developed for a set S of sensors of different types R and a bounded region A. The problem can be posed as determining the best locations for each sensor in S such that the area covered is maximized. For the discrete case in which there exists a finite set of candidate locations denoted as I for sensors, the problem can be reduced to choose $|S|$ points out of $|I|$ locations such that the area covered is maximized. Another way of posing the coverage problem is to determine the required number of sensors of each type to ensure that all points in the region are k-covered, where k is a prespecified integer parameter. Another modeling approach named as α-coverage refers to the ratio of active (alive) sensors required to

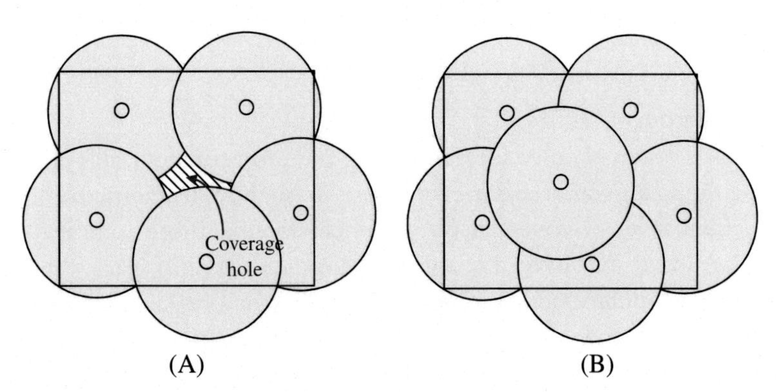

Fig. 4 (A) Coverage hole with deterministic sensing model and (B) with the deployment of sensor in the middle a degree $k = 1$ of coverage is accomplished.

ensure a full coverage of the region [63, 64]. Once the fraction of alive sensors (i.e., α) descends below a specific value the network is regarded as unsuccessful.

For a given set of sensors S and a region A, it is said that no coverage hole exists in the region, if every point in that region is covered by at least k sensors, where k represents the required degree of coverage for a particular application. Several methods and approaches have been proposed in the literature for choosing the optimal set of sensor locations. Typical solutions involve analytic geometry and geometric properties based on sensor positions. Another important metric used in area coverage for HWSNs is energy efficiency. Especially, for scenarios where the number of sensors deployed in a region is greater than the optimum required, then the problem turns out to determine the disjoint sets of sensors to be activated separately to perform area monitoring tasks individually. With this approach, the planners can conserve sensor energy and prolong the overall network lifetime by putting sensors in a low-energy sleep mode [65].

Although most of the studies dealing with area coverage problems consider homogeneous networks, there exist a few studies which utilize HWSNs. For example in [66], the authors assumed that the sensors have arbitrary-shaped nonuniform coverage regions and they can move to a specific direction. They seek to optimize the movements of mobile sensors to increase the overall network coverage. In another study, Wang and Medidi [67] developed optimal location schemes for area coverage in the presence of HWSNs with adjustable transmission and sensing radii. The authors developed algorithms, which ensure complete coverage for reliable surveillance while ensuring minimum energy consumption for extending network lifetime. Similarly, Wu and Yang [68] analyzed area coverage problem for HWSNs with variable sensing radii and developed models with the objective of maximizing coverage, while maintaining low energy.

The interested reader can also refer to the survey papers [48, 69] for more information on coverage problems with WSNs and implemented solution techniques. In addition, the review paper [70] is another important source, which discusses coverage-related problems from deployment, search area shape, sensor mobility, and objective function perspectives.

4.2.2 Point coverage

HWSNs are also commonly used for monitoring and protecting multiple critical facilities and infrastructure against different types of threats. In such applications, the objective is to provide a certain level of coverage (surveillance) around those facilities that can be considered as points to be covered in the two-dimensional (2D) space. Some examples of "points" include military facilities, high-value infrastructure, chemical plants, airports, oil platforms, energy generation facilities, high-rise buildings, city blocks, bridges, etc. [71–73]. In some applications, the points to be covered can also refer to a set of possible immobile target locations, for this reason the point coverage concept is sometimes called as "target coverage" [74]. In point coverage, different from the area coverage concept in

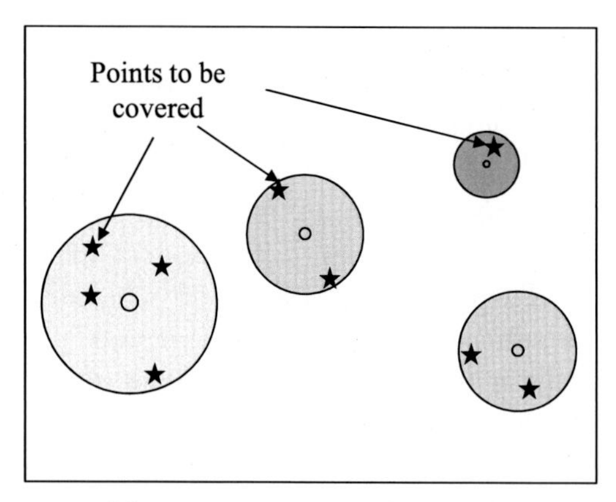

Fig. 5 Deterministic sensors used for point coverage. Each point that must be covered by a sensor is denoted by $\star$.

which the search space is defined as a bounded region, the search space consists of a set of discrete locations. Since a large field coverage is not required, solutions to the point coverage problems generally require fewer number of sensors compared to those obtained for area coverage [75]. Fig. 5 displays an example point coverage scenario, where a group of sensors of different types are used to cover a set of points each denoted as a $\star$.

The coverage status of a particular point is determined by comparing the distance between the point and each (or nearest) sensor with the detection range. A point in the field is assumed to be covered if it lies within the sensing zone of a sensor. In its general form, the point coverage with HWSNs seeks to cover a predefined set of points denoted by P in area A with a given set of sensors S of different types R. In some cases, without loss of generality, each point is also considered as a candidate sensor location. When the network is sufficiently dense, the area coverage problem can also be approximated by guaranteeing the point coverage [76]. In other words, an area coverage problem can be posed as a point coverage problem, where each point represents a discretization of the field [52, 77]. In those cases, typical solutions involve methods and algorithms to construct dominating sets and/or connected dominating sets, which are based on traditional graph theory [76, 78]. The idea of discretizing a continuous space area coverage problem is used by several scholars. As one of the foundational works, Zou and Chakrabarty [79] represented the search area as a 2D grid and proposed solution algorithms to replace randomly deployed sensors to improve coverage. Refs. [53, 76, 80–84] are some other example works which implement area discretization to model and solve an area coverage problem.

Another common way of posing the point coverage problem is to determine a selection of sensors in S and candidate locations I to site these sensors such that each point in P is covered by at least k sensors, while the total cost of sensors is minimized. This problem is also referred as the k-coverage problem.

Among the studies which tackle the point coverage problem for HWSNs, Cardei et al. [85] considered covering a set of targets with randomly deployed HWSNs that have adjustable sensing ranges. By adjusting the sensing ranges and activation times of sensors, they aimed at maximizing the lifetime of the network. In [86], the authors formulated a multiobjective mathematical model having three distinct objectives: (i) maximizing the energy consumption, (ii) maximizing coverage rate, and (iii) maximizing the equilibrium of energy consumption. Their mathematical model schedules active periods of the HWSNs to optimize these three objectives. They developed hybrid heuristic algorithms to solve the problem. Dhawan et al. [87] tackled the problem of maximizing the lifetime of an HWSN for point coverage purposes, where sensors can adjust their sensing radii. The authors proposed a linear program and a greedy algorithm to solve it. In a similar study, Cardei et al. [88] considered randomly deployed HWSNs with adjustable sensing radii. The authors attempted to determine the optimal scheduling of sensor nodes as well as transmission or sensing range of them.

4.2.3 Barrier coverage

In a barrier coverage setting, the goal is to provide coverage along a line segment or a 2D belt-shaped area that separates one region from another to prevent passages from one side to the other. Sensors are deployed along the perimeter of a region with the objective of establishing a barrier to detect mobile intruders. The effectiveness of barrier coverage along the borderline of a country is crucial since a weak coverage may result in the intrusion of nonfriendly units into critical areas. Since the problem considers detecting the intruders along their penetration paths through the region, it is sometimes called as the "path coverage" problem.

HWSNs are effectively used for barrier coverage scenarios in which some objectives include: (1) maximization of the detection probability of a moving object while crossing the deployment region of the network, (2) minimization of the number of sensors used to achieve a specific detection probability of intruders, and (3) minimizing coverage holes along a barrier with a given set of sensors. Deployment of different types of sonars and/or sonobuoys (expendable stationary sonars) to protect a channel against hostile underwater targets, for example, submarines, divers, use of guards and/or pickets in the vicinity of critical bases, and mine deployment are some of the common barrier coverage applications observed in security domain [21, 89].

In a general barrier problem, the barrier can be defined as a single line that separates one region from the other as well as a 2D belt-shaped geometric planar region with a well-defined boundary. A border region is called a "belt" if its boundary consists of two boundary parts (called banks) with equal separation distance. A belt is said to be "closed" if it is a bounded and closed region, such as ring. An open belt, on the other hand, can be regarded as a piece of region between two parallel lines [90]. Fig. 6 shows

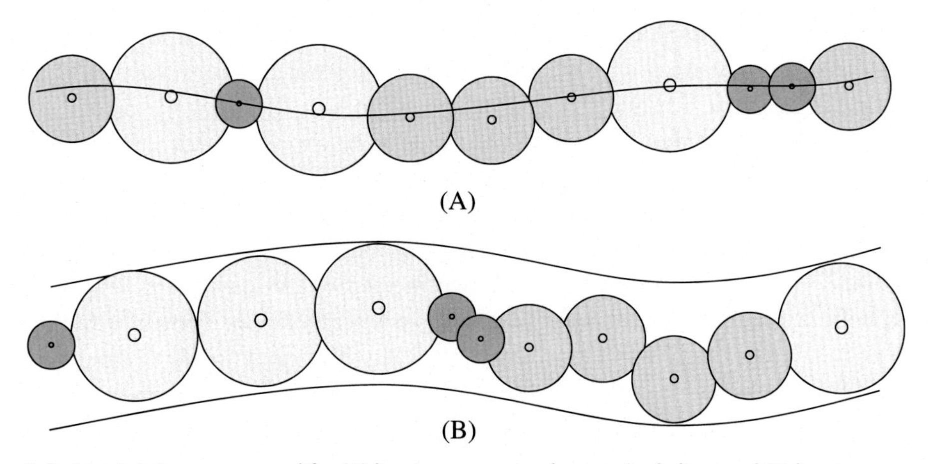

Fig. 6 Deterministic sensors used for (A) barrier coverage along a single line, and (B) barrier coverage along a belt-shaped region.

two different barrier coverage examples. In Fig. 6A, the barrier is defined as a single line, whereas in Fig. 6B, it is represented as a belt-shaped border region.

A full barrier coverage is said to be established if every point in the barrier line segment (or the perimeter of the region) is within the sensing range of a sensor. Similarly, the concept of k-barrier coverage of belt-shaped regions is introduced by Kumar et al. [91]. Chen et al. [92] define the concept of L-local barrier coverage problem, which considers maximizing the detection probability of all intruders following a path confined to a slice of length L of the belt-shaped barrier region. If the box bounding the entire intruder path has a length at most L; then this path is guaranteed to be detected by at least k sensor(s). Hence, it can be considered as a generalization of the barrier coverage problem since when L equals the length of the entire border region, L-local barrier coverage becomes the regular barrier coverage. Another problem concerning the barrier coverage is the barrier repair problem. Given a particular sensor deployment scheme over a belt-shaped border region, a required coverage quality Q^*, and quality of the actual deployment for a k-coverage barrier problem Q_k, Chen et al. [93] define a weak zone as a zone with quality such that $Q_k < Q^*$. Hence, the problem is defined as identifying all weak sections that need to be repaired. Similarly, for the case where redundant sensors exist in the network, determining the optimal sleep-wake-up schedules for the heterogeneous lifetime case (i.e., each sensor type has a different lifetime) while maintaining a specific level of coverage is another problem studied in the literature [94]. Barrier coverage performance of a network is also categorized as *weak* and *strong* [91]. In the weak barrier coverage, each target following a congruent path is guaranteed to be detected, whereas in strong barriers it is ensured that targets are always detected no matter which paths they follow. Thus, there exist some studies, for example, [95–97], that analyze barriers from this perspective and consider developing efficient algorithms for establishing weak and/or strong barriers.

Considering the problem of locating mobile HWSNs with arbitrary deterministic sensing ranges to cover a single line-shaped barrier region, Benkoczi et al. [18] aimed to minimize the total distance traveled by the sensors in the network on the barrier and developed efficient algorithms to solve it. Developing a mathematical model, Karatas and Onggo [19] aimed at finding locations of HWSNs in a barrier such that the probability of detecting an intruder is maximized. They utilized simulation to verify the solutions obtained from their model. Wang et al. [98] considered deploying both stationary and mobile HWSNs to form a barrier. Once the stationary sensors are deployed, the movements of the mobile sensors are scheduled to leverage the performance of the network. A greedy algorithm is developed for movement scheduling, while the total moving cost of the mobile sensors is minimized. Assuming a hub-spoke topology as well as a joint and probabilistic sensing model, Karatas and Onggo [20] developed two mathematical formulations and an optimization via simulation model to assess the performance and optimize the locations of HWSNs in a barrier. They also considered multiple target types, unreliable sensors, and budget constraints in their model.

There also exist a few studies, which adopt a more holistic approach and consider multiple coverage objectives simultaneously. Assuming a joint and probabilistic sensing model, Karatas [99] proposed a multiobjective mathematical formulation for a hybrid point and barrier coverage problem with HWSNs that aim at protecting critical facilities inside a belt-shaped region and detect any illegal intruders. The author developed a genetic algorithm (GA) metaheuristic to solve the multiobjective mathematical model. In another work, Karatas [21] presented a multiobjective bilevel HWSN location problem for an integrated area, point, and barrier coverage problem. They incorporated realistic requirements such as capacity constraints, network topology, and communication and interference range limitations into their model. They proposed a multiobjective mixed integer nonlinear program (MINLP) and an equivalent mixed integer linear program to solve the model. They also implemented a computational study to assess the performance of different branching and node selection strategies for solving the model. Later, this problem is also tackled by Yakıcı and Karatas [75], which proposed an NSGA-II (nondominated sorting genetic algorithm) metaheuristic to solve the problem efficiently.

Fig. 7 shows an exemplary location scheme for an HWSN developed for an integrated area, barrier, and point coverage scenario. Using gradual and joint coverage concepts, the heatmaps (A), (B), and (C) depict the overall detection probabilities for each point in the belt-shaped border region for area coverage, barrier coverage, and point coverage purposes, respectively. The example includes 10 sensors of 2 types (each type is denoted by a disk and square), connected to 3 hubs (each denoted by a star) and 10 critical facilities of 2 types (each type is represented by an upward pointing and downward pointing triangle in subfigure (C)). The dotted small circles around each critical facility represent the threat zones. In addition, the minimum interference range of each sensor is denoted by a dashed circle and the maximum communication range is denoted by a dotted circle.

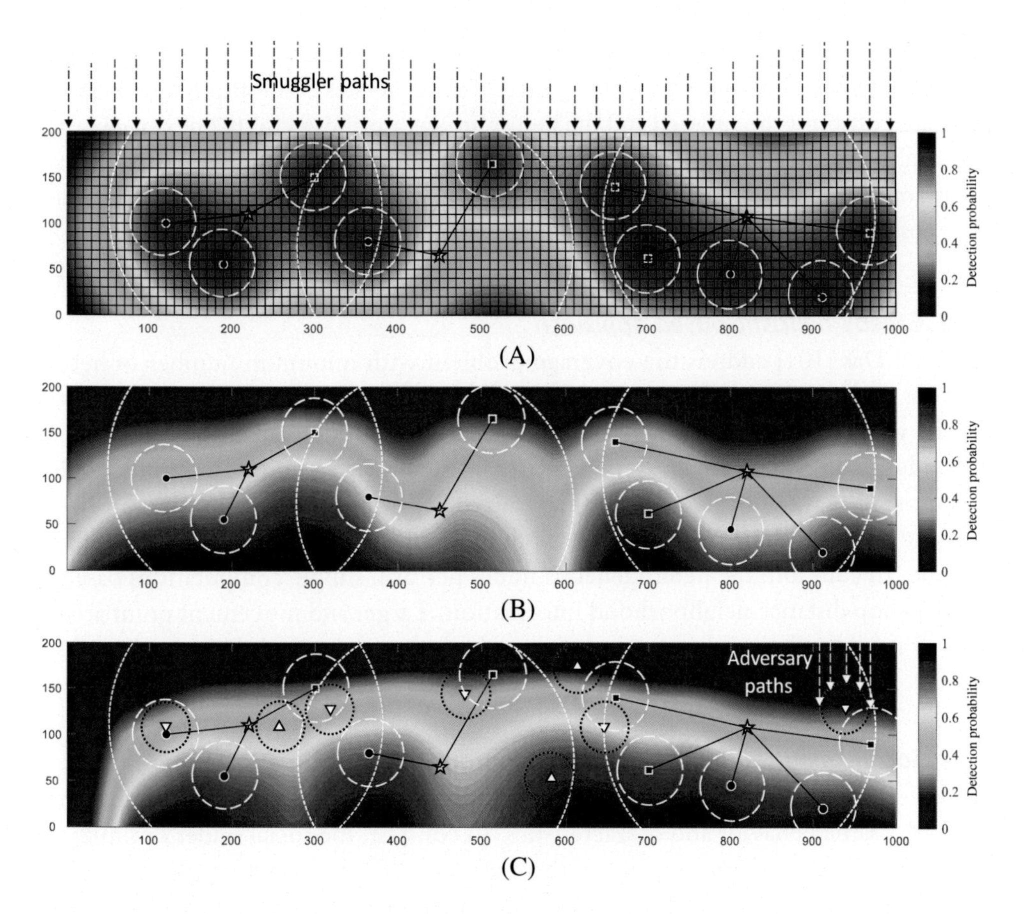

Fig. 7 Joint probabilities computed with gradual sensing sensors for (A) area coverage, (B) barrier coverage for intruders, and (C) point coverage for protecting critical facilities from attackers *(From M. Karatas, A multi-objective bi-level location problem for heterogeneous sensor networks with hub-spoke topology, Comput. Netw. 181 (2020) 107551).*

4.3 Solution approaches

In this section, we discuss the solution approaches to address the coverage maximization problem.

4.3.1 GA-based approach

In [100], the authors investigate the maximum coverage deployment problem by minimizing the number of sensors. The aim is to deploy k sensors (with different radius denoted as $r_1, r_2, \ldots, r_k$) to maximize the overall coverage. In GA, a group of solution is searched where the solutions are the positions and number of the sensor nodes (N).

These are stored in population. In each generation, $N/2$ pairs are randomly generated, and then crossover and mutation are applied. This generates $N/2$ offspring. Parents and generated offspring are ordered and the best solutions are chosen for the next generation. The evaluation is conducted using Monte Carlo method. This starts with a small number of random samples and increases the subsequent generations and then enables to reduce time cost.

4.3.2 Greedy heuristic-based approach

Saha and Das [101] address the coverage problem with minimum number of heterogeneous nodes and minimum displacement. It is assumed that heterogeneous nodes are mobile and different sensing and transmission range. Nodes can move based on the remaining energy and cover a circular area. The aim is to choose minimum subset of sensors so that the covered area is maximized. The authors also consider the energy limitations. Thus, the remaining nodes wait in the sleeping mode to improve network lifetime. In addition, the heterogeneous nodes perform simple computations based on the one-hop distance neighborhood information. Target and movement point are calculated in a distributed greedy algorithm, where the nodes find the suitable areas to cover.

4.3.3 Social spider optimization-based approach

In most of the swarm intelligence (SI) algorithms, all individuals are modeled according to the same behavior and characteristics. In contrast, the social spider optimization (SSO) algorithm performs operations through two different search agents: female and male. Therefore, each individual carries out different evolutionary processes according to its role in the colony. Information is exchanged with marriage and mating behavior and new individuals are produced. Then, the goal of finding the best individual is achieved [29]. In [29], the authors present an optimal coverage model to deploy an appropriate number of heterogeneous sensors using the improved SSO algorithm. For the initial phase, SSO algorithm is applied. After defining the positions and coverage areas of sensors, individual spiders are updated depending on the network coverage. The best individual of the spiders is calculated and the location and coverage areas of the other individuals are updated by learning from the best individual.

5. Conclusion and future outlook

Heterogeneous sensors are more effective and flexible with less deployment cost, energy consumption, and improved communication and coverage performance. They have been proven to be very promising in various applications. Sensor deployment and coverage problems in the presence of multiple sensor types, each having an arbitrary sensing zone, are more challenging compared to problems involving homogeneous WSN. Thus, deployment and coverage problems are fundamental issues in HWSNs.

In this chapter, we investigated the heterogeneous sensor deployment strategies and coverage models in HWSNs. We first itemized some of the typical challenges associated with HWSN applications (i.e., sensor diversity, coverage, energy efficiency, network lifetime, network dynamics, etc.) that draw attention both from practitioners and researchers. Next, we discussed the deployment strategies in terms of maximizing coverage, network lifetime, and connectivity. Coverage schemes were also categorized into three schemes as (i) area coverage, (ii) point coverage, and (iii) barrier coverage and then detailed.

There are still potential research areas related to HWSNs that can be studied. One potential future work direction may tackle coverage problems while considering several metrics and issues such as connectivity, energy management, topology, realistic sensing, etc., collaboratively. In addition, the inclusion of different requirements such as energy efficiency, fault-tolerance, and *k*-coverage, and realistic constraints such as adjustable sensor range, directional, or omni-directional sensing can be taken into account for each problem variant.

Acknowledgments

The second author was supported by the Scientific and Technological Research Council of Turkey (TÜBİTAK), Grant No. 118E694.

Conflict of interest

The authors declare that they have no conflict of interest.

Disclaimer

Conclusions contained herein are those of the authors and should not be interpreted as necessarily representing the official policies or endorsements, either expressed or implied, of any affiliated organization or government.

References

[1] P. Gupta, S. Tripathi, S. Singh, Energy efficient hotspot problem mitigation techniques using multiple mobile sink in heterogeneous wireless sensor network, Int. J. Commun. Syst. 33 (18) (2020) e4641.

[2] J. Guo, H. Jafarkhani, Sensor deployment with limited communication range in homogeneous and heterogeneous wireless sensor networks, IEEE Trans. Wirel. Commun. 15 (10) (2016) 6771–6784, https://doi.org/10.1109/TWC.2016.2590541.

[3] D. Trinugroho, Y. Baptista, Information integration platform for patient-centric healthcare services: design, prototype and dependability aspects, Future Internet 6 (1) (2014) 126–154.

[4] Y. ElSaadany, A.K.M.J.A. Majumder, D.R. Ucci, A wireless early prediction system of cardiac arrest through IoT, in: 2017 IEEE 41st Annual Computer Software and Applications Conference (COMPSAC), vol. 2, IEEE, 2017, pp. 690–695.

[5] A. Massaro, G. Ricci, S. Selicato, S. Raminelli, A. Galiano, Decisional support system with artificial intelligence oriented on health prediction using a wearable device and big data, in: 2020 IEEE International Workshop on Metrology for Industry 4.0 & IoT, IEEE, 2020, pp. 718–723.

[6] L. Catarinucci, D. De Donno, L. Mainetti, L. Palano, L. Patrono, M.L. Stefanizzi, L. Tarricone, An IoT-aware architecture for smart healthcare systems, IEEE Internet Things J. 2 (6) (2015) 515–526.

[7] S. Mattsson, J. Partini, Å. Fast-Berglund, Evaluating four devices that present operator emotions in real-time, Procedia CIRP 50 (2016) 524–528.

[8] F. Wu, T. Wu, M.R. Yuce, An internet-of-things (IoT) network system for connected safety and health monitoring applications, Sensors 19 (1) (2019) 21.

[9] A. Mavrinac, X. Chen, Modeling coverage in camera networks: a survey, Int. J. Comput. Vis. 101 (1) (2013) 205–226.

[10] A.T. Erman, L. van Hoesel, P. Havinga, J. Wu, Enabling mobility in heterogeneous wireless sensor networks cooperating with UAVs for mission-critical management, IEEE Wirel. Commun. 15 (6) (2008) 38–46.

[11] C. Tunca, S. Isik, M.Y. Donmez, C. Ersoy, Performance evaluation of heterogeneous wireless sensor networks for forest fire detection, in: 2013 21st Signal Processing and Communications Applications Conference (SIU), IEEE, 2013, pp. 1–4.

[12] S. Haran, S. Kher, V. Mehndiratta, Bridge monitoring using heterogeneous wireless sensor network, in: Health Monitoring of Structural and Biological Systems 2010, vol. 7650, International Society for Optics and Photonics, 2010, p. 765008.

[13] N. Wang, Z. Hong, An energy-efficient topology control algorithm for earth building monitoring using heterogeneous wireless sensor networks, IEEE Access 7 (2019) 76120–76130.

[14] Q. Huang, K. Rodriguez, A software framework for heterogeneous wireless sensor network towards environmental monitoring, Appl. Sci. 9 (5) (2019) 867.

[15] C.T. Vu, Distributed Energy-Efficient Solutions for Area Coverage Problems in Wireless Sensor Networks (Ph.D. dissertation), Georgia State University, 2009.

[16] S. Bhansali, G.H. Chapman, E.G. Friedman, Y. Ismail, P.R. Mukund, D. Tebbe, V.K. Jain, 3D heterogeneous sensor system on a chip for defense and security applications, in: Unattended/Unmanned Ground, Ocean, and Air Sensor Technologies and Applications VI, 5417, International Society for Optics and Photonics, 2004, pp. 413–424. vol.

[17] C.N. Abhilash, S.H. Manjula, K.R. Venugopal, L.M. Patnaik, Efficient network lifetime for barrier coverage in heterogeneous sensor network, in: 2013 Annual IEEE India Conference (INDICON), IEEE, 2013, pp. 1–4.

[18] R. Benkoczi, Z. Friggstad, D. Gaur, M. Thom, Minimizing total sensor movement for barrier coverage by non-uniform sensors on a line, in: International Symposium on Algorithms and Experiments for Wireless Sensor Networks, Springer, 2015, pp. 98–111.

[19] M. Karatas, B.S. Onggo, Validating an integer non-linear program optimization model of a wireless sensor network using agent-based simulation, in: 2016 Winter Simulation Conference (WSC), IEEE, 2016, pp. 1340–1351.

[20] M. Karatas, B.S. Onggo, Optimising the barrier coverage of a wireless sensor network with hub-and-spoke topology using mathematical and simulation models, Comput. Oper. Res. 106 (2019) 36–48.

[21] M. Karatas, A multi-objective bi-level location problem for heterogeneous sensor networks with hub-spoke topology, Comput. Netw. 181 (2020) 107551.

[22] J. Brown, X. Du, Detection of selective forwarding attacks in heterogeneous sensor networks, in: 2008 IEEE International Conference on Communications, IEEE, 2008, pp. 1583–1587.

[23] K.A. Kumar, A.V.N. Krishna, K.S. Chatrapati, New secure routing protocol with elliptic curve cryptography for military heterogeneous wireless sensor networks, J. Inf. Optim. Sci. 38 (2) (2017) 341–365.

[24] T. Qiu, N. Chen, K. Li, M. Atiquzzaman, W. Zhao, How can heterogeneous internet of things build our future: a survey, IEEE Commun. Surv. Tutor. 20 (3) (2018) 2011–2027, https://doi.org/10.1109/COMST.2018.2803740.

[25] S. Mohanti, E. Bozkaya, M. Yousof Naderi, B. Canberk, K. Chowdhury, WiFED: WiFi friendly energy delivery with distributed beamforming, in: IEEE INFOCOM 2018—IEEE Conference on Computer Communications, 2018, pp. 926–934, https://doi.org/10.1109/INFOCOM.2018.8486207.
[26] S. Mohanti, E. Bozkaya, M.Y. Naderi, B. Canberk, G. Secinti, K.R. Chowdhury, WiFED mobile: WiFi friendly energy delivery with mobile distributed beamforming, IEEE/ACM Trans. Netw. Early Access (2021) 1–14, https://doi.org/10.1109/TNET.2021.3061082.
[27] C. Sergiou, A. Paphitis, C. Panayiotou, P. Ktistis, K. Christou, Wireless sensor networks for planetary exploration: issues and challenges through a specific application, in: 13th International Conference on Space Operations, 2014, pp. 1–19.
[28] M. Asif-Ur-Rahman, F. Afsana, M. Mahmud, M.S. Kaiser, M.R. Ahmed, O. Kaiwartya, A. James-Taylor, Toward a heterogeneous mist, fog, and cloud-based framework for the internet of healthcare things, IEEE Internet Things J. 6 (3) (2019) 4049–4062, https://doi.org/10.1109/JIOT.2018.2876088.
[29] L. Cao, Y. Yue, Y. Cai, Y. Zhang, A novel coverage optimization strategy for heterogeneous wireless sensor networks based on connectivity and reliability, IEEE Access 9 (2021) 18424–18442, https://doi.org/10.1109/ACCESS.2021.3053594.
[30] B. Cao, J. Zhao, P. Yang, P. Yang, X. Liu, Y. Zhang, 3-D deployment optimization for heterogeneous wireless directional sensor networks on smart city, IEEE Trans. Ind. Inf. 15 (3) (2019) 1798–1808, https://doi.org/10.1109/TII.2018.2884951.
[31] J. Guo, H. Jafarkhani, Movement-efficient sensor deployment in wireless sensor networks with limited communication range, IEEE Trans. Wirel. Commun. 18 (7) (2019) 3469–3484, https://doi.org/10.1109/TWC.2019.2914199.
[32] R. Balakrishnan, K. Ranganathan, A Textbook of Graph Theory, Springer, 2012.
[33] K. Xu, H. Hassanein, G. Takahara, Q. Wang, Relay node deployment strategies in heterogeneous wireless sensor networks, IEEE Trans. Mob. Comput. 9 (2) (2010) 145–159, https://doi.org/10.1109/TMC.2009.105.
[34] F.S. Hillier, G.J. Lieberman, Introduction to Operations Research, McGraw-Hill Higher Education, 2004.
[35] W.B. Heinzelman, A.P. Chandrakasan, H. Balakrishnan, An application-specific protocol architecture for wireless microsensor networks, IEEE Trans. Wirel. Commun. 1 (4) (2002) 660–670, https://doi.org/10.1109/TWC.2002.804190.
[36] J. Feng, H. Chen, Repairing confident information coverage holes for big data collection in large-scale heterogeneous wireless sensor networks, IEEE Access 7 (2019) 155347–155360, https://doi.org/10.1109/ACCESS.2019.2949136.
[37] S. Karimi-Bidhendi, J. Guo, H. Jafarkhani, Energy-efficient node deployment in heterogeneous two-tier wireless sensor networks with limited communication range, IEEE Trans. Wirel. Commun. 20 (1) (2021) 40–55, https://doi.org/10.1109/TWC.2020.3023065.
[38] R. Kuawattanaphan, P. Champrasert, S. Aramkul, A novel heterogeneous wireless sensor node deployment algorithm with parameter-free configuration, IEEE Access 6 (2018) 44951–44969, https://doi.org/10.1109/ACCESS.2018.2865279.
[39] Y. Zou, K. Chakrabarty, Sensor deployment and target localization based on virtual forces, in: IEEE INFOCOM 2003. Twenty-Second Annual Joint Conference of the IEEE Computer and Communications Societies (IEEE Cat. No. 03CH37428), vol. 2, 2003, pp. 1293–1303, https://doi.org/10.1109/INFCOM.2003.1208965.
[40] X. Liu, G. Kang, N. Zhang, A deployment scheme for lifetime enhancement of heterogeneous wireless sensor network, in: 2014 IEEE 80th Vehicular Technology Conference (VTC2014-Fall), 2014, pp. 1–5, https://doi.org/10.1109/VTCFall.2014.6965810.
[41] J. Kennedy, R. Eberhart, Particle swarm optimization, in: Proceedings of ICNN'95—International Conference on Neural Networks, vol. 4, 1995, pp. 1942–1948, https://doi.org/10.1109/ICNN.1995.488968.
[42] X. Wang, S. Wang, Hierarchical deployment optimization for wireless sensor networks, IEEE Trans. Mob. Comput. 10 (7) (2011) 1028–1041, https://doi.org/10.1109/TMC.2010.216.
[43] N. Heo, P.K. Varshney, Energy-efficient deployment of intelligent mobile sensor networks, IEEE Trans. Syst. Man Cybern. Part A Syst. Hum. 35 (1) (2005) 78–92, https://doi.org/10.1109/TSMCA.2004.838486.

[44] R. Church, C. ReVelle, The maximal covering location problem, in: Papers of the Regional Science Association, vol. 32, Springer-Verlag, 1974, pp. 101–118.
[45] O. Berman, Z. Drezner, D. Krass, The multiple gradual cover location problem, J. Oper. Res. Soc. 70 (6) (2019) 931–940.
[46] C.-F. Huang, Y.-C. Tseng, The coverage problem in a wireless sensor network, Mob. Netw. Appl. 10 (4) (2005) 519–528.
[47] Z. Xiao, M. Huang, J. Shi, J. Yang, J. Peng, Full connectivity and probabilistic coverage in random deployed 3D WSNs, in: 2009 International Conference on Wireless Communications & Signal Processing, IEEE, 2009, pp. 1–4.
[48] G. Fan, S. Jin, Coverage problem in wireless sensor network: a survey, J. Netw. 5 (9) (2010) 1033.
[49] M. Karatas, L. Eriskin, The minimal covering location and sizing problem in the presence of gradual cooperative coverage, Eur. J. Oper. Res. 295 (3) (2021) 838–856.
[50] T. Drezner, Z. Drezner, The maximin gradual cover location problem, OR Spect. 36 (4) (2014) 903–921.
[51] M. Karatas, A multi-objective facility location problem in the presence of variable gradual coverage performance and cooperative cover, Eur. J. Oper. Res. 262 (3) (2017) 1040–1051.
[52] E.M. Craparo, M. Karatas, T.U. Kuhn, Sensor placement in active multistatic sonar networks, Nav. Res. Logist. (NRL) 64 (4) (2017) 287–304.
[53] A.R. Fügenschuh, E.M. Craparo, M. Karatas, S.E. Buttrey, Solving multistatic sonar location problems with mixed-integer programming, Optim. Eng. 21 (1) (2020) 273–303.
[54] N. Ahmed, S.S. Kanhere, S. Jha, Probabilistic coverage in wireless sensor networks, in: The IEEE Conference on Local Computer Networks 30th Anniversary (LCN'05), IEEE, 2005, p. 8.
[55] P. Si, C. Wu, Y. Zhang, H. Chu, H. Teng, Probabilistic coverage in directional sensor networks, Wirel. Netw. 25 (1) (2019) 355–365.
[56] H.M. Ammari, J. Giudici, On the connected k-coverage problem in heterogeneous sensor nets: the curse of randomness and heterogeneity, in: 2009 29th IEEE International Conference on Distributed Computing Systems, IEEE, 2009, pp. 265–272.
[57] H.M. Ammari, S.K. Das, Scheduling protocols for homogeneous and heterogeneous k-covered wireless sensor networks, Pervasive Mob. Comput. 7 (1) (2011) 79–97.
[58] H.P. Gupta, S.V. Rao, T. Venkatesh, Sleep scheduling protocol for *k*-coverage of three-dimensional heterogeneous WSNs, IEEE Trans. Veh. Technol. 65 (10) (2015) 8423–8431.
[59] J. Yu, Y. Chen, L. Ma, B. Huang, X. Cheng, On connected target k-coverage in heterogeneous wireless sensor networks, Sensors 16 (1) (2016) 104.
[60] H.M. Ammari, A unified framework for k-coverage and data collection in heterogeneous wireless sensor networks, J. Parallel Distrib. Comput. 89 (2016) 37–49.
[61] S. Gabriele, P. Di Giamberardino, The area coverage problem for dynamic sensor networks, in: New Developments in Robotics Automation and Control, IntechOpen, 2008.
[62] N. Ahmed, S.S. Kanhere, S. Jha, The holes problem in wireless sensor networks: a survey, ACM SIGMOBILE Mob. Comput. Commun. Rev. 9 (2) (2005) 4–18.
[63] M. Lehsaini, H. Guyennet, M. Feham, A-coverage scheme for wireless sensor networks, in: 2008 The Fourth International Conference on Wireless and Mobile Communications, IEEE, 2008, pp. 91–96.
[64] Z. Zheng, P. Sinha, S. Kumar, Sparse WiFi deployment for vehicular internet access with bounded interconnection gap, IEEE/ACM Trans. Netw. 20 (3) (2011) 956–969.
[65] M. Cardei, J. Wu, Energy-efficient coverage problems in wireless ad-hoc sensor networks, Comput. Commun. 29 (4) (2006) 413–420.
[66] Y. Stergiopoulos, A. Tzes, Cooperative positioning/orientation control of mobile heterogeneous anisotropic sensor networks for area coverage, in: 2014 IEEE International Conference on Robotics and Automation (ICRA), IEEE, 2014, pp. 1106–1111.
[67] J. Wang, S. Medidi, Energy efficient coverage with variable sensing radii in wireless sensor networks, in: Third IEEE International Conference on Wireless and Mobile Computing, Networking and Communications (WIMOB 2007), IEEE, 2007, p. 61.
[68] J. Wu, S. Yang, Coverage issue in sensor networks with adjustable ranges, in: Workshops on Mobile and Wireless Networking/High Performance Scientific, Engineering Computing/Network Design

and Architecture/Optical Networks Control and Management/Ad Hoc and Sensor Networks/Compil, IEEE, 2004, pp. 61–68.
[69] M. Cardei, J. Wu, Coverage in wireless sensor networks, Handb. Sens. Netw. 21 (2004) 201–202.
[70] M. Kumar, V. Gupta, A review paper on sensor deployment techniques for target coverage in wireless sensor networks, in: 2016 International Conference on Control, Instrumentation, Communication and Computational Technologies (ICCICCT), IEEE, 2016, pp. 452–456.
[71] J. Wang, N. Zhong, Efficient point coverage in wireless sensor networks, J. Comb. Optim. 11 (3) (2006) 291–304.
[72] M. Karatas, E. Craparo, G. Akman, Bistatic sonobuoy deployment strategies for detecting stationary and mobile underwater targets, Nav. Res. Logist. (NRL) 65 (4) (2018) 331–346.
[73] E.M. Craparo, A. Fügenschuh, C. Hof, M. Karatas, Optimizing source and receiver placement in multistatic sonar networks to monitor fixed targets, Eur. J. Oper. Res. 272 (3) (2019) 816–831.
[74] L. Erişkin, Point coverage with heterogeneous sensor networks: a robust optimization approach under target location uncertainty, Comput. Netw. 198 (2021) 108416.
[75] E. Yakıcı, M. Karatas, Solving a multi-objective heterogeneous sensor network location problem with genetic algorithm, Comput. Netw. 192 (2021) 108041.
[76] S. Yang, F. Dai, M. Cardei, J. Wu, On multiple point coverage in wireless sensor networks, in: IEEE International Conference on Mobile Adhoc and Sensor Systems Conference, 2005, IEEE, 2005, p. 8.
[77] E.M. Craparo, M. Karatas, A method for placing sources in multistatic sonar networks, Technical Report, Naval Postgraduate School, Monterey, CA, United States NPS-OR-18-001 (2018).
[78] J. Wu, H. Li, On calculating connected dominating set for efficient routing in ad hoc wireless networks, in: Proceedings of the 3rd International Workshop on Discrete Algorithms and Methods for Mobile Computing and Communications, 1999, pp. 7–14.
[79] Y. Zou, K. Chakrabarty, Sensor deployment and target localization in distributed sensor networks, ACM Trans. Embed. Comput. Syst. 3 (1) (2004) 61–91.
[80] Q. Yang, S. He, J. Li, J. Chen, Y. Sun, Energy-efficient probabilistic area coverage in wireless sensor networks, IEEE Trans. Veh. Technol. 64 (1) (2014) 367–377.
[81] E. Craparo, M. Karatas, Optimal source placement for point coverage in active multistatic sonar networks, Nav. Res. Logist. (NRL) 67 (1) (2020) 63–74.
[82] H. Wu, M. Shahidehpour, Applications of wireless sensor networks for area coverage in microgrids, IEEE Trans. Smart Grid 9 (3) (2016) 1590–1598.
[83] T. Yan, T. He, J.A. Stankovic, Differentiated surveillance for sensor networks, in: Proceedings of the 1st International Conference on Embedded Networked Sensor Systems, ACM, 2003, pp. 51–62.
[84] M.T. Thai, F. Wang, D.H. Du, X. Jia, Coverage problems in wireless sensor networks: designs and analysis, Int. J. Sens. Netw. 3 (3) (2008) 191.
[85] M. Cardei, J. Wu, M. Lu, M.O. Pervaiz, Maximum network lifetime in wireless sensor networks with adjustable sensing ranges, in: (WiMob'2005), IEEE International Conference on Wireless and Mobile Computing, Networking and Communications, 2005, vol. 3, IEEE, 2005, pp. 438–445.
[86] Y. Xu, O. Ding, R. Qu, K. Li, Hybrid multi-objective evolutionary algorithms based on decomposition for wireless sensor network coverage optimization, Appl. Soft Comput. 68 (2018) 268–282.
[87] A. Dhawan, C.T. Vu, A. Zelikovsky, Y. Li, S.K. Prasad, Maximum lifetime of sensor networks with adjustable sensing range, in: Seventh ACIS International Conference on Software Engineering, Artificial Intelligence, Networking, and Parallel/Distributed Computing (SNPD'06), IEEE, 2006, pp. 285–289.
[88] M. Cardei, J. Wu, M. Lu, Improving network lifetime using sensors with adjustable sensing ranges, Int. J. Sens. Netw. 1 (1–2) (2006) 41–49.
[89] A. Washburn, Barrier games, Mil. Oper. Res. 15 (3) (2010) 31–41.
[90] W. Wu, Z. Zhang, W. Lee, D.-Z. Du, Barrier Coverage, Springer International Publishing, Cham, 2020, pp. 159–181, https://doi.org/10.1007/978-3-030-52824-9_10.
[91] S. Kumar, T.H. Lai, A. Arora, Barrier coverage with wireless sensors, in: Proceedings of the 11th Annual International Conference on Mobile Computing and Networking, ACM, 2005, pp. 284–298.
[92] A. Chen, S. Kumar, T.H. Lai, Local barrier coverage in wireless sensor networks, IEEE Trans. Mob. Comput. 9 (4) (2009) 491–504.

[93] A. Chen, T.H. Lai, D. Xuan, Measuring and guaranteeing quality of barrier-coverage in wireless sensor networks, in: Proceedings of the 9th ACM International Symposium on Mobile Ad Hoc Networking and Computing, 2008, pp. 421–430.
[94] S. Kumar, T.H. Lai, M.E. Posner, P. Sinha, Maximizing the lifetime of a barrier of wireless sensors, IEEE Trans. Mob. Comput. 9 (8) (2010) 1161–1172.
[95] S. Kumar, T.H. Lai, M.E. Posner, P. Sinha, Optimal sleep-wakeup algorithms for barriers of wireless sensors, in: 2007 Fourth International Conference on Broadband Communications, Networks and Systems (BROADNETS'07), IEEE, 2007, pp. 327–336.
[96] B. Liu, O. Dousse, J. Wang, A. Saipulla, Strong barrier coverage of wireless sensor networks, in: Proceedings of the 9th ACM International Symposium on Mobile Ad Hoc Networking and Computing, 2008, pp. 411–420.
[97] L. Zhang, J. Tang, W. Zhang, Strong barrier coverage with directional sensors, in: GLOBECOM 2009: 2009 IEEE Global Telecommunications Conference, IEEE, 2009, pp. 1–6.
[98] Z. Wang, Q. Cao, H. Qi, H. Chen, Q. Wang, Cost-effective barrier coverage formation in heterogeneous wireless sensor networks, Ad Hoc Netw. 64 (2017) 65–79.
[99] M. Karatas, Optimal deployment of heterogeneous sensor networks for a hybrid point and barrier coverage application, Comput. Netw. 132 (2018) 129–144.
[100] Y. Yoon, Y.-H. Kim, An efficient genetic algorithm for maximum coverage deployment in wireless sensor networks, IEEE Trans. Cybern. 43 (5) (2013) 1473–1483, https://doi.org/10.1109/TCYB.2013.2250955.
[101] D. Saha, A. Das, Coverage area maximization by heterogeneous sensor nodes with minimum displacement in mobile networks, in: 2015 IEEE International Conference on Advanced Networks and Telecommunications Systems (ANTS), 2015, pp. 1–6, https://doi.org/10.1109/ANTS.2015.7413629.

CHAPTER 2

Efficient multitasking in heterogeneous wireless sensor networks

Kiran Singh[a], Suman Avdhesh Yadav[b], and T. Poongodi[a]
[a]School of Computing Science and Engineering, Galgotias University, Greater Noida, Uttar Pradesh, India
[b]Amity School of Engineering & Technology, Amity University, Greater Noida, Uttar Pradesh, India

1. Introduction

WSN is a frameworkless wireless organization that is expressed in a specially appointed way in a large number of wireless sensors and is used to track the framework, physical or natural conditions. WSN uses sensor nodes in combination with a locally accessible processor to track and manage the climate in a specific region. They are connected to the base station, which acts as a WSN device handling unit. Military, environmental, health, entertainment, transportation, crisis management, smart spaces, and disaster prevention are just a few of the applications of wireless sensor networks (WSNs) [1].

WSNs are composed of multiple sensor nodes, each with its own battery life and sensing range. If a sensor node is within its sensing range, it will monitor the environment. Sensors may be distributed randomly or deterministically, depending on the application requirements and feasibility. In WSNs, coverage is described as "how well sensors track." Because of their numerous applications in different fields, wireless sensor networks have increased in popularity in recent years.

A large number of sensor nodes are linked to form a wireless sensor network. Temperature, humidity, ambient light, gas, and other parameters are controlled and sent to the master node by each sensor node in the network. Sensor networks have a number of weaknesses, including energy, localization, security, self-organization, fault tolerance, and many others, despite their wide range of applications and uses. As a result, several researchers around the world are working on new algorithms, protocols, and techniques to improve the performance and reliability of wireless sensor networks. It is important to thoroughly test the proven process before bringing it into service. However, providing a live sensor network environment is not always feasible. As a result, simulation [2] is the only way to validate the study before moving forward with live implementation. There are many modeling tools for WSN networks available today, some of which are unique to wireless sensor networks and others that are applicable to both wireless and wired networks.

Comprehensive Guide to Heterogeneous Networks
https://doi.org/10.1016/B978-0-323-90527-5.00008-3

1.1 Background and overview of sensor network technology

WSNs are self-configuring and self-contained [3] wireless networks that detect physical or characteristic conditions such as temperature, sound, vibration, squeezing factor, growth, or pollutants and send data to a control zone or sink where it can be displayed and inspected. A sink or base station binds consumers and the organization. By merging demands and meeting outcomes from the sink, the association will obtain the much-needed information. A large number of sensor hubs are typical in a WSN. Radio signals can be used to communicate between the sensor hubs. A remote sensor center is fitted with radio handsets and power supplies, as well as the ability to identify and prepare items. Individual WSN hubs have limited resources, such as a slow planning rate, a high growth breaking point, and limited communication bandwidth. After the sensor hubs are sent, they are responsible for self-masterminding an efficient association scheme with them using multibounce correspondence. At that point, the sensors that had been built had begun to collect data of interest. Requests from a "control point" to execute unambiguous rules or conduct recognition tests are treated similarly by remote sensor systems. The operating strategy of the sensor hubs can be either constant or event-driven. The global positioning system (GPS) and neighborhood arranging estimations can be used to obtain area and arranging information. Remote sensor systems can be fitted with actuators, allowing them to "act" in response to specific conditions. Wireless sensor is a term used to identify these organizations on occasion. WSNs engage new applications and necessitate nonstandard ideal models for show design due to a few roadblocks. Due to the need for low device unconventionality as well as low energy consumption, a fair arrangement between correspondence and sign/data planning capacities should be found (i.e., a long association lifetime). There has been a massive amount of effort put into exploration, standardization, and mechanical endeavors in this field over the last decade [4]. At this time, much of the WSN research has been focused on creating energy-efficient and computationally efficient estimations and displays, with the application domain restricted to basic data orchestrated checking and uncovering applications (Fig. 1).

WSNs have recently been used in a variety of fields owing to the fact that sensor nodes can be transported without a foundation and the network can track a large number of unsafe or remote locations that individuals cannot access. Simultaneously, coverage, routing, and localization have been the subject of a large number of recent WSN research studies [5]. Minimal effort microsensor nodes with wireless capability, low force usage for communicating information, asset requirements, and battery energy limit are all characteristics of WSNs. Since sensor nodes have limited energy, it is important to present energy-saving techniques in order to extend the lifetime of WSNs. Many experts have recently conducted numerous studies in order to find feasible routing protocols for WSNs, with the result that a strategy with order and grouping is promising in improving the flexibility and extending the lifetime of WSNs. The LEACH [6] protocol (low-energy adaptive clustering hierarchy of importance) is an old-style convention.

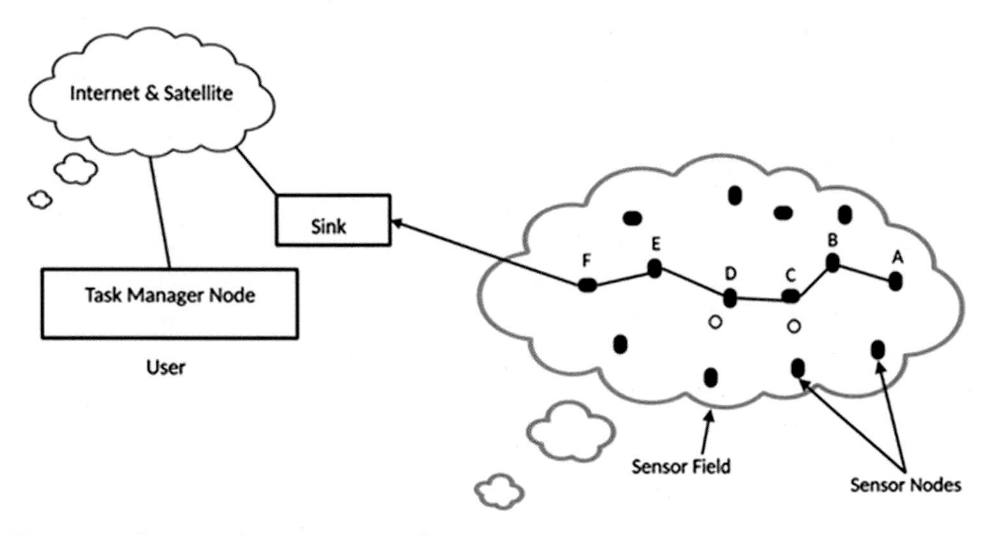

Fig. 1 An ordinary wireless sensor network.

1.2 Evolution of sensor networks

The widespread sending of extremely high-density WSNs is being propelled by recent advancements in semiconductor, framework organization, and materials science developments. Together, these developments have ushered in a new age of WSNs, one that varies dramatically from the remote organizations produced and sent just 5–10 years ago. The most recent cutting-edge WSNs are more serious, have lower affiliation and backing costs, and last longer. They are making their way into a number of environments, including our homes, workplaces, and the past, adding new outlets of knowledge, influence, and solace to our personal and professional lives. Table 1 lists a few examples of how WSNs can boost profits [7], lower costs, and even save lives.

Table 1 Benefits of WSNs in lives.

Application	Advantages
On farms, small ecosystems are evaluated	The crop yield per square kilometer is growing
Traffic detection on road frameworks	To prevent traffic delays, crashes, and improvement areas, guide traffic away from them; emergency response teams are ready
Recognize people existence in households and workplaces	Reduce the amount of energy lost in HVAC and lighting
Water/electrical metering/exhaust cloud	Improve value dispersion frameworks and lessen shortcomings

Later on, each WSN application will fuse a grounded, normalized blend of software/hardware arrangements. However, for the time being, wireless system and organization designers must sort out and comprehend the perplexing compromises among a variety of application variables, including setup costs, equipment and programming, framework dependability, security, and execution. Remotely inserted fashioners should consider these trade-offs and make a range of decisions, including transducer and battery invention, wireless operation recurrence, yield capacity, and device administration conventions. Energy storage, wireless systems administration norms, semiconductor design, and sensor/actuator creativity are just a few of the strengths needed to keep compromises in WSN arrangements at bay. The multifaceted design of WSN configuration addresses some of the most important obstacles to WSN adoption, while also offering an opportunity for equipment and programming innovation providers to add value. Governments and universities gradually began to use WSNs in applications such as air quality monitoring, forest fire tracking, catastrophic event prevention, climate stations, and primary checking. By then as designing understudies progressed into the corporate universe of advancement monsters of the day, for instance, IBM and Bell Labs, they began propelling the use of WSNs in weighty mechanical applications, for instance, power scattering, wastewater treatment, and specific handling plant computerization. Although WSNs' market revenue was high, growing beyond these restricted applications proved to be a challenge. Prior to a few years, military, science/advanced, and deep mechanical jobs were entirely focused on bulky, costly sensors and prohibitive frameworks organization shows. These WSNs placed a premium on usefulness and execution, while other considerations such as equipment and sending prices, arranging regulations, power usage, and flexibility fell by the wayside. The combination of high cost and low volume hampered the widespread gathering and transmission of WSNs for a wider range of applications. The development for large-volume modern-day and consumer applications did not exist in the 20th century, both academia and industry acknowledged the potential for such partnerships and outlined collaborative ventures to resolve the planning challenges. These academic/ mechanical exercises can be found in the following places:

- Sensors for UCLA's Wireless Optimized Network (1993)
- Pico Radio at the University of California, Berkeley (1999)
- The MIT software Adaptive Multi-zone Power Conscious Sensors (2000)
- NASA's Sensor Networks (2001)
- ZigBee Alliance (ZigBee) (2002)
- Embedded Network Sensing Hub (2002)

1.3 Convergence of heterogeneous and multitask wireless sensor networks

Unlike conventional homogeneous single-task wireless sensor networks, HWSNs permit participation among numerous different devices committed to addressing distinctive sign preparing undertakings. Notwithstanding the possible conflicts between the devices, the objective is just to allow every device to address its errand with better exhibition looked at than the situation where it would work all alone. In any case, the plan of such HWSNs is extremely testing and requires the plan of adaptable calculations that boost the exhibition of the devices without sending crude sensor signals in an unconstrained pattern. In the direction, to achieve this objective, new methods are required on signal level and on the organization correspondence stage. We have given an outline for utilizations in the area of HWSNs with a unique spotlight on the sign handling viewpoints. Also, we give an overall outline of the current calculations for circulated node explicit assessment. At long last, we talk about the principal challenges that must be handled for the plan of HWSNs. A few models are cell phones, sans hands communication packs, tablets, workstations, amplifiers, handheld cameras or much more cutting-edge devices, for example, head-mounted presentations. Logically, any of the devices can work individually for playing a specific task for signal processing of the SDST (single device for a single task) system or else performing multiple tasks for the SDMT (single device for multiple tasks) system [8]. Then again, the spatial assortment of the sensor signals developed by diverse gadgets could be used for achieving unparalleled execution. Regardless of the huge size of the data, joining together these signs needs a massive correspondence information move limit and figuring power that is consistently distant. To avoid the necessity of a central head node, all gadgets need a placement strategy to pass the collected information efficiently to the cluster head. In any case, regular WSNs typically expect a homogeneous setting wherein all the gadgets, furthermore known as a hub, are of a comparative sort and work together to address a lone association wide sign taking care of undertaking (MDST). Propelled by a variety of gadgets in emerging IoT-based networks a number of plans like SDST, SDMT, or MDST are used. But they also face a big range of limitations [8]. This general structure is suggestive of MDMT frameworks. These WSNs are formed by different gadgets that help out each other despite the way that their sensor signals rise up out of different models in view of seeing exceptional ways. Also, the gadgets of these WSNs can confuse the geographical assortment of sensors by assisting everyone despite the way that they are designed to deal with the handling variant tasks. Consequently, the utilization of each gadget in an HWSN is not stiff to its own task or alone together with a fundamental association-wide endeavor. In light of everything, the utilization of each gadget goes past its own task and premium by assisting various gadgets to handle various signal handling endeavors at the same time and achieve a preferable show as broken down over the circumstance where the devices would work isolated.

1.4 Architectural elements of a sensor network

The fundamental components and configuration focal point of sensor networks are briefly discussed in this section. These components and plan requirements should be set in light of the C1WSN sensor network environment, which is defined by many (if not all) of the following variables: a broad sensor population (e.g., the system and addressing apparatus should accommodate at least 64,000 customer units), major data spikes, insufficient/uncertain data, and high potential node d (C2WSNs have a large number of these equivalent limits, yet not all). Improvements in sensor network identification, communications, and registration are all based on technical advancements (information taking care of calculations, equipment, and programming). As previously mentioned, in order to efficiently manage the limited WSN properties, WSN guiding conventions must be energy aware. Information-driven guiding and in-network management are relevant concepts that are inextricably connected to sensor networks. The start-to-finish steering plans that have been proposed in the writing for versatile impromptu organizations are not fitting WSNs; information-driven advances are required that act in-network collection of information to yield an energy-productive spread. A sensor network is made out of countless sensor nodes that are thickly sent. To list only a couple of settings, sensor nodes might be conveyed in an open space; on a war zone before, or past, adversary lines; in the inside of a modern apparatus; at the lower part of a waterway; in an organically and additionally synthetically defiled field; in a business working; in a home; or inside or on a human body. A sensor hub customarily has embedded getting ready limits and introduced accumulating; the hub can have in any event one sensor working in the acoustic, seismic, radio (radar), infrared, optical, appealing, and manufactured or common spaces. The hub has correspondence interfaces, customarily far-off associations, to abutting spaces. The sensor node additionally regularly has area and situating information that is procured through a worldwide situating framework (GPS) or neighborhood situating calculation. (Note that GPS-based components may here and there be excessively exorbitant as well as the hardware might be excessively massive.) Sensor nodes are dispersed in an extraordinary space called a sensor field. Every one of the dispersed sensor nodes commonly has the ability to gather information, investigate them, and course them to a (assigned) sink point. Fig. 2 shows the normal WSN game plan, although in numerous conditions all WNs are assumed to have similar functionality.

1.5 The need for efficient multisensor task allocation

Multisensor frameworks can assume a significant part in checking undertakings and distinguishing targets. In any case, the ongoing designation of heterogeneous sensors to dynamic targets/errands that are obscure earlier in their areas and needs is a test. Not at all like conventional similar single-task WSNs, HWSNs permit participation among numerous different gadgets committed via addressing diverse sign preparing assignments. In spite of their heterogenous nature and the way that every device may tackle an alternate assignment, the devices could in any case profit by a coordinated effort between them to accomplish an unrivaled exhibition. In any case, the plan of such heterogeneous

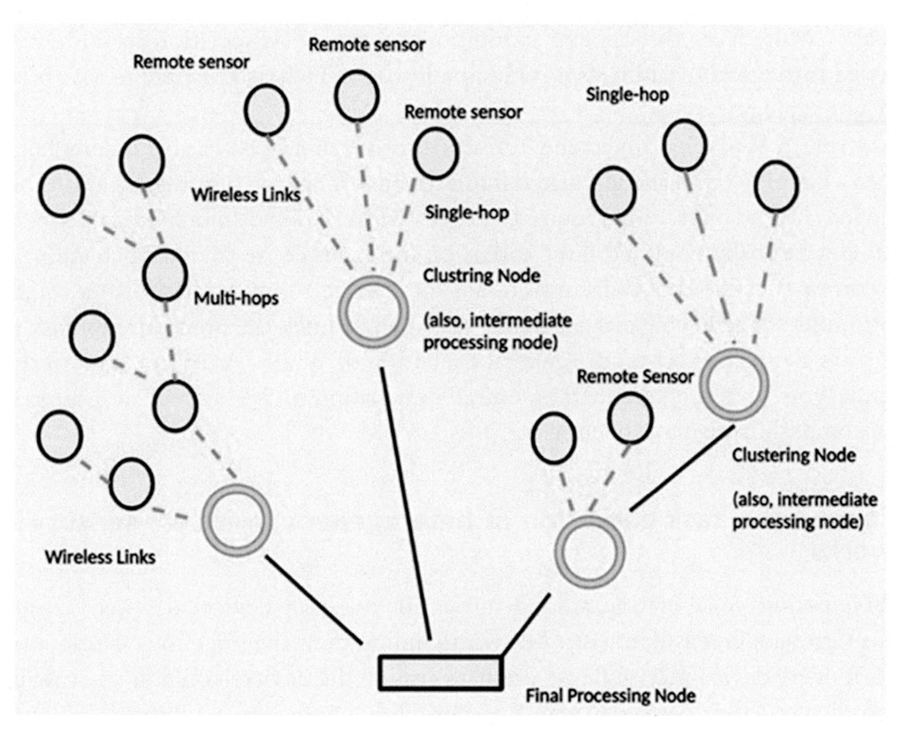

Fig. 2 Sensor network arrangement.

WSNs is extremely testing and requires the plan of versatile calculations that boost the exhibition of the gadgets without sending their crude sensor signals in an unconstrained design.

In a wireless sensor organization, sensor nodes are carefully energy and limit compelled, which makes it fundamental for them to cooperatively execute a perplexing errand. Along these lines, the task portion turns into a principal and urgent issue in wireless sensor organizations. Most past examinations created unified techniques to take care of this issue. Likewise, a typical supposition will be that all the sensor nodes are homogeneous, which is negative in numerous genuine applications. In this chapter, a disseminated task distribution technique which can deal with the issue in a heterogeneous remote sensor network is discussed. The undertaking is engendered from nodes to nodes and every node coordinates its own ability with the necessary limits until all the requested limits of the assignment are gotten. Adding to this, an efficient task allotment technique dependent on self-association is created. By using past allocating data, the nodes with legitimate limits will be chosen as up-and-comer nodes; at that point, the ways to these nodes will be enhanced. In this manner, another showing up undertaking can be assigned straightforwardly and rapidly. Reenactment results show the practicality of the proposed approach. Moreover, the general exhibition of the self-association-based system is approved through an examination with an advancement-based technique that brought the major strategy.

As a rule, a WSN is planned and conveyed for playing out a hard and complex mission. As a basically equal and circulated framework, a WSN can disintegrate a convoluted errand into little and low asset required subtasks. In any case, to an individual node, it just has an exceptionally confined limit and asset, for example, restricted calculation limit, tight correspondence data transmission, and little extra room, which further square its opportunities for achieving errands freely. Along these lines, the procedure for task allotment plays a crucial effect on the general execution of WSNs. As of late, a lot of studies and endeavors have been directed on errand designation in WSNs, and various productive accomplishments have been made.

2. Cooperative task allocation in heterogeneous wireless sensor networks

HWSNs permit collaboration among numerous different devices devoted to settling diverse sign handling assignments. Notwithstanding their heterogenous nature and the way that every device may settle an alternate errand, the devices could in any case profit by coordinated effort between them to accomplish a prevalent exhibition. In any case, the plan of such HWSNs is extremely testing and requires the plan of adaptable calculations that boosts the exhibition of the devices without sending crude sensor signals in an unconstrained pattern. In the direction to achieve this objective, new methods are required for handling signal level and on the organization correspondence stage. We have given an outline for utilizations in the area of HWSNs with a unique spotlight on the sign handling viewpoints. Also, we give an overall outline of the current calculations for conveyed node explicit assessment. At long last, we talk about the fundamental difficulties that must be handled in the planning of HWSNs.

2.1 Multitasking in WSNs

Multitasking in WSNs [9] poses a slew of concerns. What are the responsibilities? What are the activities, and where do they take place? What are device tasks, and how are they connected to application tasks? What is the timetable for the activities, and how are they communicated? Finally, how can these activities be put into effect and used across the WSN? All of these problems occur at the network level, assuming the WSN acts as a single framework. However, the existing software development strategies concentrate on individual device programming and seldom view the WSN as an interconnected framework that consolidates the capabilities of all network members into a single computing platform. Multitasking on WSNs is defined in the following sections.

Block of construction: It is a mission. Tasks are the work units that the WSN conducts. These tasks operate through the entire WSN, not just on a single network unit. This system design approach differs from previous WSN architectures in that, rather than fixing software logic for a single program or network communication, the system monitors the execution of tasks, each of which has its own computation and communication logic.

Task management: At both concept and runtime. A task is a series of physically distributed and loosely connected parallel computations. A task consists of computation and communication logic, which are combined and performed together because a WSN task cannot be completed without communication among multiple devices. A task is spread through a layered stack architecture in software, with each layer performing a specific function, such as performing local computation, establishing communication, or controlling low-level hardware output. Every task has its own stack at runtime, with its own computation and communication logic that may vary from tasks that run concurrently or at different times. From the standpoint of program implementation, Fig. 3 depicts the execution model of a WSN. The framework is constructed as a network protocol stack, with each layer containing several implementations of the same features but with different logic addressing different goals. A task runs one instance of a service implementation from each layer of the stack at runtime, selecting computation and communication resources [10] that meet the task's requirements. Any computer connected to the network runs an identical copy of the stack at the same

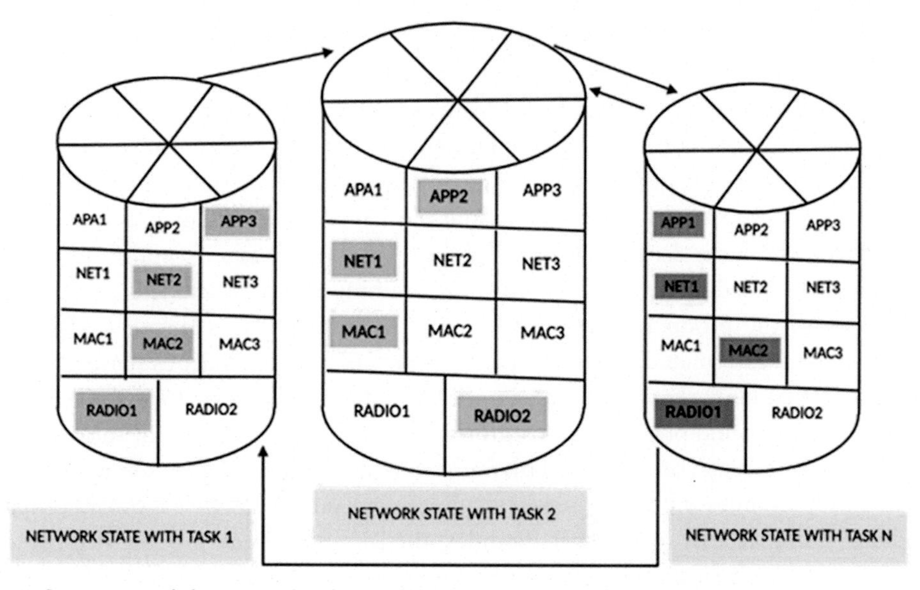

Fig. 3 Context switch between the three tasks running on the WSN at runtime.

time. The arrows between the network protocol stacks in the diagram show the WSN context transition between the tasks.

Any device's computation and communication logic are modified when the WSN changes a mission. Tasks are divided into two categories: program and framework. Tasks are capable of performing both program and system-level tasks. A stand-alone role can be included in an application. Data collection, for example, may be done as a single operation, with computation logic sampling sensors and communication logic moving sensor samples to a destination, such as a device or a network gateway. An application can also be made up of several tasks that run simultaneously or sequentially. A task can also perform work at the machine level. A job, for example, will keep track of how much energy is left on low-power devices. As a consequence, a task will perform maintenance. A device role may also provide resources and abstractions [11] that make application creation simpler.

2.2 Multiple sensors for data collection

Sensor information is the yield of a device that distinguishes and reacts to some kind of contribution from the actual climate. The yield might be utilized to give data or contribution to another framework or to manage a cycle. Sensors can be utilized to distinguish pretty much any actual component that has been planned. Furthermore, they start to be utilized for information procurement because of the different large information investigation or information mining. For instance, the action meter is incredibly helpful for our medical services with late well-being cognizance [12]. A mobile phone has a sensor unit, such as an accelerometer, a temperature sensor, and a pressing factor sensor, for another case. In this way, by utilizing these sensors, it is conceivable to give information assortment and scientific help, for instance, climate at the area, well-being, resting condition, and natural estimation, utilizing a cloud framework like AWS (Amazon Web Services). Be that as it may, the greater part of them are frequently just accessible for explicit sensors or frameworks. Also, they are frequently shut frameworks. For instance, a regularly utilized movement meter cannot gather information autonomously except if unique programming is utilized. Likewise, devoted programming is pricey and it is exceptionally hard to work the product because of complex determination. Hence, to effortlessly obtain different information in different circumstances later on, various sensor information securing framework, which can be understood as a typical stage for sensor information dissemination and processing, is required. Fig. 4 shows the design of various sensor information obtaining frameworks. This framework comprises some sensor modules, different sensor modules and simple sensor modules, for example, as well as microcomputer modules. Using at least two sensor modules, the separate sensor module is capable of gauging certain physical and compound conditions. The basic sensor module tests a few conditions near to the microcomputer module, such as temperature, stickiness,

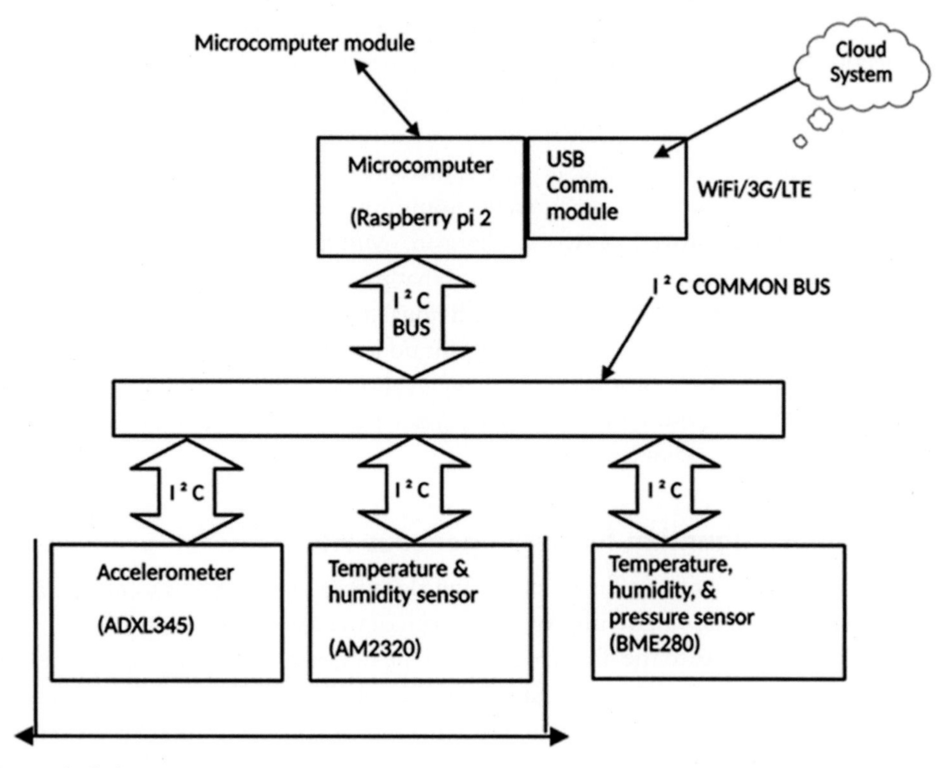

Fig. 4 Block diagram.

and pressing factor [13], using the BME280 temperature, moisture, and pressing factor sensor. The microcomputer module's capabilities involve monitoring sensors, preparing signal information (for example, information inspection and pressure), and submitting it to a cloud system.

3. Protocol and algorithms of heterogeneous wireless sensor networks

When designing a heterogeneous WSN, we must consider its energy supply limitations as well as the communication capabilities of its nodes. The nodes are normally powered by batteries. This is the reason why researchers focus on energy saving and efficient power consumption to boost the life span of the sensor nodes. In heterogeneous WSNs, it is a big challenge to design efficient algorithms because of the diversities of the harsh environment in which these nodes are placed. Several protocols and algorithms have been designed with the purpose of saving energy consumption in real-time applications. Stable election protocol (SEP) [14] is an efficient protocol that increases the stable period of the network. This protocol defines the different conditions for the nodes to be selected as a

CH. The network's average energy can be appended with the nodes' energy to dominate the selection process. The number of nodes in a cluster also plays an important role in CH selection. TEAR [15] also takes into account the repercussion of network traffic and the diversity of the sensor nodes. These nodes can sense their environment and send the data in real time to their CH or the sink; however, this data can also be transmitted on demand for particular applications like emergency monitoring, resulting in minimized energy consumption and improved network life span. Multiple aggregator multiple chain (MAMC) [16] is another clustering algorithm that partitions the total network area into 10 divisions where there is a chain within individual division deployed as PEGASIS [17]. One of the nodes is designated as the aggregator node. All the other nodes transmit their collected data to the aggregator node. This aggregator node then transmits this collected data to the sink. The division of the network area encourages even energy load distribution throughout the network.

3.1 Clustering algorithm for heterogeneous WSNs

The primary objective of an efficient clustering algorithm for HWSN is to minimize the energy consumption leading to improved life span of the network. Energy consumption is directly proportionate to the distance between the nodes in a WSN, and the placement of the selected CH also impacts the energy consumption as all the nodes transmit their collected data to the CH. The entire network is divided into zones and CHs are selected (preferably the one located in the central area) for each zone.

For transmitting a data packet of size m bits over a distance d, we can calculate the transmission energy as

$$\mathrm{ETx}(\mathrm{m}, \mathrm{d}) = \mathrm{ETx} - \mathrm{Elec}(\mathrm{m}) + \mathrm{ETx} - \mathrm{Amp}(\mathrm{m}, \mathrm{d}) \tag{1}$$

$$\mathrm{ETx}(\mathrm{m}) = \mathrm{m} \times \mathrm{Eelec} \tag{2}$$

The residual energy can be calculated as

$$\mathrm{Eres} = \mathrm{Ein} - (\mathrm{ETx}(\mathrm{m}, \mathrm{d}) + \mathrm{ETx}(\mathrm{m})) \tag{3}$$

where Ein is the node's initial energy, Eelec is the electronic energy, and ETx-Amp is the amplifier energy.

The algorithm proceeds as follows:

1. Begin.
2. If the node is placed at the center.
3. if the node has max initial energy, then the node is designated as the CH of the zone.
   Else, wait to receive the CH information.
4. Else, wait to receive the CH information.
5. The selected CH will broadcast to all nodes about its new designation.
6. All other nodes, upon receiving the information, will start transmitting their collected data to the CH.

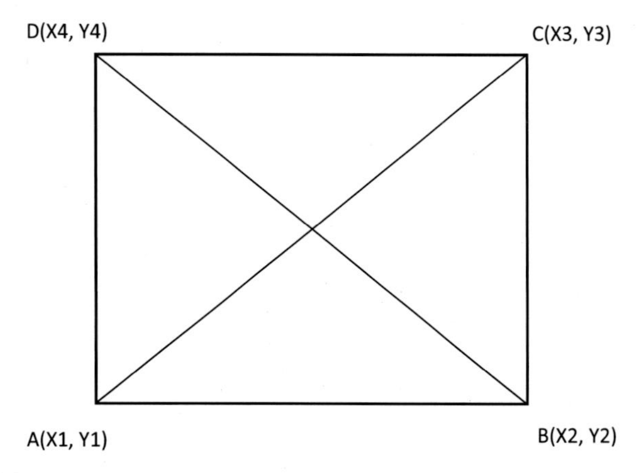

Fig. 5 Square model.

The distance can be calculated by finding out the intersection point using the square model where the diagonals represent the distance between opposite sides (Fig. 5).

The sides can be obtained using the following equations:

$$X = \sqrt{(X3 - X1)^2 + (Y3 - Y1)^2} \tag{4}$$

$$\text{Diagonal length} = 2\sqrt{(X)} \tag{5}$$

Mid-point coordinates

$$(X) = \frac{X1 + X2}{2} \cdot \frac{Y1 + Y2}{2} \tag{6}$$

The CH is selected and other nodes are clustered according to the algorithm. The CH processes its own data, merges its own information with the received data from other nodes, and then finally transmits the collected data to the sink through multiple hops of other CHs. The DEEC [18] algorithm observe the initial and residual energy of each nodes for the process of CH selection, hence, the nodes with either higher initial energy or higher residual energy have higher chances to be selected as CHs. In DEEC, the placement of the CH need not be necessarily at the center of the partition area, therefore may consume more energy, and there are chances that the CH might die soon. To improve the CH lifetime, this algorithm presents an effective way to improve the CH threshold function, $O(v_i)$ with the following equation:

$$O(v_i) = \begin{cases} \dfrac{Kp_i}{1 - p_i\left(r \bmod \dfrac{1}{p_i}\right)}, & v_i \in G \\ O, & v_i \notin G \end{cases} \tag{7}$$

where G is the nodes not serving as the CH, p_i is the probability of the common node vi becoming the CH, K is the edge degree ($0<K<1$), and R is the current round.

The probability of any node becoming the CH, i.e., p_i, can be calculated as

$$p_i = \frac{1}{1+k\lambda} \times \frac{E_i(r)}{\overline{E}(r)} \times p_{opt} \tag{8}$$

where P_{opt} is the ratio of set no. of cluster to total nodes, λ is the ratio of heterogeneous nodes to total nodes, and k is the ratio of the total initial energy of heterogeneous nodes to common nodes.

3.2 Comparative analysis of clustering algorithms

In the following section, we discuss the various clustering algorithms for heterogeneous WSNs with the objective of minimizing the energy consumption and improving the life of the network. LEACH, DEEC, and SEP are algorithms without energy supply, and PHC, EBCS, and HNS are a few algorithms with energy supply.

3.2.1 *LEACH algorithm*

The low-energy adaptive clustering hierarchy (LEACH) algorithm takes on the idea of rounds, each with two stages, namely, initialization and stabilization. In the former stage, every node initiates a number between 0 and 1 to see whether it can be chosen as a CH or not, and when this number is greater than the threshold value, the node becomes the CH and sends a message to all other nodes; otherwise, it waits for the broadcast message. The threshold value can be drawn by

$$T(n) = \begin{cases} \dfrac{p}{1-p\left(r \bmod \dfrac{1}{p}\right)} \\ 0 \end{cases} \tag{9}$$

where p is the percentage of CH nodes to total nodes, r is the round number, and n is the designated node belonging to a group of nodes that have not been chosen as CHs for the past $1/p$ rounds.

In the later stage, the nodes transmit their data to the CH, and the CH transmits the combined information to the BS. In this algorithm, each node has an equal probability to be selected as the CH, and thus the energy utilization is equally distributed among all the nodes over the network. In every round, the cluster is restructured leading to huge energy cost. The CHs that are far from the BS need more energy resulting in disturbing the energy distribution over the network [19].

3.2.2 DEEC algorithm

The distributed energy-efficient clustering (DEEC) algorithm works on multilevel heterogeneous networks where there are differing values of initial energy of the nodes. The probability of any node to be selected as the CH depends on the ratio of residual energy to average energy of the WSN. This ratio also determines the distribution of energy consumption and the life span of the nodes in the cluster.

DEEC gives the average probability, p_i, as

$$p_i = \begin{cases} \dfrac{p_{opt} E_i(r)}{(1+am)\overline{E}(r)} & \text{where the node is a normal node} \\ \dfrac{p_{opt}(1+a)E_i(r)}{(1+am)\overline{E}(r)} & \text{where the node is an advanced node} \end{cases} \quad (10)$$

Now, if we substitute Eq. (10) for p in Eq. (9), we can obtain the threshold value and select the CH. Here $E_i(r)$ represents the residual energy, $\overline{E}(r)$ represents the average energy at round r, p_{opt} is the reference value, and m is the ratio of advanced nodes that owns a times more energy as compared to the normal nodes in a cluster. The threshold is given by

$$T(n) = \begin{cases} \dfrac{p}{1 - p\left(r \; mod \; \dfrac{1}{p}\right)} \\ 0 \end{cases} \quad (11)$$

3.2.3 SEP algorithm

SEP (stable election protocol) is a clustering algorithm which works with networks with varying energy. There are two types of nodes in this algorithm, namely, advanced and ordinary. The initial energy decides the probability of each node to be selected as the Cluster Head. The number of advanced nodes is always greater than the number of ordinary nodes and their energy is also greater than that of the ordinary nodes.

Now, in each cycle, the node will choose a number between 0 and 1 and if this chosen number is less than the threshold value, the specified node will be chosen as the CH. The threshold value is specified as

$$T(n) = \begin{cases} \dfrac{p_{elec}}{1 - p_{elec} \times mod\left(r, \dfrac{1}{p_{elec}}\right)}, n \in G \\ 0, n \notin G \end{cases} \quad (12)$$

The possibility of being selected as the CH is higher for advanced nodes than normal nodes.

3.2.4 PHC algorithm

The power-harvesting clustering (PHC) is an efficient energy optimization algorithm for WSNs. It performs the CH selection considering the energy supply and is like LEACH in transmission of information among non-CH nodes and CH nodes. It prefers the high-energy nodes for CH selection so that they do not die soon and thus improves the network life. The probability of any node to be chosen as the CH increases with the node's energy level in this algorithm. The threshold for selecting the CH is given by

$$T(n) = \begin{cases} \dfrac{P}{1 - P\left[r \bmod \dfrac{1}{P}\right]} \bullet \dfrac{E_{res}(n) + E_{hrv}(n, r)}{E_{init}}; n \epsilon G \\ 0; \text{otherwise} \end{cases} \tag{13}$$

Here, P is the percentage of CH to the total nodes, $E_{res}(n)$ is the remaining energy, E_{init} is the initial energy, and E_{hrv} is the harvesting energy of the nth node during the rth round. This algorithm takes into account the change of energy of the nodes. Each node sends its information to the sink directly. The threshold value is dependent on the initial, residual, and harvesting energy levels of each node for the CH selection process.

3.2.5 EBCS algorithm

The energy-balanced clustering with self-energization (EBCS) algorithm is another clustering algorithm with the objective of uniform energy distribution over the network. It takes into account the supply given to each node throughout their lifetime for selecting the cluster head out of them. The information is shared among all the nodes about their energy supplies, thus involving all the nodes in the process of CH selection in numerous rounds. In this algorithm, the current energy level of any node does not impact its probability of being selected as a CH node. EBCS adopts the residual energy, $\widehat{E}_{rst}(r)$, as the main criteria for CH selection. $\widehat{E}_{rst}(r)$ in the rth round is derived as

$$\widehat{E}_{rst} = E_{rst}(r) + \widehat{E}_{hrv}(r) - \widehat{E}_{CH}(r) \tag{14}$$

Here, $E_{rst}(r)$ is the residual energy of the head, $\widehat{E}_{hrv}(r)$ is the harvesting energy in the rth round, $\widehat{E}_{CH}(r)$ is the consuming power in the rth round.

The threshold in EBCS is calculated as follows:

$$T(i) = \begin{cases} P\left[\dfrac{\widehat{E}_{rst}(i) + \sum\limits_{r=n-1}^{r=1} \widehat{E}_{hrv}(i, r-1)}{E_{init}} + n\text{div}\dfrac{1}{P}\left(1 - \dfrac{\widehat{E}_{rst}(i) + \sum\limits_{r=n-1}^{r=1} \widehat{E}_{hrv}(i, r-1)}{E_{init}}\right)\right] \\ 0; \text{otherwise} \end{cases} \tag{15}$$

3.2.6 HNS algorithm

The HNS (heterogeneous network includes self-supplying nodes) employs vibration energy from improving the energy balance in a WSN. A unique threshold is associated with each node within a cluster that discriminates advanced nodes with ordinary nodes, and the CH selection process depends on the residual energy and current energy supply of each individual node. In this algorithm, the advanced nodes have a higher probability to be selected as the CH against the ordinary nodes [20].

HNS calculates the average probability of any node to be selected as the CH as

$$p_i = \frac{p_{opt} \cdot N \cdot (1 + \alpha_i)}{N + \sum_{i=1}^{N} \alpha_i} \cdot \frac{E_i(r)}{E_{ave(r)}} \quad (16)$$

Here, p_{opt} is the optimal to normal node ratio in the network, $E_i(r)$ is the residual energy, and $E_{ave}(r)$ represents the avg. energy of all the nodes.

Now, these algorithms can be compared based on important parameters like their contribution in the lifetimes of the networks, network stability cycle, network energy consumption balance and the residual energy of the heterogeneous networks in which they are applied. For efficient comparison, these algorithms have to be applied under the same environment.

When the network lifetime is taken into account, it is inversely proportional to the number of rounds the algorithm is run. More than half the nodes die as the algorithm reaches 2000 rounds in case of LEACH and SEP, and the number of active nodes in PHC, HNS, and DEEC is even less than 10% after 1000 rounds. But the death of the first node is encountered after the algorithm crosses its 1000th round in case of LEACH and SEP, and thus these algorithms enhance the network lifetime [21].

When the stability cycle of the network is considered, DEEC has the longest stable cycle out of these algorithms and LEACH has the shortest cycle. The stability cycle is defined as the complete time starting from the network initialization up to the time when the death of the first node is encountered. The stability cycle is an important parameter to consider while selecting an optimum algorithm and predicting the life span of the network during the design phase. This can be shown in Fig. 6 where the algorithms are compared against the number of rounds.

The next important parameter is distributing energy consumption over the network. The ratio of CH to normal nodes plays a vital role in this energy balance. DEEC has been proved to be the best out of these algorithms as the no. of CH nodes increases with the number of rounds of the algorithm. The drawback is that a few nodes may observe a heavy load in this algorithm and may die soon.

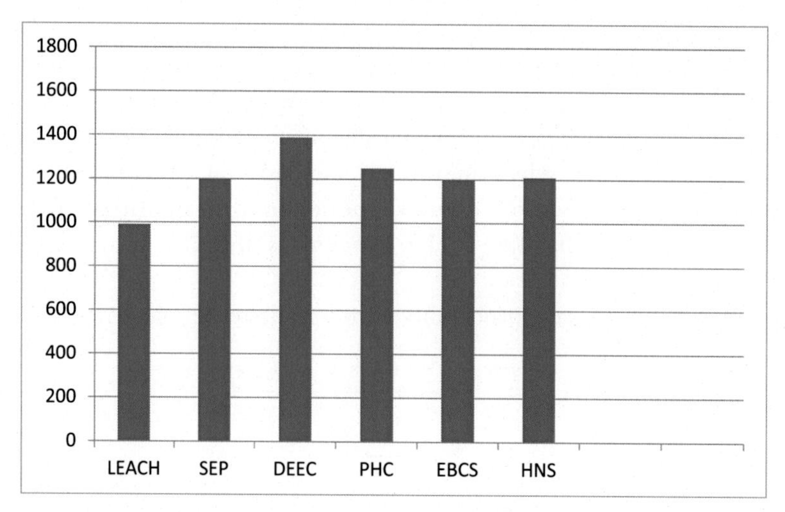

Fig. 6 Comparison of algorithms against the number of rounds.

The analysis of residual energy under different algorithms depicts that in LEACH, SEP, and DEEC the residual energy level keeps on decreasing with the increasing number of rounds. Thus, the continuous energy supply to the nodes can improve the network lifetime. Throughput in WSNs is the performance of transmission of data from nodes to their CH and from CH to the BS. This is almost the same for all the above algorithms initially. But after a certain no. of rounds, it varies for each algorithm. DEEC maintains the highest throughput among all. After DEEC, HNS, SEP, and PHC are relatively equally efficient. LEACH has the minimum throughput out of these six algorithms. This is a result of the varying stability cycles of every algorithm that we have discussed earlier.

4. Wireless sensor networks formation

For real-time applications like military surveillance, weather forecasting and other industrial applications of heterogeneous wireless sensor networks are used. In each implementation, the design objectives remain the same, i.e., improve the lifetime and manage energy distribution over the networks. There are two types of topologies, namely, central and distributed. In central formation, the control lies centrally in one device, whereas in distributed formation nodes control the flow of information through message passing (Figs. 7–9).

The centralized network formation is further classified as follows:

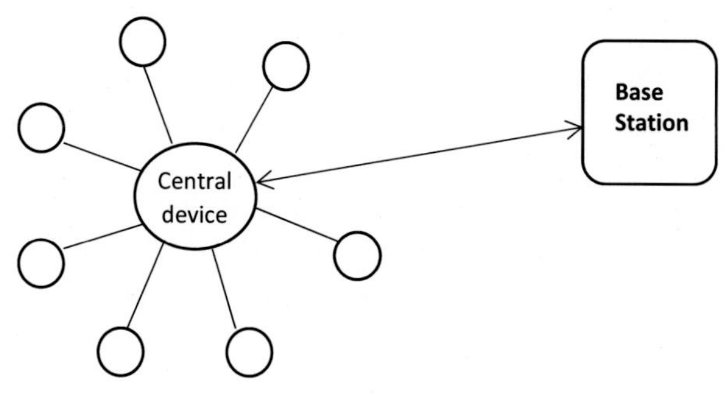

Fig. 7 Centralized formation.

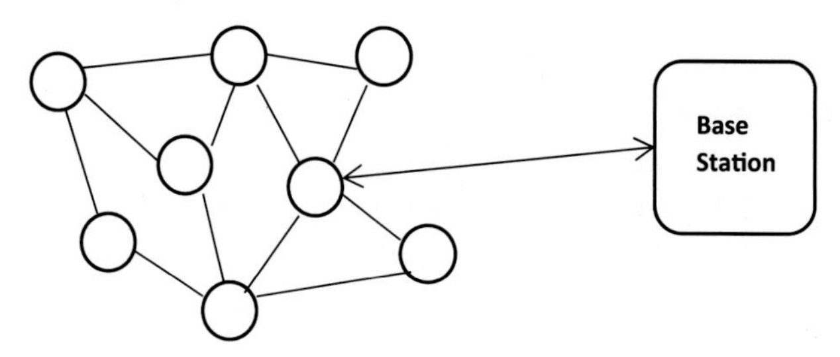

Fig. 8 Distributed formation.

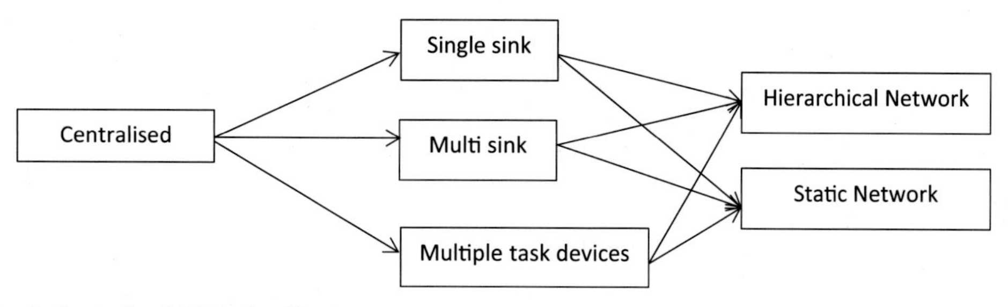

Fig. 9 Centralized WSN classification.

4.1 Centralized wireless sensor networks

In centralized formation, the network communication is controlled by one device. This centralized device is used to supervise the node localization and routing. Star topology is the most common topology applied for centralized formations of WSNs. Centralized WSNs are further classified as

i. Single Sink
ii. Multisink
iii. Multiple Task Devices

In single sink, one designated sink is functioning and the objective of this topology is to forward all the information to the designated sink. In multisink, multiple sinks are placed within the network that reduce the traffic congestion, improve the lifetime, and cover larger areas effectively. In the third category, there are multiple additional devices installed with the network to manage the diversity of the heterogeneous WSNs. All these additional devices are assigned unique jobs. Some of them are assigned the job of routing, a few are responsible for evaluating the energy level of each node, and a few are responsible for checking the status of alive nodes in the network. Further, these three classifications can be extended as

i. Hierarchical Networks
ii. Static Networks

4.1.1 Hierarchical networks

The control of the network proceeds in a hierarchical manner using the 802.15.4 [22] protocol. These networks deploy the multihop packet forwarding strategy. This formation assures high mobility among the connected sinks following a self-configuration approach to identify the suitable sink and share the collected information with the base station. The sink will broadcast messages to all other nodes. An improved performance is observed when there is minimal or no interference, but the broadcasting creates unnecessary traffic in the network. This also affects the energy consumption at each node and impacts the stability cycle of the network. The sensory nodes are free to move within the network; therefore, the routing algorithm has to address the network failures, the energy level of each node, the formation of tree through broadcast messages and transmit the collected information using TDMA [23]. Tree is constructed depending on the time and energy required for reconfiguration against the expected quality of service (Fig. 10).

4.1.2 Static networks

Static networks implement a multisink optimal solution [24] that guarantees energy balance in the network with enhanced data sharing and data processing speed. In these networks, sink is placed at the center of the cluster following the particle swarm optimization [25]. The network can be represented as a graph with defined sink locations. This formation aims at identifying the suitable location for each sink that is at an optimum distance from every other CH in the network. The reduced distance results in reduced energy consumption leading to the improved lifetime of the network. The location of sinks is defined using PSO and the *k*-mean algorithm, but this algorithm keeps on identifying the location of all nodes continuously, and the time involved in this process increases with the increasing number of sinks in the cluster. The topology is reconfigured whenever a failure is encountered. Once the clusters are configured, their CH is designated and they do not change frequently. They are only changed whenever there are issues in energy distribution.

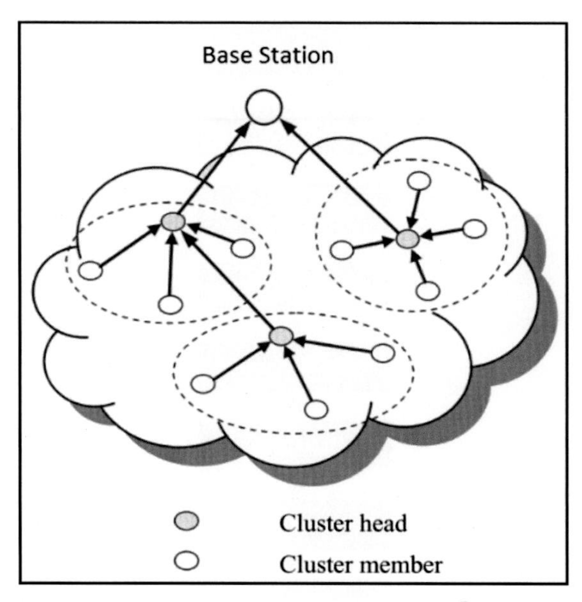

Fig. 10 Hierarchical network.

5. Open research challenges and future directions

Normally, the homogeneous network cannot handle heterogeneous data transmission and it is highly complex to satisfy the service quality requirements. Heterogeneous clustering protocols manage cluster nodes and their members by balancing the energy consumption in the complete network. Balancing the energy consumption of sensor nodes that are mobile in nature is highly challenging in HWSN. Robust damage resistance with heterogeneous data is the strongest property of the tree topological structure.

Some of the reasons for incorporating the heterogeneity feature in WSN are: improves scalability, security; reduces energy consumption, performance upgradation, cost-effectiveness, and maintains consistent network functionality. It also supports several broadband applications [26] and complex routing protocols. With big data, the massive amount of data that is generated in the heterogeneous WSN is managed. In addition, it also takes care of high-speed transmission with heterogeneous data that are eventually hard to gather, store, and process. The data generated by different nodes may not be significant and produces a huge volume of data. Managing and securing such data is extremely challenging in HWSN.

HWSN is more vulnerable to various kinds of threats or attacks and risks because of its intrinsic features such as transmission medium, remote deployment, and security breach [27]. If the sensor node is compromised, it leads to injection of fake messages, eavesdropping, data integrity breach, and network resource consumption. The resources in HWSN are limited, which includes

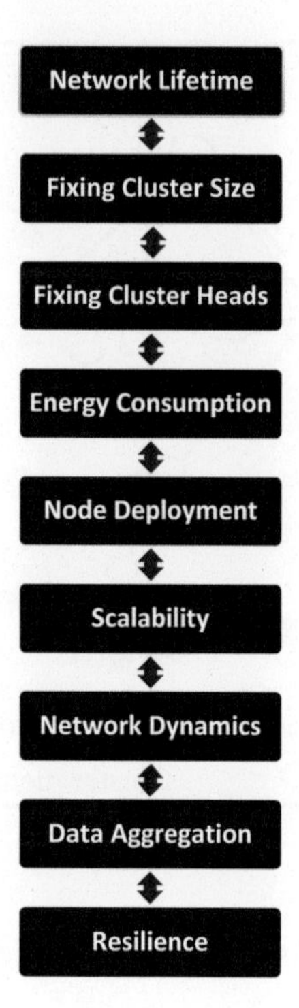

Fig. 11 Research challenges in HWSN.

Battery power: If the sensors are deployed once, it is very hard to recharge or replace.

Processing capacity: Less or limited capacity is available for processing cryptographic functions.

Storage space: Due to the limited capacity of memory, use a smaller cryptographic key size and algorithms.

In HWSN, high-end sensors have more energy, computational power, and memory and low-end sensors have limited resources which can be resolved easily using heterogeneity features [28]. Some of the research challenges and issues that need to be focused on in the future are described below and are shown in Fig. 11:

Network lifetime:

The sensors in HWSN are completely dependent on batteries and they are resource constraint based on the energy source, which maximizes the network lifetime.

Cluster size:

The load can be assigned to the cluster head evenly by the cluster members. In case the cluster head is overloaded, it leads to a lot of energy consumption and node failure needs to be addressed and resolved [29].

Fixing the no. of cluster heads:

Cluster heads are chosen using a systematic approach; however, concluding the fixed number of cluster heads guarantees the minimum energy indulgence. Moreover, it reduces the power consumption and the network lifetime is increased.

Geographical location:

Maintaining consistent information of geographical location and residual energy optimizes the clustering processes by significantly enhancing the energy utilization.

Energy conservation in the idle state:

The data is collected from sensor nodes when they are active. Thus, sensor nodes can be designed with power-off enabled facility in the concern of minimizing energy consumption. The sensor nodes can switch over periodically from active to inactive mode in order to conserve energy during the idle state.

Node deployment:

The deployment of nodes is completely dependent on the application and it has an impact on the routing protocol performance [30]. Moreover, the deployment can be either self-organizing or deterministic. In the self-organizing approach, the sensor nodes are located randomly and the ad hoc method is followed for creating a route. In the deterministic approach, the sensor nodes are located manually and the data is transmitted via predetermined paths. However, it faces a problem in optimal clustering since there is no uniform distribution of nodes.

Network dynamics:

Routing protocols in the sensor network should support both cluster heads and sink nodes. Forwarding messages among mobile nodes is highly challenging due to factors such as bandwidth, energy, and route stability. Depending on the application, the event handling may be either static or dynamic. For static events, take action based on triggers and for dynamic events, it is handled periodically.

Energy consumption:

The process of establishing routes for data transmission consumes more energy. Normally, the transmission power is proportional to twice of its distance or even more in the presence of impediments. Rather, it functions well in direct routing if all the nodes are nearer to the sink node, whereas excess memory is consumed.

Scalability:

If the performance of the system is enhanced when the nodes are getting increased, it is said to be scalable. Routing strategies must be appropriate to handle the huge number of sensor nodes in a wireless sensor network.

Resilience:

Because of battery power or any other environmental issues, sometimes sensors stop functioning unexpectedly. It could be resolved by choosing an alternative path. The robust algorithms and protocols are required to handle such kinds of situations, which in turn lead to fault tolerance.

Data aggregation:

It is the process of gathering data retrieved from various heterogeneous source nodes. Built-in functions are used for computation; it consumes less amount of energy when compared to the energy used for communication. Hence, traffic optimization and energy efficiency are achieved using this technique.

Data delivery models:

There are three kinds of data delivery models followed in HWSN; they are continuous, query-/event-driven, and hybrid. Data is transmitted periodically in the continuous data-driven model. In the query-/event-driven model, data is transmitted if a query or event is triggered. The hybrid model combines the functionality of both data-driven models. It ensures route stability and less energy consumption.

Node capability:

Load balancing and data aggregation are handled by the powerful nodes in HWSN. Deployment of sensor nodes creates several technical issues in terms of data routing. Moreover, some applications require different sensors to monitor temperature, humidity, and pressure for deploying in the surrounding environment. Such kinds of sensors can be deployed which will function independently or the sensors can be embedded in other sensors based on the requirement. The sensors deliver different quality of service (QoS), data rates, where the various data delivery models are exploited. Thus, the heterogeneous nodes in WSN make the routing process highly challenging.

6. Comparison of various algorithms

If 100 nodes are randomly deployed in an area of $100 \times 100\,\text{m}^2$ area, and we compare the different algorithms in MATLAB simulation software in real time, we get the following outcomes.

Parameters	LEACH	SEP	PHC	HNS	DEEC
Network lifetime (No. of live nodes after 1500 rounds)	Less than 20%	Less than 40%	Less than 70%	Less than 80%	More than 90%
Stability cycle (Approx. no. of rounds at which first death of node is encountered)	950	1000	1100	1200	1350
Energy consumption (No. of CHs after 15 rounds)	11	15	10	7	13
Residual energy (After 500 rounds)	Decreases	Decrease	Almost constant	Almost constant	Decreases
Scalability (As compared to each other)	Least	Average	Better than SEP	Approximately equal to PHC	Highest

7. Conclusions

The utilization of heterogeneous nodes results in successful data transmission and provides adequate service quality in HWSN. Hence, it obtains a good accuracy rate at the time of data transmission. The data transmission in a heterogeneous network includes several kinds of data such as audio, video, image, and 2-D data. It requires an extremely well-organized topological structure in HWSN for handling a large size of data packets. HWSN is more popular nowadays due to its predominant features in a huge number of applications such as health monitoring, region surveillance, environmental conditions, etc. The most significant challenge in HWSN is improving the network lifetime; energy utilization is also a major factor that directly impacts the network lifetime. Moreover, energy utilization can be optimized by clustering the nodes exploiting efficient clustering algorithms. Various state-of-the-art techniques for implementing robust clustering strategies are discussed. The open research challenges and future directions are highlighted that would be beneficial for incorporating different clustering algorithms in HWSN.

References

[1] A. Nayyar, R. Singh, A comprehensive review of simulation tools for wireless sensor networks (WSNs), J. Wirel. Netw. Commun. 5 (1) (2015) 19–47.

[2] S. Singh, S. Kumar, A. Nayyar, F. Al-Turjman, L. Mostarda, Proficient QoS-based target coverage problem in wireless sensor networks, IEEE Access 8 (2020) 74315–74325.

[3] C. Buratti, A. Conti, D. Dardari, R. Verdone, An overview on wireless sensor networks technology and evolution, Sensors 9 (9) (2009) 6869–6896.

[4] M.A. Matin, M.M. Islam, Overview of wireless sensor network, in: Wireless Sensor Networks-Technology and Protocols, Intech Publisher, 2012, pp. 1–3.

[5] A.J. Al-Mousawi, Evolutionary intelligence in wireless sensor network: routing, clustering, localization and coverage, Wirel. Netw (2019) 1–27.

[6] T. Jamal, S.A. Butt, Low-energy adaptive clustering hierarchy (LEACH) enhancement for military security operations, J. Basic Appl. Sci. Res. (2017) 2090–4304. ISSN.

[7] A. Benefit, The Evolution of Wireless Sensor Networks, vol. 1, Silicon Lab. Inc., Austin, TX, USA, 2013, pp. 1–5. White Paper Rev.

[8] J. Plata-Chaves, A. Bertrand, M. Moonen, S. Theodoridis, A.M. Zoubir, Heterogeneous and multitask wireless sensor networks—algorithms, applications, and challenges, IEEE J. Sel. Top. Signal. Process. 11 (3) (2017) 450–465.

[9] M.K. Szczodrak, Multitasking on Wireless Sensor Networks, (Doctoral dissertation, Columbia University, 2015.

[10] K. Sohraby, D. Minoli, T. Znati, Wireless Sensor Networks: Technology, Protocols, and Applications, John Wiley & Sons, 2007.

[11] A. Boukerche, X. Cheng, J. Linus, Energy-aware data-centric routing in microsensor networks, in: Proceedings of the 6th ACM International Workshop on Modeling Analysis and Simulation of Wireless and Mobile Systems, 2003, pp. 42–49.

[12] T. Harada, M. Yokoyama, S. Cho, A. Tanaka, M. Yasuda, Multiple Sensor Data Acquisition System using Commonly Available Sensor Devices for Sleep and Car Conditions, 2017.

[13] Z. Yan, L. Sun, T. Krajnik, Y. Ruichek, EU long-term dataset with multiple sensors for autonomous driving, 2019. arXiv preprint arXiv:1909.03330.

[14] G. Smaragdakis, I. Matta, A. Bestavros, SEP: A Stable Election Protocol for Clustered Heterogeneous Wireless Sensor Networks, Boston University Computer Science Department, 2004.

[15] D. Sharma, A.P. Bhondekar, Traffic and energy aware routing for heterogeneous wireless sensor networks, IEEE Commun. Lett. 22 (8) (2018) 1608–1611.
[16] P. Harichandan, A. Jaiswal, S. Kumar, Multiple aggregator multiple chain routing protocol for heterogeneous wireless sensor networks, in: 2013 International Conference on Signal Processing and Communication (ICSC), IEEE, 2013, pp. 127–131.
[17] R.K. Yadav, A. Singh, Comparative study of PEGASIS based protocols in wireless sensor netwroks, in: 2016 1st India International Conference on Information Processing (IICIP), IEEE, 2016, pp. 1–5.
[18] W.R. Heinzelman, A. Chandrakasan, H. Balakrishnan, Energy-efficient communication protocol for wireless microsensor networks, in: Proceedings of the 33rd Annual Hawaii International Conference on System Sciences, IEEE, 2000, p. 10.
[19] L. Qing, Q. Zhu, M. Wang, Design of a distributed energy-efficient clustering algorithm for heterogeneous wireless sensor networks, Comput. Commun. 29 (12) (2006) 2230–2237.
[20] X.P. Fan, X. Yang, S.Q. Liu, Z. Qu, Clustering routing algorithm for wireless sensor networks with power harvesting, Comput. Eng. 34 (11) (2008) 120–122.
[21] X.L. Xu, Q. Lv, W.L. Wang, X.J. Huangfu, Clustering routing algorithm for heterogeneous wireless sensor networks with self-supplying nodes, Comput. Sci. 44 (2017) 134–139.
[22] IEEE Standard Association, IEEE STD. 802.15.4: Wireless Personal Area Networks (PANs), 2005.
[23] M. Singh, M. Sethi, N. Lal, S. Poonia, A tree-based routing protocol for mobile sensor networks (MSNs), Int. J. Comput. Sci. Eng. 2 (1S) (2010) 55–60.
[24] D.R. Dandekar, P.R. Deshmukh, Energy balancing multiple sink optimal deployment in multi-hop wireless sensor networks, in: 2013 3rd IEEE International Advance Computing Conference (IACC), IEEE, 2013, pp. 408–412.
[25] G. Venter, J. Sobieszczanski-Sobieski, Particle swarm optimization, AIAA J. 41 (8) (2003) 1583–1589.
[26] J.M. Corchado, J. Bajo, D.I. Tapia, A. Abraham, Using heterogeneous wireless sensor networks in a telemonitoring system for healthcare, IEEE Trans. Inf. Technol. Biomed. 14 (2) (2010) 234–240.
[27] T. Poongodi, M. Karthikeyan, Localized secure routing architecture against cooperative black hole attack in mobile ad hoc networks, Wirel. Pers. Commun. 90 (2) (2016) 1039–1050. 2016.
[28] D. Kumar, T.C. Aseri, R.B. Patel, EEHC: energy efficient heterogeneous clustered scheme for wireless sensor networks, Comput. Commun. 32 (4) (2009) 662–667.
[29] S. Singh, S. Chand, R. Kumar, A. Malik, B. Kumar, NEECP: novel energy efficient clustering protocol for prolonging lifetime of WSNs, IET Wirel. Sens. Syst. 6 (2016) 151–157.
[30] T. Poongodi, A. Rathee, R. Indrakumari, P. Suresh, Principles of Internet of Things (IoT) Ecosystems: Insight Paradigm, IoT Sensing capabilities: Sensor deployment & Node discovery, Wearable sensors, Wireless Body Area Network (WBAN), Data acquisition, Springer series, 2020, pp. 127–151. ISBN: 9783030335953.

CHAPTER 3

Network selection in a heterogeneous wireless environment based on path prediction and user mobility

Fayssal Bendaoud
LabRI-SBA Lab, Ecole supérieure en Informatique, Sidi Bel Abbes, Algeria

1. Introduction

The convergence and simultaneity of numerous wireless access networks, such as the well-known WiFi or IEEE 802.11, and cellular networks such as UMTS, HSPA, LTE, and 5G, characterize the next generation of wireless networks [1]. The goal is to realize the always best connected (ABC) concept [2] by allowing users to easily connect to diverse networks without impacting their QoS, assuming that they have multimode devices. Radio access networks were primarily homogeneous prior to the emergence of 3G networks. Individuals are now outfitted with cell phones looking for the ABC vision on a constant basis, thanks to the enormous rise of Internet apps and services, the development of network technology, and the mobile users' industry. The diversity of RATs creates a heterogeneous wireless environment, which is at the heart of next-generation networks. As illustrated in Fig. 1, a heterogeneous wireless program enables mobile users to select a RAT out of a set of possibilities, depending on a multitude of factors. Network selection technique is the name for this procedure, and it is the subject of this chapter [3,4]. A mobile user in an area with overlapping RAT coverage can choose which RAT to utilize in a heterogeneous wireless scenario. In theory, the mobile terminal should recognize this and dynamically and smoothly choose and attach to the best possible network according to its current requirements. In order to deliver an effective autonomous network selection decision in this multiuser, multitechnological, and multiprovider context, new methodologies and standards must be developed. To make this judgment, the majority of past solutions in this context have relied on multicriteria decision-making algorithms. Game theory, fuzzy logic, and artificial intelligence can all be utilized to investigate the interaction of competitive and cooperative behavior among service providers and users in this domain.

Network selection is the operation that permits us to pick the best RAT from the accessible ones [5]. Because of the different models involved, such as the cost of employing such a RAT, QoS, and the amount of energy utilized, the network selection

Comprehensive Guide to Heterogeneous Networks
https://doi.org/10.1016/B978-0-323-90527-5.00002-2

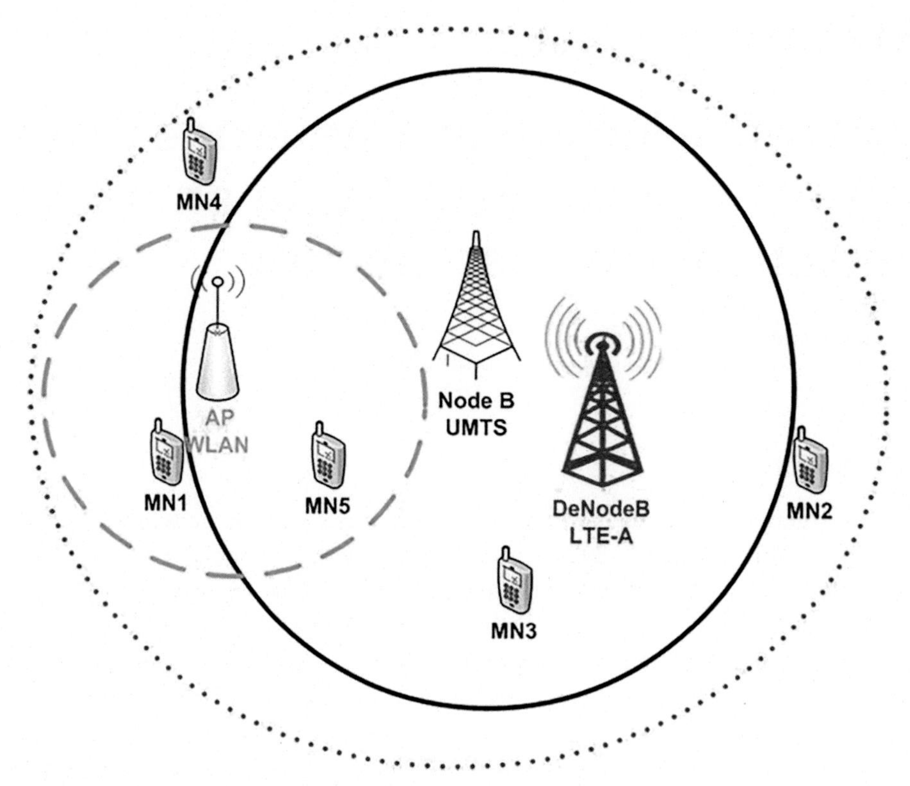

Fig. 1 A heterogeneous wireless environment.

procedure is usually mind-boggling and difficult to figure out. To expand on this concept, we can state that the network selection problem can be defined as the dynamic and automatic selection of the best RAT while taking into account various constraints. The network selection mechanism in older cell technologies was based on the received signal strength, and the mobile device was typically routed to the preeminent-available cell tower. Users may prefer to attach to a less charged RAT situated at a larger distance with a lower RSS instead of a more highly congested cell with a stronger RSS; therefore, such a choice strategy is clearly not appropriate for heterogeneous networks. Because of the numerous criteria and parameters involved, network selection in a heterogeneous environment can be viewed as a multiple attribute decision-making MADM problem [6]. Other methods for dealing with the network selection problem include game theory, fuzzy logic [7], and utility functions (Fig. 2). Supplementary techniques linked to multi-criteria optimization are also used to manage the network selection issues [3,5], including neural networks, artificial intelligence, and genetic algorithms; however, the last two will not be covered in this chapter due to their rarity in this case study.

The goal of this chapter is to give a thorough examination of the current research on various methods for network selection solutions. The most important contributions are as follows:

- An overview of the network selection decision problem is provided, as well as a discussion of supporting standards and some distinct approaches in the literature, which are classified according to the mathematical technique used for multiple criterion decision-making.
- A new paradigm for network selection is proposed, with a focus on user path prediction and mobility; the findings are assessed, discussed, and compared to previous work.

The fundamental steps of this work are as follows:

- To begin, we suggest a network selection mechanism in which people are as portable as possible. The idea is to foresee the paths that the customer will take with precise probabilities.
- The operator then performs network selection activities, taking into account the zones that make up the predicted routes.
- Following the network selection activity, the next stage entails locating the optimum RAT for each expected path's configuration.
- Finally, the operator selects the RAT configuration with the highest mean value in each direction. This list of RAT is probably the suited solution for meeting customers' QoS requirements throughout the entire call session.

The rest of the chapter is organized as follows: in Section 2, we present the network selection formulation with some mathematical descriptions of the well-known methods used with this aim. Section 3 presents our framework for network selection; all steps of the proposal as presented and explained. In Section 4, an evaluation of the performance of our proposal is provided; the results are presented and discussed with a comparison with other works. Section 5 concludes the chapter and presents some future works.

2. Network selection formulation

In the forthcoming generation of networks, wireless access that is heterogeneous is an attractive feature where clients are smugly adaptable to choose the most appropriate network, as indicated by their necessities. In these situations, the determination of the best RAT has a significant duty for the smooth working of the entire framework; in fact, the network selection cycle comprises switching between RATs to provide the optimal network for the user [3,8]. When a multiinterface user detects the presence of multiple RATs in a nearby neighborhood, he ought to be capable of choosing the optimal RAT based on throughput, jitter, and packet loss rate. The following activities make up the network selection task:

- The monitoring process that includes determining which RAT is available, gathering network radio circumstances, and other features. A portion of the criteria is assessed and others are determined at this stage.

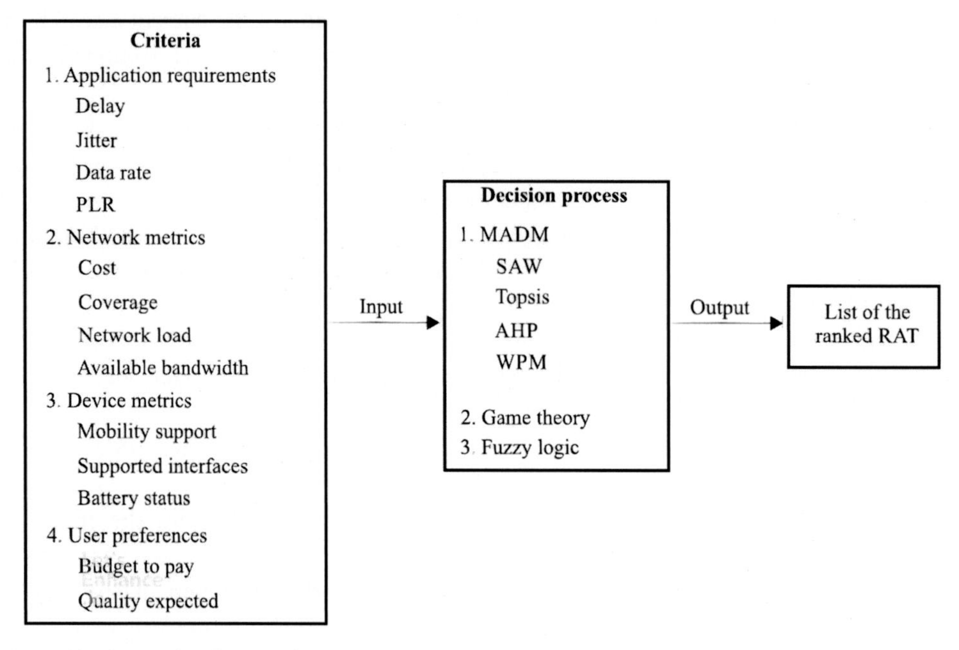

Fig. 2 The network selection diagram [3].

- The network selection task begins with the decision phase; the best RAT is chosen based on the monitoring step and the many parameters offered by the mobile user, such as client willingness. At this stage, the choosing method is used to rank the RATs.
- The completing stage assigns users to the best-positioned RAT.

When an application's demand is established, such as video streaming, VoIP call, or a file download, as well as when the RSS falls under a fixed threshold and the user's radio link deteriorates, particularly when the user is moving, the network selection procedure [5] begins. Furthermore, the network selection strategy relates to the nature of the application requested; for VoIP, latency and packet loss rate are vital factors; for video, data rate and latency are important factors; and for best-effort applications, just data rate is important.

2.1 MADM: Multiple attribute decision-making methods

MADM is a mathematical method for problem-solving that focuses on individual decisions. It deals with challenges using numerous choice-related models and is widely used for a variety of topics, such as the economics [9]. This characteristics of this method are grouped into three categories:

- Alternatives: A group of entertainers that will be placed in the network selection circumstance; the choices are a list of RATs.

- Criteria for the network selection condition: They include data rate, jitter, packet loss, and latency, and the set of attributes addresses the rules employed in the dynamic cycle.
- Weights: The meaning of provided models or behaviors is dependent on the activity of choice.

The framework is represented as a grid, with columns representing criteria and lines representing networks. There have been a number of MADM strategies proposed [10–13], but we will use the most well-known, SAW, AHP, TOPSIS, and WPM.

2.1.1 Simple additive weighting (SAW)

Multiple-criteria optimization methods include simple additive weighting [3,6,14]. The inputs are first normalized in SAW, and then the option with the highest/lowest (depending on the system) value is chosen. SAW is mathematically described as follows:

$$S_{SAW} = \sum_{i=}^{n} p_i{}^{*} s_{ij}$$

S_{SAW} represents the value of each option, p_i represents the weight value for the criteria i, and s_{ij} represents the normalized value for each criterion.

2.1.2 Technique to order preference by similarity to ideal solution (TOPSIS)

TOPSIS is a compensating accumulating approach based on the assumption that the arrangement should have the least geometric distance from the positive ideal solution and the greatest geometric distance from the negative ideal solution [3,6,15–17]. All parameters' standardized inputs are weighted, and the mathematical distance between each alternative and the best alternative is calculated. The following is how the TOPSIS calculation is completed:

$$S_{TOPSIS} = \frac{A_P}{A_P + A_N}$$

where

$$A_P = \sqrt{\sum_{i=0}^{N} \left(p_j^{2*} \left(s_{ij} - b_j^P \right)^2 \right)}$$

$$A_N = \sqrt{\sum_{i=0}^{N} \left(p_j^{2*} \left(s_{ij} - a_j^N \right)^2 \right)}$$

The variable p_j represents the weight value and s_{ij} is the normalized value of the parameter j and the network i. The variable b^P represents the best alternative, a^N is the worst alternative, and N is the number of the decision criteria.

2.1.3 Weighted product model WPM (multiplicative exponential weighting (MEW))

Similarly, the WPM technique is analogous to the SAW technique [3,6], with the exception that the add process in SAW is replaced by the product operation in MEW. Every choice in WPM is assimilated with the others by performing several ratio multiplications, one for each criterion. Every ratio is multiplied by the weight of the criterion it is related to. The numerical depiction is as the following:

$$s_{WPM} = \prod_{i=1}^{N} \left(p_j{}^* s_{ij}\right)$$

WSM is an alternative to WPM. WSM is similar to WPM, the key distinction being that the approach uses a sum rather than a product. WSM is based on the assumption that the total value of all options is equal to the total value of all goods. WSM is simply applied to problems with the same unit ranges across criteria; however, when the unit ranges differ, such as when qualitative and quantitative attributes are used, the problem becomes difficult to handle, and normalization procedures should be used. Because of WSM's straightforward nature, it is common to combine it with other methodologies, such as AHP. As a result, for n criteria and m choices, the best solution is found using this mathematical model:

$$U_{WSM}^* = max \sum_{i=}^{n} w_i{}^* r_{ij}$$

U_{WSM}^* is the weighted sum score.

2.1.4 Analytic hierarchy process (AHP)

The decomposition of a complex issue into several hierarchical simple subproblems is considered by the AHP. [3,6,18–20] The AHP steps are as follows:

- First, break down the problem into a hierarchy of subproblems, with the top node representing the eventual aim and the lower nodes representing the problem's solutions.
- Second, Determine the importance of the criterion in relation to the objective in each level by comparing the choice factors pair-wise based on their levels of influence on the scale shown.
- Third, calculate the weights at all levels of the hierarchy.
- Four, the AHP approach is combined with the GRA method of gray relational analysis, with the AHP being used for weighting and the GRA method being used for ranking (ee Table 1).

The problem of network selection has been addressed in a number of articles in the literature. These studies focus on improving users' network selection decisions in order to support a large number of services with the best QoS in a timely way. The following is a list of works in the network selection that one should know about.

Table 1 AHP level of importance.

Definition	Importance
Same importance	1
Fair importance	3
Strong importance	5
Intense importance	7
excessive importance	9
in-between importance	2,4,6,8

A. Sgora et al. used SAW to generate a list of grouped networks in Ref. [9], whereas the authors in Ref. [21] used game theory and SAW approaches. SAW's main advantages are its simplicity and ease of use. It has two major drawbacks: one criterion may be surpassed by another, and the rank reversal addresses the MADM method as a whole. A similar report was discussed in Ref. [9], in which the authors used SAW and TOPSIS to examine the performance of the VHO. The TOPSIS technique, according to the authors, is superior to the SAW strategy. As a rule, pay approaches such as TOPSIS avoid the problem of one criterion being surpassed by another; this is accomplished by the use of various criteria concessions. This means that a poor value in one criterion is neglected in favor of an acceptable value in another. This notion provides greater faith in the entire NS process as compared to noncompensation systems that use thresholds and do not allow any compensation. K. Savitha et al. proposed a TOPSIS-based approach that used the entropy formula to calculate the weight vector in Ref. [21]. They claim that the mathematical results show that the suggested method can select the best available networks in a diverse environment based on user preferences and service requirements. In terms of the vertical handover, a comparison was made between SAW and WPM [22]. To conclude that WPM is superior to SAW, Q.T. Nguyen Vuong et al. used relative standard deviation as a correlation measure. D.L. Olson used the MANP approach to deal with heterogeneous systems in Ref. [23]. They believe that combining MANP with TOPSIS creates a more powerful unique dynamic technique that can avoid penalizing poor-quality attributes more effectively. B. Bakmaz et al. used the analytic hierarchy process to classify the importance of the different standards used and analyzed the materiality of various Internet promoting networks in Ref. [24]. The proposed model provides a level-headed and effective way for sponsors to employ when selecting Internet promoting networks. The original analytic hierarchy process was compared by M. Lahby et al. in Ref. [25] to a modified version called the fuzzy analytic hierarchy process. The author employs a fuzzy corresponding grid and a fuzzy consistent grid to modify the classic AHP's consistency requirement. The numerical findings demonstrate that the fuzzy analytical hierarchy method provided is superior to the standard analytical hierarchy approach.

2.2 Game theory

Game theory allows for the visualization of competitive scenarios that show intelligence among competing actors with mutual or possibly contradicting interests [9,21,22]. It provides a scientific tool for predicting the outcome of intricate connections between plausible but incompatible pieces. Three sets are used to illustrate the fundamentals of a game:

- A group of players that includes the players who are competing for a larger prize.
- A range of actions depending on the information available in the system. Clearly, the players are looking for ways to increase their own earnings.
- The reward that all participants aim for can be defined as the total advantage the user obtains by using the existing system. The game is rehashed till players are no longer capable of additional gains, at which point the game is said to have arrived at the Nash equilibrium (NE).

Nash equilibrium is a key notion in game theory, and we will define it in the next section. Consider the game (S, R), where S represents the set of strategy profiles and R represents the set of reward profiles. When a player i selects the s_i strategy, he or she earns a profit equal to $R(s_i)$. It is vital to note that the pay-off is determined by player is strategy as well as the strategies of all other participants.

If no change in strategy by any single player is profitable, a strategy profile $s^* \in S$ is a Nash equilibrium, that is, if for all i,

$$R(s^*) \geq R\left(s_1^*, s_2^*, \ldots s_{i-1}^*, s_i, s_{i+1}^*, \ldots, s_n^*\right)$$

We talk about strict Nash equilibrium when the inequality is strict. If not, s^* is referred as a weak Nash equilibrium. Some games may or may not have a Nash equilibrium, and some games may have multiple Nash equilibriums.

The Pareto optimality is another key notion. It is a situation where no individual or preference criterion can be made better off without making at least one individual or preference criterion worse off.

Following that, we will look at various works that examine network selection using game theory. E.H. Watanabe et al. demonstrated a contending situation between players for one AP as an evolutionary game in Ref. [26]. Mobile users act as adversaries in order to increase the transmission rate. The transmission rate is the game's strategy, and the payout is determined by the objective function. This capacity computes the mean opinion score MOS, which is an audio call quality assessment, by combining delay and packet loss rate. In Ref. [26], free users are employed, as well as VoIP as the sole application, and the creators claim that the resulting harmony is optimum. The issue of less packed passage guarantee is highlighted in Ref. [27]. The class of techniques includes a variety of AP options, with the result being a trade-off between necessary throughput and the work required to switch to a new AP. The authors demonstrated that this game results in a dispersion of users on the AP. M. Cesana et al. address the issue with only WiFi networks and several APs in Ref. [28]. For the time being, people can only join one AP; the prize function is determined by the AP's blockage level and the cost of using it. The Nash equilibrium is reached, according to the authors. The players are the networks, and the reward sorts out which radio transmitter will deliver the requirement stated by the user, which includes the repartition of service needs among the networks, according to M. S. Z. Khan et al. [29]. D. E. Charilas et al. [30] used a multiple-stage game between two 802.11 networks to investigate the admission control problem. The two networks are the game's players, while the strategies are the users' service requests. The dispersion of service requests between the warring RATs is the game's outcome. Various works, such as in Refs. [31–35], attempted to solve the problem of network selection using game theory.

2.3 Fuzzy logic

The artificial intelligence subsystem is employed in a variety of fields [36]. There are different degrees of content of a condition in fuzzy logic [37–40]. Unlike classical logic, which assumes that all propositions are correct or incorrect, fuzzy logic adds a [0,1] level of credibility to decide from. Fuzzy logic is an extension of classical logic that is based on the concept of fuzzy sets. Fuzzy logic propels the advancement of midway truth, despite the fact that the reality's value can range from completely trustworthy to completely fraudulent. Employing fuzzy logic as an ordering system, not many works have looked at the network selection problem. Fuzzy logic is used in network selection in one of two ways: as a mixture of MADM or as a ranking tool.

F. Bendaoud et al. suggested a standard to deal with the multicriteria network selection problem in Ref. [7]. The multirules network selection goal is achieved in their proposal taking into account the customers' requirements and QoS. The proposal is evolving, and it is well suited for dealing with large numbers of RATs that meet a variety of requirements. Simulations demonstrate that the suggested technique outperforms the reference solution in terms of performance and S. Kher et al. show new fuzzy logic plans in Ref. [41]. Users are involved in these plans to assess the appropriateness of various P2P-based grid network combinations. Cost, limit, and reliability are all examples of regularly utilized attributes. The suggested sorting method uses a natural rule improvement configuration with Boolean logic to catch input combinations. S.B. Cho and J.H. Kim proposed a combination strategy-based fuzzy logic approach for various network plans in Ref. [42]. The most significant benefit is the overall perspective on the value of various networks. The authors demonstrated that the proposal considerably improves ability. The vast bulk of recent fuzzy logic works [43,44] incorporate MADM techniques. Regarding network selection, the usage of fuzzy logic as the center of the ranking technique is not widely accepted. MADM, on the other hand, has long been associated with fuzzy logic.

2.4 Utility functions

Utility is the value we gain from utilizing, possessing, or making an action. It is what gives us the ability to pick between options [45–47]. The utility of a preference function gives properties to the ordering of a collection of options. This is important for inspecting customer sentiments while attempting to maximize customer happiness. When dealing with a variety of choices and restricted funds, we will choose the option that best meets our needs. Utility functions are commonly written as U(y1; y2; y3), which signifies that for monotonic functions, U (function) is an element of the amounts of y1; y2, etc., and if y1 > y2, then U (y1) > U (y2). Utility implies the degree of enjoyment that goods or services provided to the decision-maker if "y1" is favored over "y2" for reaching a conclusion [48]. A utility function is a term that relates to the value copied by a user from goods or services. Various users with distinct client desires will have variable utility qualities for a comparable object. Personal preferences should be considered when measuring utility in this way.

Bari, F. and Leung, V. C. show in their paper [49] that a large number of frequently used MADM techniques, such as SAW, MEW, and TOPSIS, are unsuitable in their model structure due to flat growth or reduction of attribute utilities. They insist on taking into account both linear and non-linear utilities, making them better adapted to attaining their objectives. In Ref. [50], M.A. Senouci et al. presented a user-based RAN selection technique based on increasing economic rent while staying within user-defined transfer completion time constraints. A review of many prospective utility capacities is offered, based on the risk attitudes of various users. They claim that the outcomes of experiences matched the utility depictions of user inputs. The risk-taker spends more but has a shorter wait time. In Ref. [51], Ormond, O. et al. proposed a device-driven RAT selection technique focused on maximizing client excess while conforming to mobile-defined transfer completion time limits. A study of a variety of alternative utility functions based on different risk attitudes is presented. The simulations, they claim, produced results that were consistent with the user utility descriptions. The risk-taker spends more but has a shorter wait time. SUTIL is a network selection instrument for heterogeneous networks that was introduced by L. Pirmez et al. in Ref. [52]. It focuses on the networks that are most critical to the application and use the least amount of energy, allowing for complete and consistent availability regarding cell phones and apps. They also suggest some future study initiatives, such as examining the dealings among several occurrences of SUTIL and the concept of habitats, particularly the unique environment in which SUTIL operates.

3. Mobility-based network selection

This is the most crucial section of the chapter. Indeed, our strategy aims to provide the best QoS to the mobile user during the call. User movement is one of the most prevalent causes of QoS deterioration; previous network selection papers sought to quickly assign users with the given QoS without establishing and maintaining this QoS during the conversation.

The first stage in achieving our aim [53,54] is to estimate the trail with expectations. These are the possibilities that are used to determine the mobile user's geographic itinerary (in terms of cells). The second stage is to apply network selection to the anticipated paths, allowing the RATs to be sorted at each zone of the paths. The third stage includes getting all of the RAT configurations that represent all of the anticipated trajectories (both initial and alternative; more information will be provided later). The final step is to select the configuration that will be applied to all of the paths, and then use the mean value formula (Fig. 3).

Our solution's initial two components are the most crucial: route discovery and network selection on projected trajectories. The RATs that have been ranked are listed at the end of the second phase. In each zone, a configuration is made of a sequence of RATs in the same ranking order. We added stages 3 and 4 to the vertical handover procedure to improve it.

3.1 First step: The prediction of paths

Our goal is to develop real-time predictors that take into account a user's movement map, extort the present place, and forecast the user's upcoming position. When the next place is determined, the user's location/position is actualized, and subsequently the location (which is now the actual place) is added to the history, and so on. The LZ-based family [55,56] and the order-k Markov predictors [57–59] are the two most well-known families of predictors in the literature. In Ref. [57], the authors evaluate the two types of predictors and reach the following assertions:

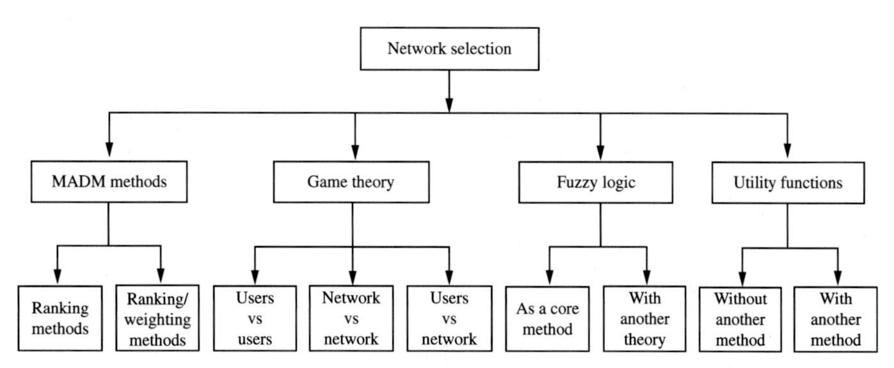

Fig. 3 Summary of the network selection approaches [3].

- The order-1 Markov model and the LZ-based family both perform similarly.
- In terms of performance, the order-2 Markov process outperforms the order-1 Markov process.
- Order-2 Markov predictors beat higher-order Markov forecasters and more sophisticated predictors. For example, O(3) (or higher) Markov predictors do not give upturn over O(2) Markov predictors since their complexity impairs accuracy by causing them to fail to generate accurate predictions.

So, in the next section, we will go over the basics of the Markov process and present a mathematical description with an example to assist in understanding the entire process and how we apply our technique in a real-world scenario. The Markov process is a stochastic process based on the Markov property, which states that all useful information for future prediction is inserted solely in the current position of the map route, with no reliance on prior states. As a result, the system is called memoryless [60], Eq. (1). As proven in Refs. [57, 61], order 2 Markov models make more precise precipitation compared to the order 1 Markov models. In this study, the path calculation is performed via the Markov model order 2.

$$P = P(R\_next/R\_actual) \tag{1}$$

Prediction models are utilized in a wide range of circumstances, including weather forecasting, chess games, and gambling. In this article, this model will be used to forecast the increase of mobile users. The Markov model has been used to predict user pathways in a number of earlier studies [62,63]. In the meanwhile, our approach includes three independent processes in addition to path forecasting to give users an effective and productive network selection, ensuring that customers receive the best possible service during the entire call. The order-k Markov algorithms are based on the assumption that the succession of the k latest zones in the area map can predict the next emplacement [57]. At any given time, a user has a separate zone or is associated with a lone access point. The location is represented by a string that illustrates the historical context of the places where the client worked. The group of all feasible sites is labeled 'A,' and one place from that group is labeled a. Every user has a list of previously visited (antecedent) places, $R_n = R_1 \ldots R_n$. The order of the stochastic Markov process formula is

$$P(R\_(i = a)/R\_1 \ldots R\_(i-1)\,) = P(R\_(i = a)/R\_(i-k) \ldots R\_(i-1)\,) \tag{2}$$
$$\forall a \in A;\, i, k < n$$

The area at instant is represented by R_i. To study the potential following zones online is simple. The forecast specialist is supposed to be updated regularly by the mobile user shift, and the forecast's efficiency is linked to the warning periods. R^m is the position map of the device m; the evaluation calculation of the order-n Markov representation is as follows:

$$P(R\_(i = a)/R\_1 \ldots R\_(i-1)) = \frac{O\left(R^m_{i-n}, \ldots, R^m_{i-1}, a; R^m\right)}{O\left(R^m_{i-n}, \ldots, R^m_{i-1}; R^m\right)} \tag{3}$$

where $O(r; C)$ is the number of times the substring r appears in the string C. The transition matrix is made up of these probabilities. The user's position is logged once every 10s in the position map, which is a series of annotations. The Markov model treats the R area history as a set of undefined theoretical symbols.

To show how the Markov approach works, we present a basic model: We take into account a territory with a long history in the area L="A-D-G-H-E-F-C-B-A-B-C-F-H-G-E-D-A-B-C-F-H-G-D-A-B-C-E-H-G-D-A-D-G-D-E-H-F-C-B-A" (see Fig. 4). Suppose that the user starts from zone "A" at time t; the user was in a random place previously (called the past area). Then we will look at how the order-2 Markov functions, as well as how to read out the probability of a road. With a chance of 0.1, we also expect that the mobile user will be able to stay in the actual zone.

The Markov technique component of the equation $R^m_{i-n}, \ldots, R^m_{i-1}$ is identical to "A" in this sample at the initial step, and the function will attempt to track down the "a" part, which is the following zone, according to formula (3). The set-up in Fig. 5 illustrates the

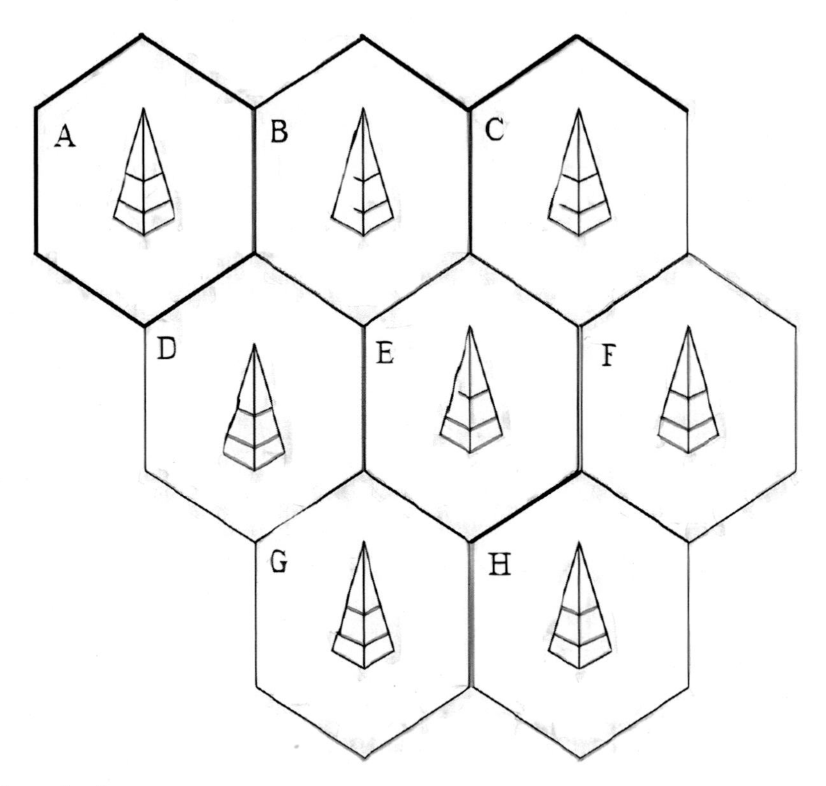

Fig. 4 Example of a zone map.

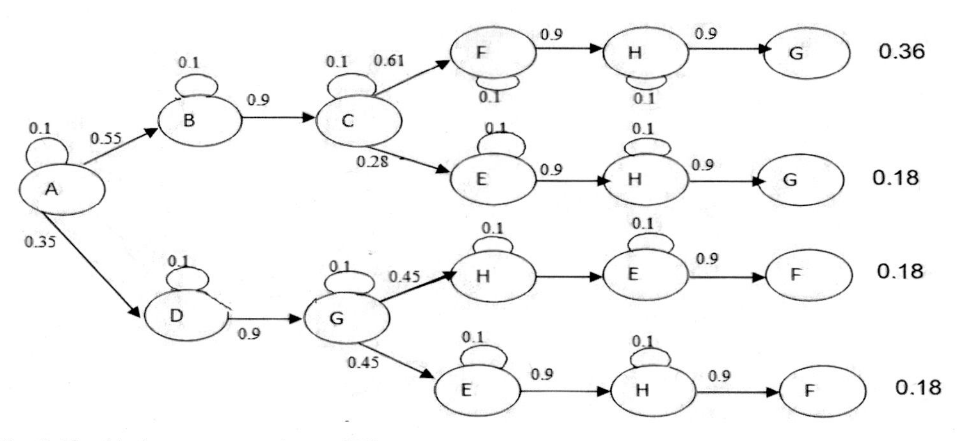

Fig. 5 The Markov process scheme [53].

Markov process shown in Fig. 4. It begins by a regular position and finds the following positions by looking at the zone's map. The transition costs represent the probabilities of moving from one state to the next. We started with a subjective state and computed the shift probabilities using the area history "L" with formula (3). At last, the probability of the way is calculated by multiplying the probabilities of the path's transitions.

3.2 Second step: Network selection

At this point, the network selection stage is introduced, and the implicated parameters of the NS approach are discussed. As a result, we provide a method for classifying the available RATs. The purpose of this project is to keep the customer optimally provided with the highest rank accessible RAT in the situation of frequent motion. In order to reduce the latency and PLR produced by this phenomenon, it is vital to limit the amount of vertical handovers, i.e., lowering the needless vertical hand-offs. A vertical handover is an interaction in which the itinerant user alters the antenna or the access point used to give the aforementioned sorts of assistance while preserving a predetermined QoS. A mobile user in a heterogeneous radio system hears various signals from numerous RATs of varying quality: access medium, connect layer, and so on [64]. As a result, vertical handover selection, when combined with network selection, has become an even more sophisticated and important phase in attaining seamless mobility and implementing ABC's goal. The vertical hand-off process is integrated with the network selection process to get the most out of all available networks; rather, the vertical handover process is combined with the network selection process to get the most out of all available networks. Vertical handover that is not necessary increases signaling overhead and delays. It also forces the mobile user to utilize more energy in order to achieve a good VHO [65]. To minimize these unconstructive effects on QoS, we intend to maintain the frequency of vertical handovers as low as possible.

Our algorithm's backbone is made up of the equations below:

$$S_i = \sum_{j=1}^{m} rank_{ij} \tag{4}$$

$$rank_{ij} = (\alpha - k_{ij}) * p[j] \tag{5}$$

$$k_{ij} = Min\,(Mat_{ij}) \tag{6}$$

where α is a constant number equal to the number of alternatives, k_{ij} represents the ranking order of the network i for the criterion j, Mat_{ij} is the column vector of the input matrix, $p[j]$ is the weight of the alternative i associated to the criteria j.

Algorithm 1 is as follows:

To begin, we divide the input matrix into a column vector with the same number of criteria as the number of criteria. Because we are working with a minimization system, the best situation for a network I is when R I in the order vector is the smallest. So, a lesser order[i] means the highest local rank with $k_{ij}=0$, $S_{ij}=\alpha * p[j]$. The bad scenario is when the network has the highest value for the criterion j, i.e., $k_{ij}=\alpha-1$ and $S_{ij}=p[j]$. Based on the application demands, we use the weighting approach such as MADM in our proposal to highlight the crucial criteria and distinguish them from the less significant criteria. As a result, the scores of the related networks are higher. Some of the advantages of this method are as follows:

- It avoids a network with good values in several criteria from being damaged by a low value in one.
- We may describe application requirements in the system using the weight concept, which is how we differentiate among apps because all of them have distinct needs.

ALGORITHM 1 The modified SAW function [14].

```
Data: The matrix mat [m][n] and the weight vector
          W[], α=m
Result: Income vector for each network
For (j=0; j<m; j++) do
     For (i=0;i<n;i++) do
            tab[i]=mat[i][j]
     End
     tabind=Sort(tab)
     For (k=0; k<tabind.length; k++) do
           Income[tabind[k]]=income[tabind[k]]+(α − k)*w[k]
     End
     return income[]
End
```

Voice over IP (VoIP) necessitates the smallest latency and packet loss for video applications. Furthermore, a high throughput is required for VoIP service. For best-effort applications, there are no special QoS requirements, but the fee criterion is critical. To map this information into numerical calculations, the eigenvector method is utilized.
In steps three and four, vertical hand-off optimization is used. A VHO occurs when the user is not connected to any RAT for a short time; this absence of connectivity results in some missing packets, which is known as packet loss. In addition, packet losses increase the latency of receiving packets. To prevent these problems, we will use these two processes to try to come up with a configuration (a set of geographical zones) that provides good QoS while avoiding a lot of unnecessary vertical handovers. The third and fourth rounds of our technique are used to evaluate C1's performance against that of C2.

3.3 Step three: Selecting the RAT's configuration

In this stage, we will go over configurations in greater depth. A configuration is a collection of RATs, each of which is the best fit for the zone. Assume we have three geographical zones A, B, and C, with three RATs, R1, R2, and R3, available for each of them. On the surface, the optimum RAT arrangement is one in which each RAT is best suited for the zone in question. One of the available set-ups is to utilize the same RAT for all zones (R1 for all or R2 or R3). This eliminates the need for a vertical hand-off for a mobile user; however, this is not always the best option in terms of overall QoS for the mobile user. As a result, the goal of this stage is to collect all of the alternative configurations and show how we might choose one above the others. A network selection operation is performed (step 2) for each path from the list of paths that the user can choose, and it results in an ideal configuration called C1 that has a list of the best-matched RATs on each zone. Because the RATs that make up C1 may introduce a large number of vertical hand-offs, resulting in increased latency and packet loss, we also develop alternative configurations. The alternate arrangements essentially try to limit the number of vertical hand-offs, particularly those that are superfluous. As our system only has three RATs (WiFi, 3G, and LTE), there are only two potential configurations because each RAT can only support three technologies. We must evaluate the three configurations, the ideal and the two alternatives, based on packet loss, latency, and jitter created by the VHOs; the comparison is also dependent on the type of application; for example, with a VoIP service, packet loss is less relevant than delay. These criteria are equally important for the video service.

The vertical hand-off delay is measured as the period between disconnecting from the old RAT and receiving the first packet from the new RAT. According to Ref. [66], communication latency is the time it takes for a packet to travel from the access point to the mobile user. It is the sum of transmission, propagation, and processing delays; these delays are added together to get the end-to-end latency (e2e). In this paper, the hand-off latency is modeled as jitter; in fact, the VHO represents an augmentation of the transmission latency for the mobile user, which is usually referred to as jitter. The goal of this stage is to determine whether to operate with the optimal configuration, which has the best latency and PLR values for each RAT individually but is prone to VHOs. Alternatively, choose an alternative set-up with increased latency and PLR for a RAT but no VHO resulting in zero jitters. Because these metrics (latency, PLR, and jitter) have an impact on the user's QoS, it is necessary to provide the user with a configuration that results in less QoS degradation. In comparison to the optimal arrangement, alternative configurations must lower the number of VHOs. Formulas (7)–(9) are used to select a configuration:

$$C2_i.latency \leq 150 \text{ ms} \tag{7}$$

$$C1_i.latency + VHO.latency > 150 \text{ ms} \tag{8}$$

$$C2_i.latency - C2_{i-1}.latency < |30| \text{ ms} \tag{9}$$

The geographical zone is represented by the letter i in the formulae. The highest values for the delay and jitter acceptable in which mobile services such as video and VoIP can be provided without QoS deterioration are "150 ms" and "20 ms." The aforementioned Eqs. (7)–(9) only involve latency and jitter since, according to various studies [66–68], the PLR produced by VHO is only approximately 2%, which is negligible. This stage concludes by generating only one configuration for all paths available to the user; the other configurations are omitted, and the next step proposes a system for selecting only one configuration for all paths.

3.4 Step four: Choosing an appropriate configuration

The framework comes to a close with this stage, which involves selecting only one configuration to be utilized for all conceivable paths. We will use the following function to accomplish this. Algorithm 2 involves taking into account all places that are likely to be passed by the mobile user and associating them with the RAT that has been chosen by the greatest number of pathways.

The aim is to take all of the zones from Z and select the suitable RAT for each one of them using Algorithm 2. For all trails p_i we have a configuration C_{pi}. The projected pathways and the selected configuration for each path are inputs to Algorithm 2, and a zone with a RAT as an output is produced.

Algorithm 3 concludes with a description of our framework

1- The path prediction is the initial stage. The order-2 Markov process was applied in this step. The method takes two parameters as inputs: "Mob-position" and "Loc-history," which represent the mobile user's current location and location history. The pairs of pathways and their probabilities are the results of this stage.

ALGORITHM 2 The choice function [53].

```
Inputs: pi, Cpi Local variables
For (z in L) do
        For (i<n) do
               If (z in pi)
                         pi [z]++
               End if
        End for
End for
Return max (p[z])
```

ALGORITHM 3 The proposed framework [53]

```
Inputs: Mob-position, Loc-history
Local variables: chi, probi, mat[][], Rki[][], C_init[], C_alt []
Step 1 : Path prediction
O(2) Markov model (Mob-position, Loc-history)
Results: (chi, probi)
Step 2: Network selection
Foreach chi do
   Foreach zone in chi do
       Construct the input matrix mat[][]
       Gain function (mat[][])
   End
   Results: RKi[][]
Step 3: RAT configuration and evaluation
Foreach RKi[][] do
   If(nb_VHO_(C_init[])==0 or (nb_VHO_(C_alt[])>= nb_VHO_(C_init[]))) then
     Choose C_init[]
   Else
       For (i < C_alt[].length and C_alt[i]!=C_init[i])
         If formula 7, formula 8 and formula 9 then
                  Choose C_init[]
         End
       End
   Choose C_alt[]
End
Step 4: Selection of the configuration
Do Algorithm2
```

2- The second stage is to create a network selection based on the expected pathways. We perform a network selection on all places for all projected routes in this step. For each zone, we begin by establishing a matrix system in which the rows represent RATs and the columns represent factors such as bandwidth, latency, PLR, energy usage, and cost. This phase yields a matrix with the RAT ranking order for each zone along the course.

3- We collect all of the configurations in the third stage and choose only one for each path. To begin with, we use the RATs in the order of precedence; therefore, the optimal configuration we design is simple. The alternate configurations are a mix of first- and second-ranking RATs, with the latter minimizing the frequency of vertical handovers. Following the collection of these configurations, we apply Eqs. (7)–(9) to choose simply one configuration for each path and discard the others.

4- Finally, we create a new configuration that we refer to as the appropriate configuration, which has the RAT that has been selected by the greatest number of pathways on each zone.

4. Performance evaluation

Before we get into how to validate our plan, it is vital to go over the simulation parameters in depth. As a result, we will use the QoS values in Table 2, which were obtained using the NS3 simulator [69]. In our simulations, we used three sorts of traffic apps: VoIP, video service, and best effort. We also used a different number of users each time: 40% VoIP users, 40% video service users, and the remainder is best-effort users. The RSS's range was set at 1 km. The simulations were run multiple times, and the average value for the QoS parameters was used. The goal is to evaluate our idea to MADM methods on the basis of QoS (latency, PLR, jitter, and data rate) provided to the user.

In terms of the proposal, we use the same example as in Fig. 4. Assuming the mobile user is in the zone "zoneA," we start by utilizing an O(2) Markov model to predict the various routes the mobile device follows and their likelihood. One of the anticipated trajectories is $ch_i = zoneA - zoneB - zoneC - zoneE - zoneH$ with a predetermined probability; the next step is to apply a network selection algorithm to all of the zones that comprise the path ch_i.

Table 2 The data model.

	Throughput (mb/s)	Latency (ms)	Packet loss (%)	Cost
Wi-Fi	1–12	90–160	0.3–2.9	1
HSPA	1–15	35–80	0.3–2.9	5
LTE	1–90	50–95	0.3–2.9	2

4.1 Testing the proposed framework

We will utilize the same example to show how our framework compares to other alternatives in terms of performance and efficiency. So, on the path ch_i=zoneA – zoneB – zoneC – zoneE – zoneH, a simple network selection yields the optimum configuration C1 = [Rat1, Rat2, Rat2, Rat1, Rat1]. At each zone of the path ch_i, this arrangement provides the best QoS for the user (individually). The alternate configuration, as outlined above, is one in which VHO recurrence is minimized while a high degree of QoS is maintained. Based on Table 4, which shows the outcomes of the network selection procedure on the path, the alternative configuration will include RAT from the second row of the table, which means RAT is classified as second in each zone, giving us C2 = [Rat2, Rat2, Rat2, Rat2, Rat2]. The VHOs are no longer needed in this set-up. According to sources [70,71], a vertical hand-off imposes a delay of 50–100 ms and a PLR of 1%– 3%. We will employ a VHO delay of 50 ms and a PLR of 2% in our investigation, which are lower values than those used in the previous experiments.

As previously stated, the proposal results in an ideal configuration of C1 = [Rat1, Rat2, Rat2, Rat1, Rat1] (see Table 4); other configurations are C2 = [Rat2, Rat2, Rat2, Rat2, Rat2] or C2 = [Rat1, Rat1, Rat1, Rat1, Rat1]. "Rat1" is clearly better suited for zone "A" than "Rat2" (see Table 3); thus, we will explore C2 = [Rat1, Rat1, Rat1, Rat1, Rat1] as an alternative design. When comparing C1 and C2, the goal is to lower the amount of VHOs while maintaining a high value of QoS; if this is the case, C2 is chosen; otherwise, C1 is used. In C2, we will examine if the first condition

Table 3 Input data.

Zones	RATs	Throughput (mb/s)	Latency (ms)	Packet loss (%)	Cost
"zoneA"	Rat0	10.87	115.54	1.57	0.2
	Rat1	4.43	31.25	1.22	1
	Rat2	35.47	80.23	2.53	0.4
"zoneB"	Rat0	9.62	103.89	1.73	0.2
	Rat1	6.08	30.65	2.64	1
	Rat2	32.47	82.86	1.47	0.4
"zoneC"	Rat0	8.48	120.22	4.16	0.2
	Rat1	1.83	40.84	2.46	1
	Rat2	50.02	79.98	1.65	0.4
"zoneE"	Rat0	4.54	120.24	2.73	0.2
	Rat1	4.79	40.17	1.15	1
	Rat2	41.05	68.14	2.86	0.4
"zoneE"	Rat0	10.55	102.85	2.87	0.2
	Rat1	8.15	41.57	2.15	1
	Rat2	30.52	69.12	2.05	0.4

(lowering the amount of VHOs) is met and if the RATs in the second and third locations in C1 cause QoS degradation or not.

A comparison of QoS between the two configurations C1 and C2 is shown in Table 4. Only the difference between the delays of subsequent RATs is used to determine the jitter in Table 4; we do not include the time delay induced by the VHO, as seen in Fig. 5. Figs. 6 and 7 show a comparison of C1, the ideal configuration, and C2, the alternate design; as previously stated, the network selection process outputs the C1 configuration. C1 has two VHOs, which means more delays and PLR. As seen in Figs. 6 and 7, the added delays and PLR have a significant impact on the users' QoS. In terms of user

Table 4 C1 versus C2.

Zones	Configurations	Throughput (mb/s)	Latency (ms)	Packet loss (%)	Jitter (ms)
"zoneA"	C1[1]=Rat1	4.42	31.23	1.23	0
	C2[1]=Rat1	4.42	31.23	1.23	0
"zoneB"	C1[2]=Rat2	32.46	82.80	1.44	51.57
	C2[2]=Rat1	6.09	30.63	2.62	0.6
"zoneC"	C1[3]=Rat2	50	79.95	1.63	2.85
	C2[3]=Rat1	1.82	40.81	2.48	10.18
"zoneE"	C1[4]=Rat1	4.78	40.19	1.12	40.76
	C2[4]=Rat1	4.78	40.19	1.12	0.62
"zoneH"	C1[5]=Rat1	8.13	41.54	2.13	1.35
	C2[5]=Rat1	8.13	41.54	2.13	1.35

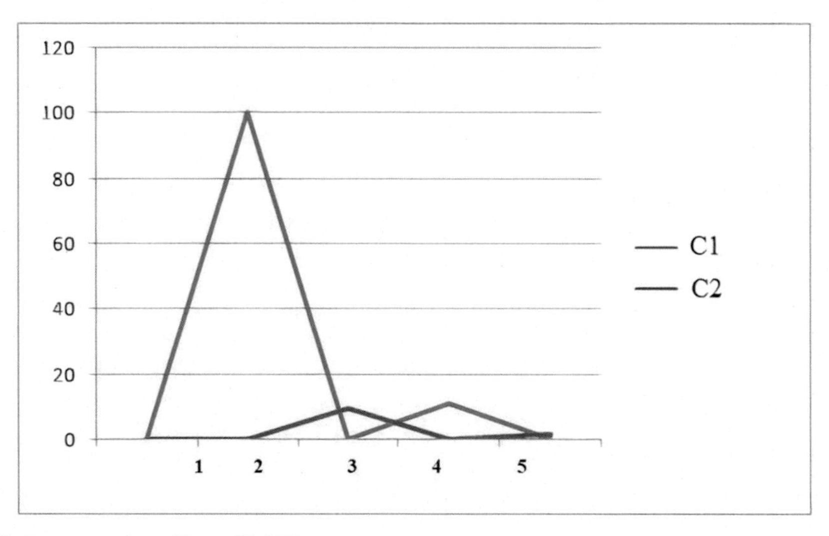

Fig. 6 Jitter comparison C1 vs C2 [53].

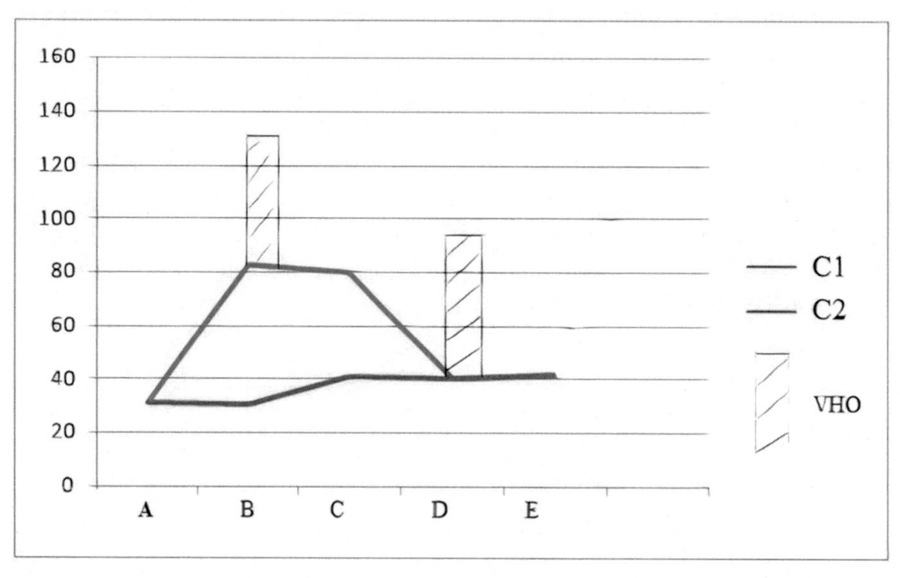

Fig. 7 Instant delay comparison C1 vs C2 [53].

benefit, C2 outperforms C1. For the remaining paths, the same method must be followed. Indeed, we use the proposed procedure to find the optimal configurations for each selected configuration, and we use it for each chosen path. This example's findings clearly illustrate that our selection technique enhances the ranking order. In fact, for all ways, we find the suited configuration that reduces the frequency of vertical hand-off although maintaining a high quality of QoS.

4.2 Comparison with other works

From zone "zoneA" to zone "zoneE," Table 5 illustrates the ranking order of RATs using various methodologies; we will compare these results in the following section. Because our approach differs from the other techniques in terms of rank, we will compare the ranked RATs in each zone for all methods to discover which one is the most efficient.

We may conclude from Figs. 8 and 9, and Table 4 that our solution is more user-friendly and provides a more exact rating of RATs, resulting in a higher QoS. For example, in "zoneA," our suggestion chooses "Rat1" as the appropriate RAT. WPM and AHP chose R2, whereas TOPSIS chosed "Rat0." Figs. 6 and 7 show that "Rat1" is more suitable for "zoneA," and the same for the rest of the places. We may deduce from these figures that our method of selection produces the most precise RAT rating order. Several works [22,24,71] have employed these methodologies, such as TOPSIS, WMP, and AHP. Table 6 shows the benefits and drawbacks of some of the most prominent works in this sector.

Table 5 Network selection on the path ch_i [53].

	zoneA	**zoneB**	**zoneC**	**zoneE**	**zoneF**
Proposal	Rat1	Rat2	Rat2	Rat1	Rat1
	Rat2	Rat1	Rat1	Rat2	Rat2
	Rat0	Rat0	Rat0	Rat0	Rat0
TOPSIS	Rat0	Rat0	Rat0	Rat0	Rat0
	Rat2	Rat2	Rat2	Rat2	Rat2
	Rat1	Rat1	Rat1	Rat1	Rat1
WPM	Rat2	Rat0	Rat0	Rat0	Rat0
	Rat0	Rat2	Rat2	Rat2	Rat2
	Rat1	Rat1	Rat1	Rat1	Rat1
AHP	Rat2	Rat0	Rat0	Rat0	Rat0
	Rat0	Rat2	Rat2	Rat2	Rat2
	Rat1	Rat1	Rat1	Rat1	Rat1

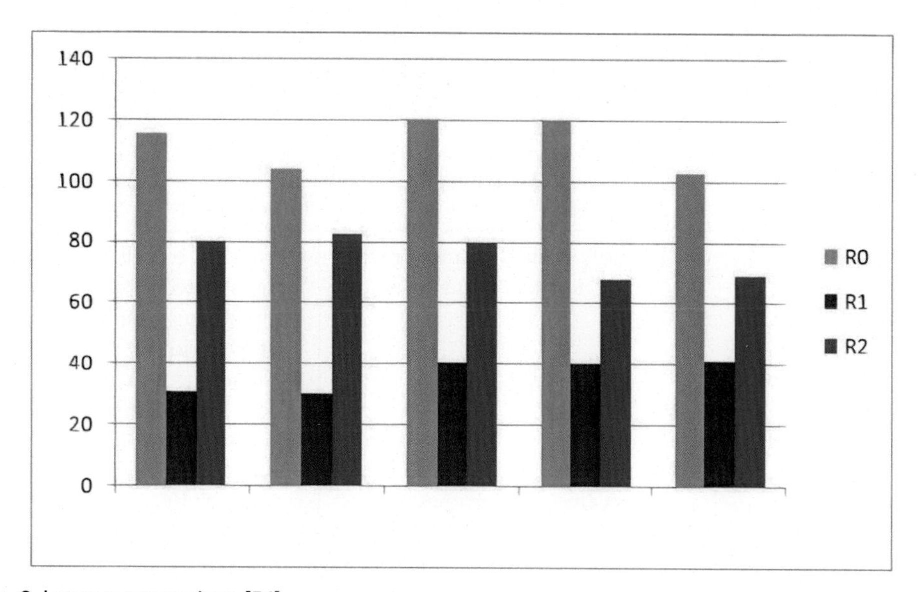

Fig. 8 Latency comparison [54].

In this study, we used an example to demonstrate the uniqueness of our concept. In this case, whichever other QoS-based approach would select the C1 configuration, it would result in additional unwanted vertical handovers. Our intelligent solution, on the other hand, lets the user enjoy a good QoS while reducing the number of vertical hand-offs. We have not found any other works that take the same approach as this one. In Table 5, we compare and contrast our suggestion with other works.

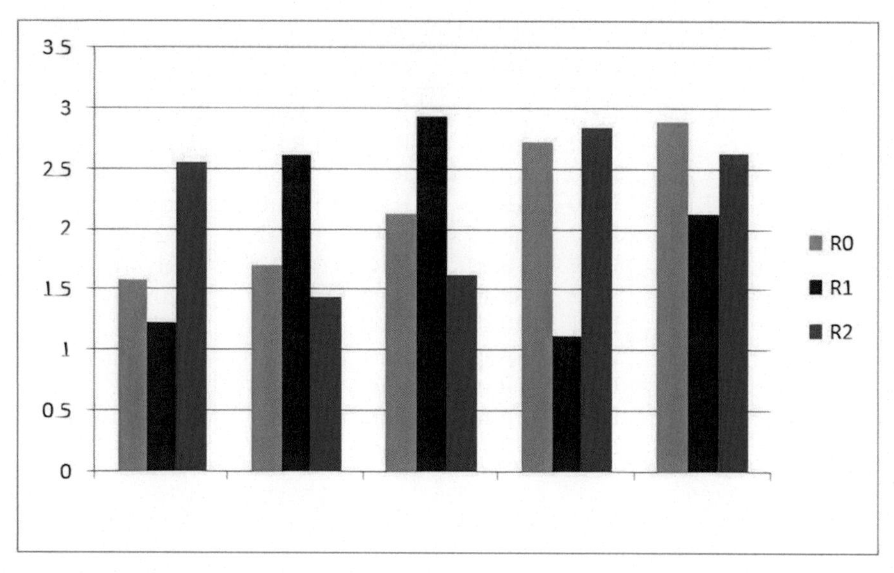

Fig. 9 PLR comparison [54].

Table 6 Theoretical comparison [53].

Author name	Advantages	Inconvenient
Hasib, M	The study enables consumers to experience a less end-to-end delay	Only 802.11n networks are investigated. The research is focused on a queue with packet priority, which is not as efficient as an approach and does not take into account vertical hand-off
Nguyen-Duc, T., & Kamioka, E	The research enables consumers to receive the appropriate QoS	It employs a large number of correlated parameters, which can result in overlapping situations. The study makes use of the RSS to simulate mobility, which isn't entirely accurate. The vertical hand-off is not taken into account
Wang, L., & Binet, D.	The study uses a Markov model similar to ours to predict mobility and takes into account vertical hand-off	There are only WLAN and UMTS options, therefore there is no network selection available
Proposal	The Markov model is used in our approach to predict mobility. The vertical handoff is the study's focal point. It also provides for a reduction in the end-to-end delay	There are a lot of processes, which means it will take longer to acquire the best RAT configuration. It relies on a probabilistic model that isn't always accurate

5. Conclusions

This chapter provides both a case study and a solution to the network selection problem. For various types of services, the performance of a network selection based on path prediction and user mobility is evaluated (VoIP, video services, and best-effort ones). The concept permits the user to construct a practical use case for network selection in a heterogeneous environment; the study's strength is that it considers mobility as a user's environment, which we believe is the correct approach to this difficulty.

Others, on the other hand, handled mobility as if it were a simple data in the vector selection procedure, which is actually wrong. We feel that this research is more thorough, and that it is the only method to choose the best network in a complex situation. In the meantime, the consequences are contingent on the correctness of the path prediction, which is a downside of our approach.

Future work should focus on reducing the overall computing time and improving the calculation model to convert it from a forecasting approach to a certainty approach; we also want to be capable to put the concept into practice.

References

[1] F. Bendaoud, Management of Joint Radio Resources in Heterogeneous Networks beyond 3G, 2018. Doctoral dissertation, 07-06-2018.

[2] E. Gustafsson, A. Jonsson, Always best connected, IEEE Wirel. Commun. 10 (1) (2003) 49–55.

[3] F. Bendaoud, M. Abdennebi, F. Didi, Network selection in wireless heterogeneous networks: a survey, Res. J. Telecommun. Inf. Technol. 4 (2018) 64–74.

[4] L. Wang, G.S.G. Kuo, Mathematical modeling for network selection in heterogeneous wireless networks—a tutorial, IEEE Commun. Surv. Tutorials 15 (1) (2012) 271–292.

[5] R. Trestian, O. Ormond, G.M. Muntean, Game theory-based network selection: solutions and challenges, IEEE Commun. Surv. Tutorials 14 (4) (2012) 1212–1231.

[6] F. Bendaoud, Multicriteria Optimization Methods for Network Selection in a Heterogeneous Environment. Intelligent Network Management and Control: Intelligent Security, Multi-criteria Optimization, Cloud Computing, Internet of Vehicles, Intelligent Radio, Wiley Online Library, 2021, pp. 89–115.

[7] F. Bendaoud, M. Abdennebi, F. Didi, Network selection using game theory, in: 2015 3rd International Conference on Control, Engineering & Information Technology (CEIT), IEEE, 2015, pp. 1–6.

[8] F. Bendaoud, M. Abdennebi, F. Didi, Network selection schemes in heteregenous wireless netwoks, in: Proc of International Conference on Information Processing and Electrical Engineering icipee, vol. 14, 2014.

[9] A. Sgora, D.D. Vergados, P. Chatzimisios, An access network selection algorithm for heterogeneous wireless environments, in: The IEEE symposium on Computers and Communications, IEEE, 2010, pp. 890–892.

[10] F. Bendaoud, Méthodes d'optimisation multicritères pour la sélection de réseaux dans un environnement hétérogène, Sécurité intelligente, optimisation multicritères, Cloud Computing, Internet of Vehicles, radio intelligente, Gestion et contrôle intelligents des réseaux, 2020, p. 95.

[11] R.V. Rao, Introduction to multiple attribute decision-making (MADM) methods, in: Decision Making in the Manufacturing Environment: Using Graph Theory and Fuzzy Multiple Attribute Decision Making Methods, Springer, 2007, pp. 27–41.

[12] E. Stevens-Navarro, J.D. Martinez-Morales, U. Pineda-Rico, Evaluation of vertical handoff decision algorightms based on madm methods for heterogeneous wireless networks, J. Appl. Res. Technol. 10 (4) (2012) 534–548.
[13] E. Triantaphyllou, Multi-criteria decision making methods, in: Multi-Criteria Decision Making Methods: A Comparative Study, Springer, Boston, MA, 2000, pp. 5–21.
[14] F. Bendaoud, A modified-SAW for network selection in heterogeneous wireless networks, ECTI Trans. Electr. Eng. Electron Commun. 15 (2) (2017) 8–17.
[15] L. Ren, Y. Zhang, Y. Wang, Z. Sun, Comparative analysis of a novel M-TOPSIS method and TOPSIS, Appl. Math. Res. Express 2007 (2007).
[16] Z. Pavić, V. Novoselac, Notes on TOPSIS method, Int. J. Res. Eng. Sci. 1 (2) (2013) 5–12.
[17] L. Dymova, P. Sevastjanov, A. Tikhonenko, A direct interval extension of TOPSIS method, Expert Syst. Appl. 40 (12) (2013) 4841–4847.
[18] C.S. Yu, A GP-AHP method for solving group decision-making fuzzy AHP problems, Comput. Oper. Res. 29 (14) (2002) 1969–2001.
[19] Z.C. Lin, C.B. Yang, Evaluation of machine selection by the AHP method, J. Mater. Process. Technol. 57 (3–4) (1996) 253–258.
[20] J. Pérez, Some comments on Saaty's AHP, Manag. Sci. 41 (6) (1995) 1091–1095.
[21] K. Savitha, C. Chandrasekar, Trusted network selection using SAW and TOPSIS algorithms for heterogeneous wireless networks, Int. J. Comput. Appl. 975 (2011) 8887.
[22] Q.T. Nguyen-Vuong, Y. Ghamri-Doudane, N. Agoulmine, On utility models for access network selection in wireless heterogeneous networks, in: NOMS 2008–2008 IEEE Network Operations and Management Symposium, IEEE, 2008, pp. 144–151.
[23] D.L. Olson, Comparison of weights in TOPSIS models, Math. Comput. Model. 40 (7–8) (2004) 721–727.
[24] B. Bakmaz, Z. Bojkovic, M. Bakmaz, Network selection algorithm for heterogeneous wireless environment, in: 2007 IEEE 18th International Symposium on Personal, Indoor and Mobile Radio Communications, IEEE, 2007, pp. 1–4.
[25] M. Lahby, L. Cherkaoui, A. Adib, An intelligent network selection strategy based on MADM methods in heterogeneous networks, arXiv (2012). preprint arXiv:1204.1383.
[26] E.H. Watanabe, D.S. Menasché, E.D.S. e Silva, R.M.M. Leao, Modeling resource sharing dynamics of VoIP users over a WLAN using a game-theoretic approach, in: IEEE INFOCOM 2008-The 27th Conference on Computer Communications, IEEE, 2008, pp. 915–923.
[27] K. Mittal, E.M. Belding, S. Suri, A game-theoretic analysis of wireless access point selection by mobile users, Comput. Commun. 31 (10) (2008) 2049–2062.
[28] M. Cesana, N. Gatti, I. Malanchini, Game theoretic analysis of wireless access network selection: models, inefficiency bounds, and algorithms, in: Proceedings of the 3rd International Conference on Performance Evaluation Methodologies and Tools, 2008, pp. 1–10.
[29] M.S.Z. Khan, S. Alam, M.R.H. Khan, A network selection mechanism for fourth generation communication networks, J. Adv. Inf. Technol. 1 (4) (2010) 189–196.
[30] D.E. Charilas, O.I. Markaki, P.T. Vlacheas, Admission control as a non-cooperative multi-stage game between wireless networks, in: 2009 16th International Conference on Systems, Signals and Image Processing, IEEE, 2009, pp. 1–5.
[31] R. Trestian, O. Ormond, G.M. Muntean, Reputation-based network selection mechanism using game theory, Phys. Commun. 4 (3) (2011) 156–171.
[32] C. Yang, X. Yubin, X. Rongqing, S. Xuejun, A heterogeneous wireless network selection algorithm based on non-cooperative game theory, in: 2011 6th International ICST Conference on Communications and Networking in China (CHINACOM), IEEE, 2011, pp. 720–724.
[33] D. Charilas, O. Markaki, E. Tragos, A theoretical scheme for applying game theory and network selection mechanisms in access admission control, in: 2008 3rd International Symposium on Wireless Pervasive Computing, IEEE, 2008, pp. 303–307.
[34] M.A. Khan, U. Toseef, S. Marx, C. Goerg, Game-theory based user centric network selection with media independent handover services and flow management, in: 2010 8th Annual Communication Networks and Services Research Conference, IEEE, 2010, pp. 248–255.

[35] N. Sui, D. Zhang, W. Zhong, C. Wang, Network selection for heterogeneous wireless networks based on multiple attribute decision making and evolutionary game theory, in: 2016 25th Wireless and Optical Communication Conference (WOCC), IEEE, 2016, pp. 1–5.
[36] J. Godjevac, Idées nettes sur la logique floue, PPUR presses polytechniques, 1999.
[37] L.A. Zadeh, Fuzzy sets and information granularity, in: Advances in Fuzzy Set Theory and Applications, vol. 11, World Scientific, Singapore, 1979, pp. 3–18.
[38] T. Aouam, S.I. Chang, E.S. Lee, Fuzzy MADM: an outranking method, Eur. J. Oper. Res. 145 (2) (2003) 317–328.
[39] S. Zolfani, M. Yazdani, D. Pamucar, P. Zarate, A VIKOR and TOPSIS focused reanalysis of the MADM methods based on logarithmic normalization, arXiv 18 (2020) 341–355.
[40] M.K. Hasan, M. Shahjalal, M.Z. Chowdhury, M.T. Hossan, Y.M. Jang, Fuzzy logic based network selection in hybrid OCC/Li-Fi communication system, in: 2018 Tenth International Conference on Ubiquitous and Future Networks (ICUFN), IEEE, 2018, pp. 95–99.
[41] S. Kher, A.K. Somani, R. Gupta, Network selection using fuzzy logic, in: 2nd International Conference on Broadband Networks, 2005, IEEE, 2005, pp. 876–885.
[42] S.B. Cho, J.H. Kim, Multiple network fusion using fuzzy logic, IEEE Trans. Neural Netw. 6 (2) (1995) 497–501.
[43] C. Kahraman, U. Cebeci, Z. Ulukan, Multi-criteria supplier selection using fuzzy AHP, Logist. Inf. Manag. 16 (2003) 382–394.
[44] Y.M. Wang, T.M. Elhag, Fuzzy TOPSIS method based on alpha level sets with an application to bridge risk assessment, Expert Syst. Appl. 31 (2) (2006) 309–319.
[45] D. Jiang, L. Huo, Z. Lv, H. Song, W. Qin, A joint multi-criteria utility-based network selection approach for vehicle-to-infrastructure networking, IEEE Trans. Intell. Transp. Syst. 19 (10) (2018) 3305–3319.
[46] C.J. Chang, T.L. Tsai, Y.H. Chen, Utility and game-theory based network selection scheme in heterogeneous wireless networks, in: 2009 IEEE Wireless Communications and Networking Conference, IEEE, 2009, pp. 1–5.
[47] H. Chan, P. Fan, Z. Cao, A utility-based network selection scheme for multiple services in heterogeneous networks, in: 2005 International Conference on Wireless Networks, Communications and Mobile Computing, vol. 2, IEEE, 2005, pp. 1175–1180.
[48] P.C. Fishburn, Utility Theory for Decision Making, Research analysis corp McLean VA, 1970.
[49] F. Bari, V.C. Leung, Use of non-monotonic utility in multi-attribute network selection, in: Wireless Technology, Springer, Boston, MA, 2009, pp. 21–39.
[50] M.A. Senouci, S. Hoceini, A. Mellouk, Utility function-based TOPSIS for network interface selection in heterogeneous wireless networks, in: 2016 IEEE International Conference on Communications (ICC), IEEE, 2016, pp. 1–6.
[51] O. Ormond, J. Murphy, G.M. Muntean, Utility-based intelligent network selection in beyond 3G systems, in: 2006 IEEE international conference on communications, Vol. 4, IEEE, 2006, pp. 1831–1836.
[52] L. Pirmez, J.C. Carvalho Jr., F.C. Delicato, F. Protti, L.F. Carmo, P.F. Pires, M. Pirmez, Sutil–network selection based on utility function and integer linear programming, Comput. Netw. 54 (13) (2010) 2117–2136.
[53] F. Bendaoud, M. Abdennebi, F. Didi, Mobility aware network selection in a heterogeneous wireless environment, Transp. Telecommun. 21 (1) (2020) 32–46.
[54] F. Bendaoud, Markov model and a modified-SAW for network selection in a heterogeneous wireless environment, in: 2020 Second International Conference on Embedded & Distributed Systems (EDiS), IEEE, 2020, pp. 147–151.
[55] A. Rodriguez-Carrion, C. Garcia-Rubio, C. Campo, Performance evaluation of LZ-based location prediction algorithms in cellular networks, IEEE Commun. Lett. 14 (8) (2010) 707–709.
[56] A. Rodriguez-Carrion, C. Garcia-Rubio, C. Campo, A. Cortés-Martín, E. Garcia-Lozano, P. Noriega-Vivas, Study of lz-based location prediction and its application to transportation recommender systems, Sensors 12 (6) (2012) 7496–7517.
[57] L. Song, D. Kotz, R. Jain, X. He, Evaluating location predictors with extensive Wi-Fi mobility data, in: Ieee Infocom 2004, vol. 2, IEEE, 2004, pp. 1414–1424.

[58] D. Joseph, D. Grunwald, Prefetching using markov predictors, IEEE Trans. Comput. 48 (2) (1999) 121–133.
[59] M.A. Teixeira, G. Zaverucha, Fuzzy Bayes and fuzzy Markov predictors, J. Intell. Fuzzy Syst. 13 (2–4) (2002) 155–165.
[60] C. Roman, R. Liao, P. Ball, S. Ou, Mobility and network selection in heterogeneous wireless networks: user approach and implementation, Netw. Protoc. Algorithms 8 (2) (2016) 107–122.
[61] S. Gambs, M.O. Killijian, M.N. del Prado Cortez, Next place prediction using mobility markov chains, in: Proceedings of the First Workshop on Measurement, Privacy, and Mobility, 2012, pp. 1–6.
[62] L. Wang, D. Binet, MADM-based network selection in heterogeneous wireless networks: a simulation study, in: 2009 1st International Conference on Wireless Communication, Vehicular Technology, Information Theory and Aerospace & Electronic Systems Technology, IEEE, 2009, pp. 559–564.
[63] J.M. François, G. Leduc, Prédiction de mobilité par le mobile ou par le point d'accès: comparaison sur base de traces réelles, in: Colloque Francophone sur l'Ingénierie des Protocoles-CFIP 2006, Hermès, 2006, p. 12.
[64] B. Benmammar, F. Krief, Gestion dynamique du handover horizontal et vertical basée sur le profil de mobilité de l'utilisateur, in: Colloque GRES 2005: Gestion de REseaux et de Services, 2005.
[65] Q. Song, A. Jamalipour, An adaptive quality-of-service network selection mechanism for heterogeneous mobile networks, Wirel. Commun. Mob. Comput. 5 (6) (2005) 697–708.
[66] P. Payaswini, D.H. Manjaiah, Simulation and performance analysis of vertical handoff between WiFi and WiMAX using media independent handover services, Int. J. Comput. Appl. 87 (4) (2014).
[67] A. Fettouh, N. El Kamoun, A. El Fazziki, Improving vertical handover performance of real time applications over heterogeneous wireless networks, Int. J. Comput. Appl. 62 (21) (2013).
[68] L. Nithyanandan, I. Parthiban, Vertical handoff in WLAN-WIMAX-LTE heterogeneous networks through gateway relocation, Int. J. Wirel. Mob. Netw. 4 (4) (2012) 203.
[69] https://www.nsnam.org/.
[70] L.A. Magagula, H.A. Chan, O.E. Falowo, PMIPv6-HC: handover mechanism for reducing handover delay and packet loss in NGWN, in: 2010 IEEE Global Telecommunications Conference GLOBECOM 2010, IEEE, 2010, pp. 1–5.
[71] D.M. Shen, H. Tian, L. Sun, The QoE-oriented heterogeneous network selection based on fuzzy AHP methodology, in: Proc. of UBICOMM2010 (The Forth International Conference on Mobile Ubiquitous Computing, Systems, Services and Technologies), 2010, pp. 275–280.

CHAPTER 4

Reducing control packets using covering rough set for route selection in mobile ad hoc networks

T. Sudhakar[a] and H. Hannah Inbarani[b]
[a]School of Computer Application, Lovely Professional University, Phagwara, India
[b]Department of Computer Science, Periyar University, Salem, India

1. Introduction

Routing is a challenging and active research problem in wireless networks due to high mobility, stipulation resources, and no fixed links among nodes. In this chapter, a CRS approach is proposed for route selection in wireless ad hoc networks. In a wireless network, mobile nodes (MNs) are deployed randomly in a simulation region. A mobile ad hoc network (MANET) is one type of wireless network which has many challenges to be addressed [1]. This work addresses the problem of reducing redundant broadcast packets [route request (RREQ) packets].

The rough set is a mathematical model to deal with vagueness and uncertainty problems. Since the network parameters such as bandwidth, delay, packet byte rate, and packet loss rate change due to the frequent mobility of nodes, it leads to uncertainty in wireless networks [1]. This type of uncertainty can be very well handled using the rough set concept. The traditional RST algorithm and the proposed CRS for route selection algorithm have been compared with conventional routing protocols such as AODV and DSR [2]. The proposed covering rough set-based route selection approach reveals better accuracy than traditional algorithms. The comprehensive simulation study substantiates that the proposed covering rough set shows a promising outcome.

In recent times, wireless networks have become a challenging and active research field in academia and industry [3]. In the fast-growing domain of wireless technologies, more and more businesses comprehend the flexibility of using wireless networks. Depending on the area and needs, a small number of devices such as nodes can create local area networks (LANs) or wireless groups. Wireless networks are either infrastructure-dependent or infrastructure-less. Ad hoc networks are infrastructure-less and enable users to spontaneously form a dynamic communication system [4].

Conversely, to offer high-quality and low-cost services to nodes, several technical challenges need to be addressed by researchers. To mention a few, first, wireless networks

Comprehensive Guide to Heterogeneous Networks
https://doi.org/10.1016/B978-0-323-90527-5.00006-X

are plagued by a scarcity of communication bandwidth; therefore, a key issue is to satisfy user requests with minimum service delay [5]. Second, if network nodes provide minimal energy capacity, it is important to reduce the energy spent on transmitting information across the network [6].

The wireless network is a general term that refers to different types of systems that connect without the need for wires [7]. Wireless networks can be classified broadly into two classes based on network structures: wireless ad hoc networks and cellular networks [8]. No fixed infrastructure is required for wireless ad hoc networks; so, setting up and implementing a wireless ad hoc network is reasonably straightforward. Without expanded infrastructure, the topology of a wireless ad hoc network is unpredictable, and often varies [9]. The concept of static or a particular topology of a wireless ad hoc network is not practical [10]. Two unique styles should be listed, i.e., wireless sensor networks and wireless mesh networks, in addition to traditional wireless ad hoc networks. Wireless sensor networks (WSNs) are ad hoc wireless networks, most of which are sensors tracking a specific scene through network nodes [11]. In terms of computing capacity, power storage, bandwidth, and other processing parameters, wireless sensors are often weak devices [12]. Wireless mesh networks are wireless networks with either a total topology of mesh or a limited topology of mesh in which any or all nodes are linked directly to all other nodes. Redundancy in the connectivity of wireless networks provides excellent reliability and excellent flexibility in network packet delivery [13]. The main idea of reducing control packets using the rough set concept was introduced by Nagaraju et al. [14]. In this work, the researchers further replace the traditional rough set with the extension of rough set concepts, namely CRS and NRS for the existing MANET [15] routing protocols. Fig. 1 shows the types of wireless networks.

In the cellular network, a fixed base station is present while the wireless ad hoc network does not require any fixed infrastructure-based network [9]. It is not possible to

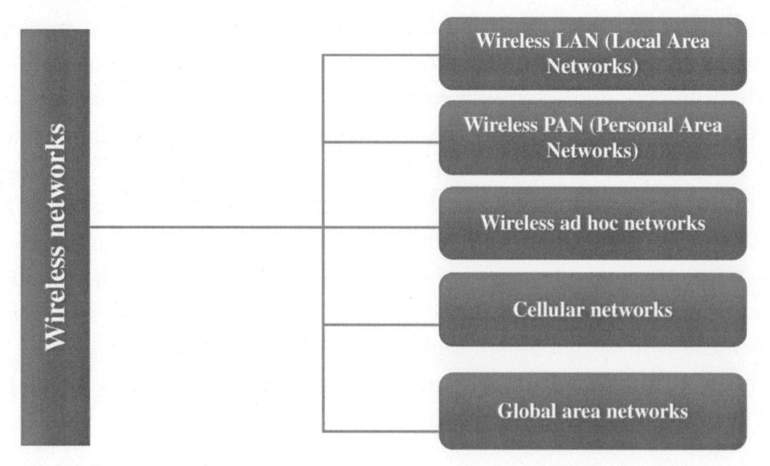

Fig. 1 Types of wireless networks.

expect a static or unique topology for an ad hoc wireless network. Two specific forms are wireless sensor networks and wireless mesh networks, in addition to traditional wireless ad hoc networks. A wireless sensor network is a type of wireless ad hoc network, in which most network nodes track the target scene [11]. WSNs are mostly battery-based devices in the form of bandwidth usage, computational energy, and other backing resources [16]. Redundancy in the communication of wireless networks offers strong stability and outstanding versatility in the distribution of network packets [13].

1.1 Overview of mobile ad hoc networks

The mobile ad hoc network (MANET) [15] is an autonomous and noncentralized wireless technique. MANETs involve mobile nodes, which are unstable in the network. Mobile nodes (MNs) are gadgets, cell phones, laptop computers, individual electronic devices, MP3 players, and personal computers (PC) that form the network. These nodes can work like a host/router or both at the same time. They can form irrelevant topologies based on their connection with each other in the network. Because of their self-setting capabilities, they can be implemented rapidly without the need for any preexisting infrastructure-based network [17]. The nodes that are within each other's transmission range communicate explicitly through intermediary nodes that can forward packets; so, these networks are also called multihop networks [18]. The known routing protocols define how routers connect and communicate in a computer network that can have any two nodes [19]. In general, routing methods are one of the complicated and stimulating analysis phases. Many routing methods have been designed for MANETs [15], such as, ad hoc on-demand distance vector (AODV) [15], optimized link-state routing (OLSR) [15], and dynamic source routing (DSR), etc. [17]. The two types of wireless networks are infrastructure-based and infrastructure-less networks as shown in Figs. 2 and 3. In an

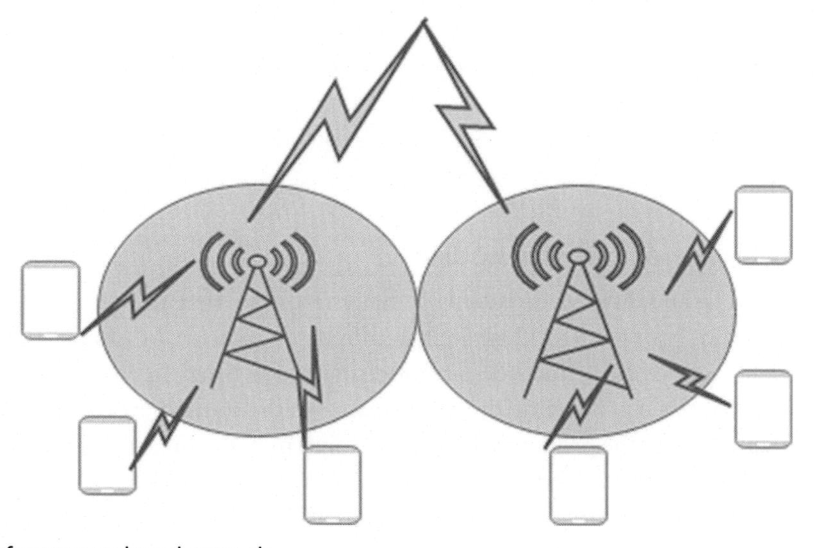

Fig. 2 Infrastructure-based network.

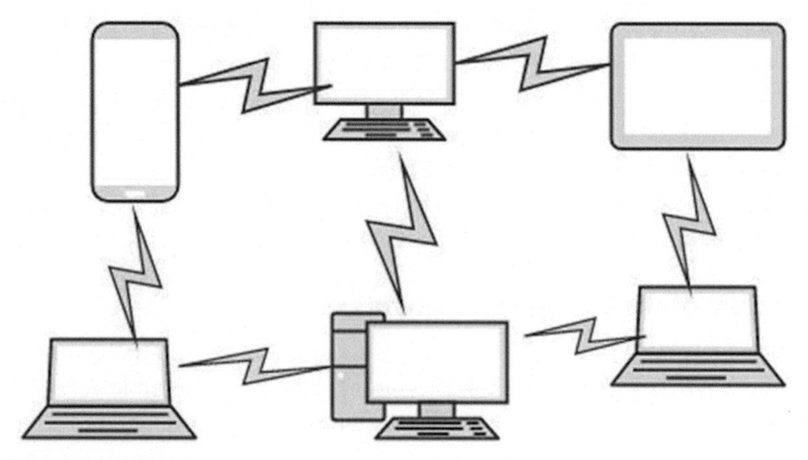

Fig. 3 Infrastructure-less network.

infrastructure-based network [9], MNs can move while communicating, and the base stations are permanent. If the MN goes out of the range of base stations, instantly, the MN may enter another base station coverage, for example, a cellular network. A MANET is also called a spontaneous network, and it works without a base station network. It is a collection of autonomous nodes, which is based on the infrastructure-less network.

Each mobile node in MANET serves as a router and host. There are drawbacks and downsides to both wired and cellular networks. The wired network is much more reliable than a wireless network because the user can easily fix the problem of identification and resolve it. However, the problem of identification is easy in the wired network but it cannot reach the entire region [20]. To overcome the problem of the wired network, wireless technology has been rolled out. In this research work, route selection and route classification have been addressed based on soft-computing approaches in mobile ad hoc networks. Multipath routing is one way to improve the reliability of the information conveyed [21].

1.2 An overview of mobility nodes

Nodes are mobile and distributed equally initially across the network [22]. The wireless mobility model is proposed to illustrate the movement pattern of mobile nodes, and their location, acceleration, and velocity of change over time. In MANET, mobility plays an important role in deciding the performance of the routing protocol. Therefore, it is important to select the proper applicable mobility model when evaluating MANET protocols. The MNs in the random waypoint (RWP) model behave in a distinctly

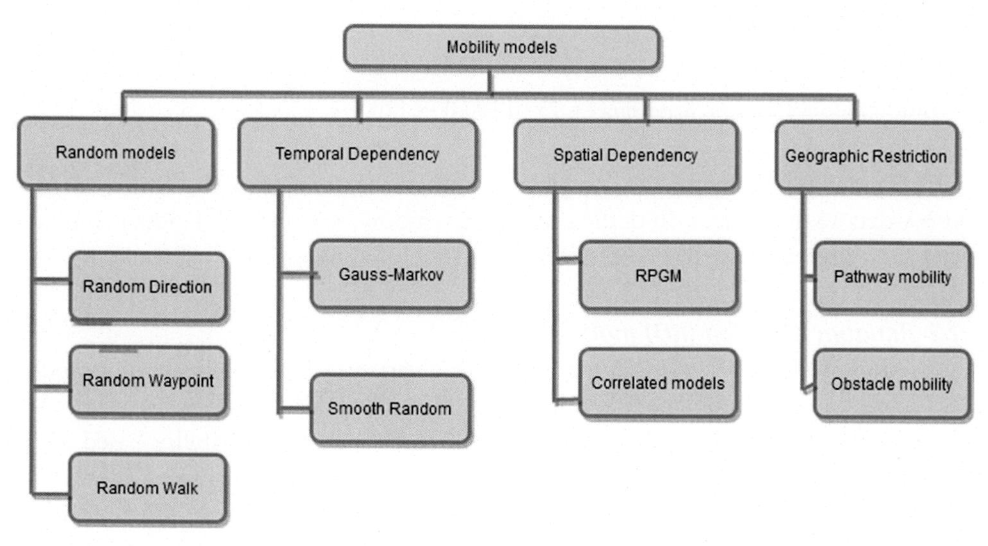

Fig. 4 Taxonomy of mobility models.

different manner from other mobility models such as the random direction (RD) model, reference point group mobility (RPGM) model, pathway mobility, and obstacle mobility [3].

In this thesis, two mobility models, namely RWP and RD, were used to evaluate the performance of the routing protocol. Both the mobility models are created with the aid of the Bonn motion tool. Fig. 4 shows the taxonomy of mobility models.

1.2.1 Random waypoint (RWP) mobility model

It was first introduced by Broch et al. [23] and, later, it became a benchmark mobility model to evaluate wireless routing protocols, because of its wide availability and simplicity. The implementation of the RWP model is as follows: When the simulation starts, the MNs randomly select one location in the simulation region as the destination point. An MN moves continuously and arbitrarily to this destination at a constant speed from $[0, V_{max}]$where V_{max} is the maximum allowable velocity for each MN. The pause time defines the stop duration of the mobile node using the T_{pause} parameter. If the T_{pause} is equal to 0, then it leads to continuous mobility. The entire process is continued until the simulation end; otherwise, the new destination is selected in the simulation region [3]. Fig. 4 gives different types of mobility models in the ad hoc network.

Mobility models allow mimicking of the mobile node's behavior when simulating network performance. The findings of the simulation are directly linked to the paradigm for mobility.

```
./bm -f indoor RandomWayPoint -n 15 -d 100.0 -x 120.0 -y 120.0 -e 0.25 -h 0.5 -l 0.3
```

where −n denotes the number of nodes, −d denotes the duration of the simulation, −x and −y denote the dimension of the simulation window, −h and −l denote maximum speed and minimum speed, −e denotes pause time of a node.

1.2.2 Random direction (RD) mobility model

The random walk (RW) and RD models are two variants of random waypoint models [3]. The mobile node movement in RD is quite similar to RWP, but it was developed to overcome the flaws of the RW model. The RW model uses a probability-based destination selection in the simulation field. In the RD model, the mobile node chooses random velocity and direction from a specified range $[0, 2 * \pi]$ and $[0, V_{max}]$. When the node reaches the border of the simulation region, the node selects a new direction from the $[0, \pi]$ range. This process is repeated until the end of the simulation; otherwise, a new direction is selected in the specified range [3].

```
./bm -f indoor RandomDirection -n 15 -d 100.0 -x 120.0 -y 120.0 -e 0.25 -h 0.5 -l 0.3
```

1.2.3 Random walk (RW) mobility model

The random walk (RW) mobility is an independent mobility model, and it has a relationship to Brownian motion (BM), which was first mathematically defined by Einstein. Brownian motion, however, is the boundary of both simpler stochastic processes, such as random walking, more suited for mobile ad hoc networks and more difficult stochastic processes depicted in physics to study the mobility sequence of the moving objects in a highly unpredictable manner [24,25].

The model was also used, particularly in MANET, due to the simplicity of implementation and analysis, although there are many problems with realistic scenarios. It can be an efficient model for aggregate node movement in a massive ad hoc network, despite the apparent constraints of the Brownian model. A large number of derivatives of this model, including 1D, 2D, 3D, and d-dimensional walks, have been developed [26].

```
./bm -f indoor RandomWalk -n 15 -d 100.0 -x 120.0 -y 120.0 -e 0.25 -h 0.5 -l 0.3
```

Some of the characteristics of the RW mobility model:

- The random walk model can have retaining barriers, which imply that mobile nodes cannot move past them.

- The random walk model may have obstacles.
- Random walking can also have hurdles that capture mobile phones from where they cannot exit. That is, breaking an absorbing barrier is unlikely.
- From the predefined ranges [V_{min}, V_{max}], speed $v(t)$ can be selected by each node following a uniform distribution or Gaussian distribution at each new interval t, where V_{min} and V_{max} are the minimum and maximum speed, respectively.

```
/* Random model formation in ad hoc networks */
   Parameter Initialization
       Channel allocation //wireless channel
       Propagation model       // Shadowing
       Antenna type         //Omni directional
       Scheduler creation
              Create new Simulator
              Create Trace and nam files
              Create node()
                 //set no.of nodes
      Node Configuration
              Assign initial node position
              Assign mobility

                      for i = 0 to n no. of nodes
                      {
                           //set no. of nodes
                      } end for
   Establish Connection
                 //set no.of UDP connections
   Stop procedure
                Procedure stop()
   End define()
   /*End of Simulation Scenarios Procedure*/
```

1.3 Types of wireless routing protocols

The routing protocol is a method to choose the correct route from source to destination for the data to travel [27,28]. Nodes in these networks switch arbitrarily; hence, the topology of the network is constantly and unpredictably changing. Besides, there is minimal bandwidth and battery power. Such limitations, together with the complex topology of the network, make routing and multicasting extremely challenging in ad hoc networks [28,29].

In this section, the different routing protocols of the wireless network are summarized [7] as shown in Fig. 5. A variety of routing protocols has been suggested for MANETs to solve the issues associated with both distance vector and link-state

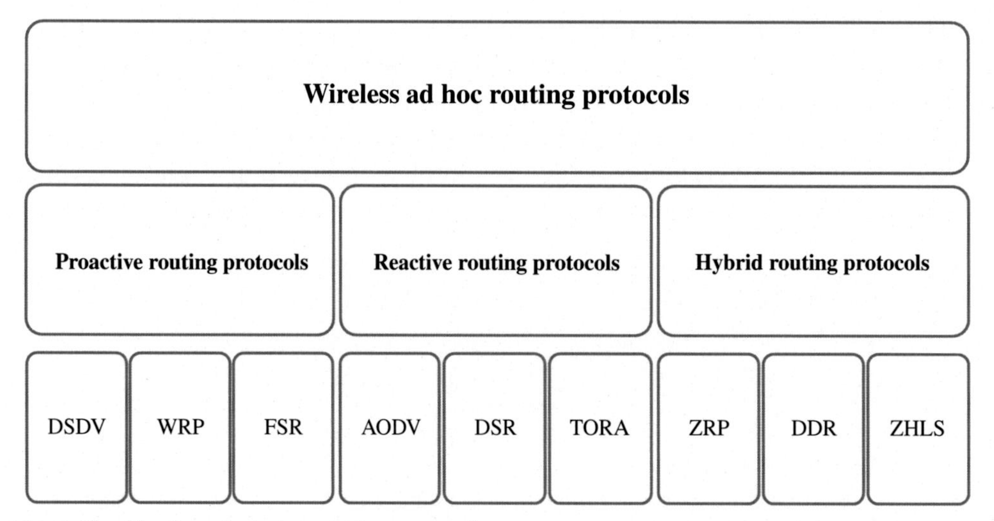

Fig. 5 Classification of wireless routing protocols.

algorithms. It is important to divide such protocols into three different groups: proactive, reactive, and hybrid [30]. The routes to all the endpoints are defined at start-up in constructive routing protocols and they are achieved using a standard route updating method. Routes are defined in adaptive protocols when they are needed by the origin using a method of route discovery [31]. Hybrid routing protocols incorporate the basic characteristics of the first two routing protocols. That is, by essence, they are both responsive and constructive. Every group has several routing strategies using a flat or hierarchical routing structure [7].

1.3.1 DSR (dynamic source routing)

It is a simple and efficient protocol for multihop ad hoc networks [32]. The route discovery process initiates from the source on an on-demand basis [32]. This route discovery is based on the link-state algorithm. The destination address, including the address of the intermediate nodes, will be loaded into each packet. This protocol creates multiple routes for multiple senders to any destination. Moreover, it allows each sender to select and control their routes [33].

Figs. 6 and 7 show the working concept of the DSR protocol. In DSR [34], the source node floods route request (RREQ) with the sequence number in the network by the broadcast method. The sequence number is used to avoid the recursive looping in the RREQ. The corresponding receiver will reply to the sender, and this reply is called route reply (RREP). The same RREQ path is used by the receiver to send the RREP, and the same path might be saved in the source's cache for maintaining the route.

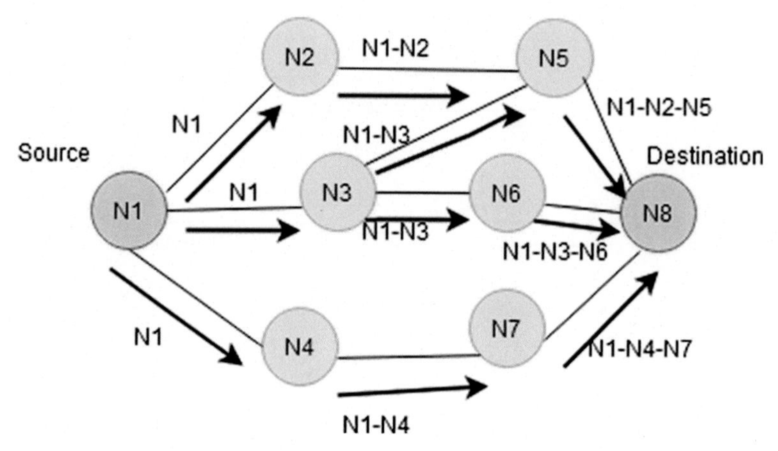

Fig. 6 Route request in DSR.

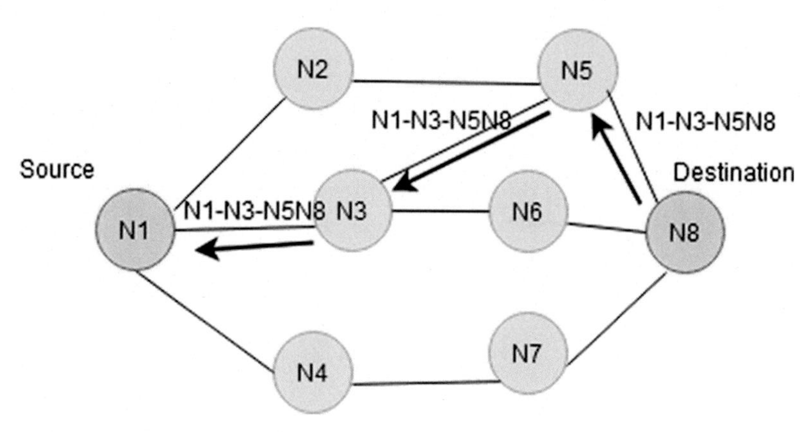

Fig. 7 Route reply in DSR.

1.3.2 AODV (ad hoc on-demand distance vector routing)

AODV is a purely on-demand routing protocol [30]. The nodes need not maintain the neighboring node information specified, nodes communicate with each other [35]. The selected route is maintained until the end of the session only. It supports both unicast and multicast routing in the MANET. Figs. 8 and 9 show the working concept of the AODV protocol. There are three control messages called RREQ, RREP, and RERR (route error) [30] used for unicast routing. The RREQ initiates when the sender wants to send data. The RREP [30] includes a destination address, sequence number, and hop count information. The RERR maintains the list of unreachable nodes and their sequence number in the same route. The lifetime of the route is extended by sending hello

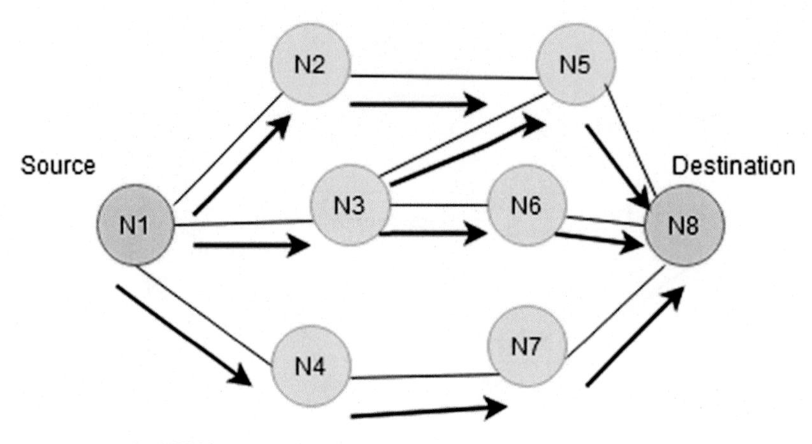

Fig. 8 Route request in AODV.

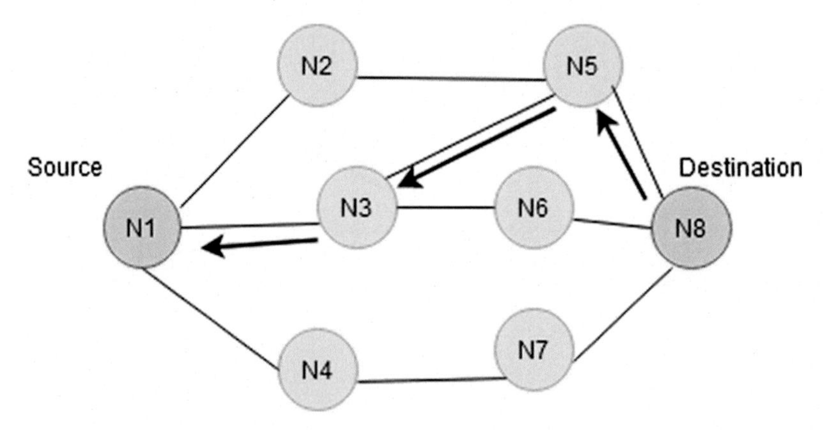

Fig. 9 Route reply to source in AODV.

messages periodically to its neighbor in the routing path [36]. A node that does not receive the hello message within the time limit is identified, and its node number is included in the RERR list.

This message is disseminated to its neighbor to update the broken node details in the table. Once the sender detects the RRER message in reverse path communication, the route reply acknowledgement (RREP-ACK) packet is sent by the sender in response to the RREP message. This status indicates the presence of a bidirectional link in the selected path [37].

1.4 Covering rough set (CRS)

The rough set theory (RST) introduced by Pawlak is a valuable mathematical tool for managing vague and uncertain data [38–40]. The rough set theory can accomplish a subset of all attributes, which preserves the clarity of original features using the data without any additional information. Pawlak's rough set model presents certain limitations when applied to complex decision tasks. However, an extension of the rough set is made available to overcome these limitations.

In this chapter, an extension of the rough set model, known as the CRS approach, is used to select the routes based on MN's resources. The RST is an important approach proposed by Pawlak [38] mainly for handling vague or imperfect data. The CRS approach can be used to reduce the attributes by redefining the approximation space of generalized rough sets [41]. In rough set theory, an equivalence relation is a fundamental concept. Several authors have proposed a modification of generalized rough sets such as modifying upper and lower approximations. In the fuzzy set theory, the selection of a member function itself is an uncertain problem. Therefore, the fuzzy set theory fails to deal with uncertainty. In the rough set concept, the lower and upper approximations are used to deal with imprecise knowledge. Therefore, the rough set theory can be applied to solve some uncertain problems [42–44].

Soft computing techniques are used for the decision-making process in many areas such as health care, mobility prediction, and route prediction [45]. In the health-care field, diseases are predicted using supervised classification models such as MLP [46], decision tree [47], and probabilistic neural network [47].

1.4.1 Motivation and contributions

In general, selecting a route between nodes is always a rigorous task in MANET due to high mobility, i.e., nodes move rottenly in a simulation area. Also, a review of authors who have already applied route selection algorithms such as RST and weighted rough set (WRS) to reduce redundant control packets is provided [48,49]. In wireless networks,

1. Mobility affects routing.
2. Mobility affects signal transmission; it leads to path failure between source and destination nodes.
3. Bandwidth consumption is very high.

Research contributions:

1. To the reduce the problem of redundant control packets for route selection, a CRS is proposed and compared with existing routing protocols.
2. Increasing the packet delivery ratio.
3. Reducing the number of RREQ packets.

This chapter is organized as follows: The contributions and motivations are presented in Section 1. Section 3 explains the RST algorithm. The CRS for the route selection algorithm is described in Section 4. In Section 5, the experimental results are discussed. The salient features of the proposed algorithms are summarized in Section 6.

2. A review of rough set-based techniques for reducing redundant packets

This literature review seeks to clarify the different approaches to rough set studies in the literature. Authors have just begun designing RST-based routing algorithms for MANETs. This survey shows how these different approaches have particular strengths and drawbacks in their work. The review concludes that the standard rough set and extension of rough set approaches are distinct but share certain features. The following summary of kinds of literature addresses the problems of mobile ad hoc networks using rough set and extension of rough set concepts such as variable precision rough set (VPRS) and weighted rough set (WRS).

Stojmenovic et al. [50] implemented a dominating and neighbor elimination-based broadcasting algorithm for wireless networks. In this work, the authors have proposed to reduce the communication overhead or eliminate it by using the dominating sets. The significant feature of the proposed work is reliability. In addition to preserving the location of adjacent nodes, their management does not require any overhead contact. For efficient broadcasting, only retransmission by internal nodes in a dominant system is necessary. Current dominant sets are strengthened by using node degrees as the primary keys rather than their ids. They exclude neighbors that have already obtained the message and retransmit it only if the list of neighbors would require that the message is nonempty. Retransmission is often defined after the scheme of negative acknowledgements [50].

Khabbazian and Bhargava [51] investigated reducing redundant broadcast in wireless ad hoc networks. Reducing the number of unnecessary communications is one of the critical goals of practical wireless ad hoc network-transmitting algorithms. There are several possible regional broadcast algorithms to reduce transmitting numbers. In the worst-case scenario, though, they do not enforce a reasonable cap on the number of transmissions and the first localized broadcast algorithm guarantees maximum distribution; and a constant approximation to the minimum number of communications is needed in the worst-case scenario. The proposed broadcast algorithm is a self-pruning algorithm focused on knowledge regarding neighbors in one hop. The experimental results validate the algorithm's analytical analysis and show a substantial decrease in the number of transmissions; and end-to-end latency is compared to one of the better broadcast algorithms based on neighboring one-hop details [51].

Nagaraju et al. [52] applied a heuristic technique to the AODV protocol to reduce redundant broadcasts. This approach is used to boost the efficiency of standard protocols for wireless routing. The heuristic analysis technique was adopted in the cycle of seeking a route from the source node to the destination node to prevent the issue of the flood of wireless transmission. The heuristic method takes into account the MANET characteristics such as bandwidth and the number of nodes inside the given range. When a node source (S) tries to submit a packet to destination (D) in a flooding system, S sends a packet to all its neighboring nodes; however, the suggested scheme aims to limit broadcasting by seeking the neighborhood heuristic steps of S. The heuristic steps are measured with the implementation of a function to all neighbors. A heuristic is a technique that allows a search method to be more effective, likely by sacrificing claims of completeness [52].

Rajam et al. [53] proposed temporary information systems, which are then applied to ad hoc mobile routing. Each mobile node keeps a cache of known routes. In this article, it is seen that the latest directions in the route cache need to be given further attention. This has contributed to the notion of weighted elementary sets in temporal information systems, where more focus was given to more recent primary sets. Recent routes (DSRrecent) were found to increase the packet delivery ratio and standardize overhead control, and temporary information has been introduced into the information systems. New elementary sets in the two proposed methods, TIME_WT and variable precision rough set (VPRS), were been given more importance. The VPRS method makes use of VPRS notions. The control overhead was seen to be much higher, while the packet transmission ratio, average hop duration, and average end-to-end delay were slightly better than DSRs. It has also been seen that efficiency enhancement increases with the number of connections [53].

Aitha et al. [54] proposed a technique to reduce broadcasting by finding equivalence classes using RST. This protocol aims to create a consistent long-term route from the source node to the destination node. This ensures that the format is more commonly used in various implementations such as digital and video conferencing. In our suggested process, rough set details accessible at each node would be used instead of locating the next node to create the route request step. This form of raw set method manages the overhead of the AODV routing algorithm route request (RREQ) phase by deleting the corresponding nodes [54].

Nagaraju et al. [55] examined the AODV routing protocol using the neighbor matrix method in stable wireless ad hoc networks. The established ad-hoc on-demand routing protocol (AODV) for MANETs is more suited for extremely complex environments. Still, many of the topology conditions shift gradually in typical ad hoc networks. In the literature, the proposed neighbor matrix approach is used to find the shortest path in static ad hoc networks. In the proposed method, whenever a node wants to see a way to the destination it evaluates the neighbor matrix of the existing ad hoc network, which helps in finding the path to the target using incremental neighbor searching till the goal has been reached. This method prohibits the current AODV reactive routing protocol for MANETs from submitting the path request process frequently. The proposed algorithm allows more efficient use of the neighboring matrix and evaluates the direction. This method prohibits the current AODV reactive routing protocol for MANETs from submitting the path request process frequently. The proposed algorithm allows more efficient use of the neighboring matrix and evaluates the direction [54].

Nagaraju et al. [55] had presented a model of reducing control packet load using the weighted rough set. The authors had proposed approaches to reduce the route request packet in the existing wireless ad hoc network protocol. Only a single reactive protocol has been used to control the packets using the weighted rough set model. This approach may not be suitable for both proactive and reactive routing protocols of the ad hoc network. The experimental findings indicate that due to less collision and less path split relation to the weighted rough set (WRS), the model provides more packet distribution rather than the AODV protocol. In a routing table, each MN maintains a unique ID, destination, hop count, next-hop, and routing table expiration time. Also, it improves the results of the classical rough set through the weighted rough set models. The WRS is used to reduce redundant broadcast packets in wireless networks [55].

Seethalakshmi et al. [56] introduced a path selection approach in a wireless network using fuzzy and rough set approaches. Also, the authors proposed a fuzzy multipath routing approach in a wireless network to find an effective shortest path routing based on minimum resources. The proposed work addresses the rough set concept for evaluating the best route between nodes, and it can be deployed to generate rules, and insignificant features (resources) are removed. Computer resources are bandwidth, computer efficiency, power consumption, and traffic load, and several intermediate nodes are considered as single path parameters in a network. Results showed that the proposed approach provides a much better outcome than others [56].

Nagaraju et al. [14] proposed a rough set and weighted rough set (WRS) method to determine better route request forwarding message node among multiple neighbor nodes. This approach is informed by rough set route selection based on the node's route request. This work highlights the importance of the hop count parameter in an ad-hoc wireless network. The simulation studies are made using a single hop count, two hop counts, and three hop counts. The simulation study is conducted based on the parameters such as pause time, bandwidth, relative distance, and battery power. In this work, equivalent nodes are computed using corresponding hop counts and based on equivalent nodes, lower and upper approximations are calculated. Finally, the findings of the simulation indicated that the proposed weighted rough set model provides better precision in terms of packet distribution ratio, end-to-end delay, and number of packets [14].

The following review of work confirms that the rough set-based approach can give best results compared to other prevailing traditional methods. Kumar et al. [57] have addressed a rough set calibration problem for routing protocols in mobile ad hoc networks. This approach is used for enhancing the performance of energy-efficient routing protocols. Also, the episodic-based association has been employed for every metric in a wireless network. The authors found that the point-based association function is used in the fuzzy set rather than an episodic association function, which reveals higher accuracy than others in terms of packet loss and throughput. Moreover, only a single reactive routing protocol is considered for this work. The following table shows the work of various authors, which addressed the problem of routing in a wireless network using fuzzy, rough, and soft computing approaches [57]. Table 1 summarizes the RST and different methods proposed for reducing redundant broadcast packets for route selection.

3. Rough Set Theory (RST) mechanism in manets

RST is a mathematical model proposed by Pawlak [39], which considered problems of uncertainty and vagueness. RST is an approach used to deal with vagueness. It initially has an information system, i.e., knowledge about elements are presented in the universal set. If an element in a set can be related to the same information, then the relationship can be similar or indiscernible. In Ref. [63], the author has shown the generalized rough set properties induced by some of the binary relations on a set. The lower approximation RX is defined by certain classified objects in a set X concerning R, and the upper approximation $\overline{R}X$ is defined by possibly classified objects in a set X concerning R. Moreover, the rough set theory is a calculus relation, and fuzzy sets are continuous relations of a set [64]. The difference between the lower and upper approximation is called the boundary region. Figs. 10 and 11 show the general rough set stages and concept.

Why rough set?

- To research intelligent systems that are characterized by inadequate and incomplete knowledge, rough sets are used.
- The main purpose of the rough set is to induce (learning) approximations.
- It provides mathematical tools for finding patterns hidden in knowledge.
- The indiscernibility relationship is the starting point for the theory of rough sets [65].

The generalized rough set theory has some fundamental properties that should satisfy the elements in set X, such as reflexive, symmetric, and transitive [66]. The rough set theory is an approximate description of sets, and the equivalence classes of partition R are called elementary sets [40,50,67,68].

Definition 1. [69–71]: The basic concept of a rough set is discussed in the rest of the sections. Let $I=(U, CA\cup\{da\})$, which is an information table [45,63], where U is the universe with a nonempty finite set of objects, and CA is the conditional attributes with a nonempty finite set of objects, da is the decision table which contains decision attributes, and $\forall a\in R$ is a corresponding function $f_a: U\rightarrow V_a$. An equivalence relation denoted by $R\subseteq CA$ is given in the following equation.

$$IND(R) = \{(x, y)\in U \times U|\forall a\in R, f_a(x) = f_a(y)\} \quad (1)$$

If x and y are indiscernible attributes [45,63] from R denoted by $(x, y)\epsilon IND(R)$, the equivalence class of the R-indiscernibility relation is indicated by$[x]_r$. The lower

Table 1 Summary of reducing redundant broadcast packets using RST and other techniques.

S. no	Year	Authors	Technique	Parameters	Routing protocols	Limitations
1.	2002	Ivan Stojmenovic et al. [50]	Broadcasting algorithm	Bit rate, slot time and overhead	DSR	Single hop distance
2.	2007	Majid Khabbazian and Vijay K. Bhargava [51]	Self-pruning algorithm	One-hop neighborhood information	–	Self-pruning suitable for one-hop network
3.	2007	A. Nagaraju et al. [52]	Heuristic tchnique	Bandwidth and no of nodes	AODV and DSR	Delay of the network is high
4.	2008	V. Mary Anita Rajam et al. [53]	Temporal information system for MANETs	Route Cache table parameters (hop count, source and dest id, etc.)	DSR	Static approach
5.	2009	A. Nagaraju et al. [14]	Rough set	Pause time, traffic, relative distance and battery power	DSR and AODV	Used general rough set and weighted rough set
6.	2011	Aitha Nagaraju et al. [54]	Neighbor matrix technique	Bandwidth	DSR and AODV	Single parameter not sufficient to evaluate the routing protocol
7.	2011	Nagaraju Aitha et al. [55]	To reduce control packets using the weighted rough set approach	Pause time, Traffic, relative distance and battery power	DSR and AODV	Mobility models were not used
8.	2011	Seethalakshmi et al. [56]	Efficient path selection using fuzzy and rough set techniques	Bandwidth, computer efficiency, power consumption, traffic load, and number of intermediate nodes	–	Static network

9.	2014	Nagaraju et al. [14]	Reduce redundant broadcast using rough sets	Pause time, bandwidth, relative distance and battery power	DSR and AODV	Time complexity
10.	2018	Sathish Kumar et al. [57]	The rough set calibration scheme	Energy and distance	DSR and TORA	Applied only general rough set
11.	2019	Rafia Kulsum et al. [58]	Analyze AODV protocol using soft computing	Distance and density	AODV	Applied basic soft computing approaches
12.	2020	Banghua Wu et al. [59]	Rough set approach for identifying attacks on data packets	End-to-end delay	–	Applied classification algorithm using rough set
13.	2020	Masood Ahmad et al. [60]	Cluster optimization using memetic algorithm	Distance	DGAC and EMPSO	Other conventional routing protocols not used
14.	2021	Bonu Satish Kumar et al. [61]	Enhance the performance of AODV using soft computing approaches	Active route imeout (ART)	AODV	Stuck with only one protocol
15.	2021	Dimitris and Cuomo [62]	Recent developments in MANET	Channel rate and mobility	MAC protocols	Comparison of protocols

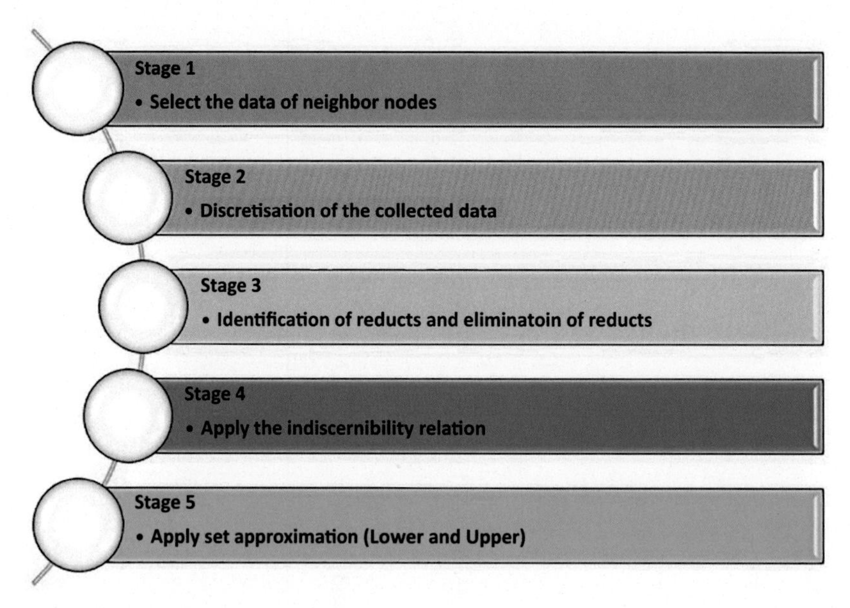

Fig. 10 The general rough set process stages.

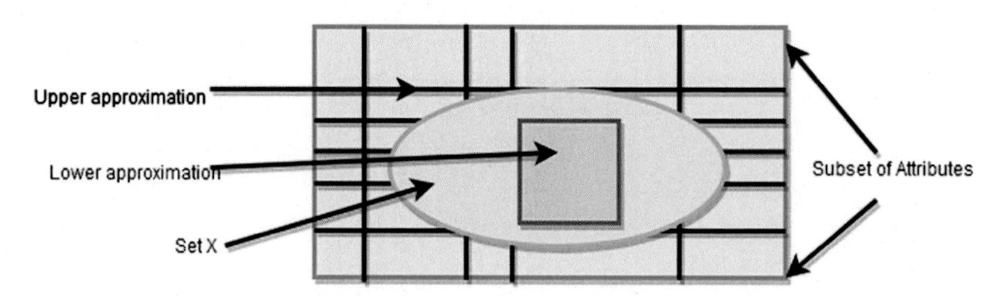

Fig. 11 Rough set theory [65].

approximation $\underline{R}X$ and the upper approximation $\overline{R}X$ of set X; $X \subseteq U$ are given in the following equations:

$$\underline{R}X = \{x \in U | [x]_r \subseteq X\} \tag{2}$$

$$\overline{R}X = x \in U | [x]_r \cap X \neq \emptyset \tag{3}$$

The positive region and the negative region are given in Eqs. (4) and (5): PBR_r and NBR_r where Q is denoted by $Q \subseteq CA$, equivalence relation over universe U. The difference between the upper approximation and the lower approximation is defined in Eq. (6).

Algorithm: Using 1-hop nodes to find equivalent node

Input: Matrix with parameters of 1-hop MNs in terms of Bandwidth and Hop count

Output: 1-hop equivalent class nodes

Step1: Discretize the 1-hop node matrix using binning algorithm

Step 2: Find the indiscernibility by matching the discretized inputs using equation (1)

Step 3: Step 2 results are denoted as [x]. For each subset, obtain equivalence class with 1-hop nodes.

Fig. 12 Using 1-hop nodes to find equivalent nodes.

Algorithm: Using 1-hop and 2-hop nodes

Input: *Matrix with parameters of 1-hop and 2-hop nodes*

Output: *Lower and upper approximations of 1-hop and 2-hop nodes*

Step 1: *Discretize the 1-hop and 2-hop matrix using the binning algorithm.*

Step 2: *Apply the indiscernibility condition on 1-hop and 2-hop node inputs and compute the equivalent classes.*

Step 3: *Evaluate the membership value using the below Equation.*

$$\mu_x^B(X): U \rightarrow [0,1] \text{ and}$$

$$\mu_x^B(X) = \frac{|[x]_B \cap X|}{|[x]_B|}$$

Step 4: *Apply lower and upper approximation using the Equations (2 and 3) and get the 1-hop and 2-hop lower upper approximations.*

Fig. 13 1-hop and 2-hop MNs in lower and upper approximations [14].

$$PBR_r(Q) = \bigcup_{X \in U/Q^{-}} RX \tag{4}$$

$$NBR_r(Q) = U - \bigcup_{X \in U/Q} \overline{R}X \tag{5}$$

$$BDR_r(Q) = \bigcup_{X \in U/Q} \overline{R}X - \bigcup_{X \in U/Q^{-}} RX \tag{6}$$

Figs. 12 and 13 explain how the lower and upper 1-hop and 2-hop MNs are generated based on the conventional rough set [14].

4. Reducing redundant control packets using covering rough set in manet for route selection

4.1 Covering rough set in manets

CRS is an extension of the classical rough set to manage more unpredictable issues [70] in an information system. Route selection is a critical decision for sending information to a reliable path. In this work, in the first step, collect the data through a network simulator (NS2) [72] and create a dataset, i.e., information gain table, which comprises attributes and instances. Node attributes are pause time, bandwidth, relative distance, and battery power [14]. Pawlak's classical rough set works are based on equivalence relation or partition of U, where the equivalence relation is replaced by the covering of U in CRS. The key point of the rough set is lower and upper approximations of set X, which are defined as Eqs. (7) and (8). The primary goal of the CRS is to solve two important problems such as reasonable set approximations and attribute reduction. The covering created by the symmetric relation and reflexive is distinctive [70]. Firstly, it creates the equivalence relation of the covering and its property that is the intersection of two elements of the union of finite elements of a covering [69,71–77]. The classical rough set is used to extract the rules and reduce attributes by partitions; although, in the CRS theory, the same attribute is characterized by covers [78].

Definition 2 (Covering). [69–71]: Let U be a universe of discourse and C be a lineage of the subset of U. The covering approximation is $\cup C = U$; C is called a covering of U,which is called the covering approximation.

Let (U, C) be an ordered pair; U is called the universe, which should be a nonempty set. C is a covering finite element, i.e., none of the C subsets of U are empty. Let (U, C) be the approximation space, where C is $\{K(x)/x \in U\}$ for any set $X \subseteq U$.

$$\underline{C}(X) = \{K(x) \epsilon C : K(x) \subseteq X\} \tag{7}$$

Eq. (7) is defined as the approximating set X. $\underline{X} = \cup \underline{C}(X)$ is called the lower approximation of the set X.

$$\overline{C}(X) = \{\cup\{\cap\{K \in C/x \in K(x)\}/x \in X\}\} \tag{8}$$

Eq. (8) defined as the approximation of set $\overline{X}$. $\overline{X} = \cup\overline{C}(X)$ is called the upper approximation of the set X. Table 2 explains the properties of covering the lower approximation $\underline{C}(X)$ and the upper approximation $\overline{C}(X)$ [74].

4.2 CRS-based route selection

The aim of CRS is to minimize the packets for path requests in the conventional routing protocols for AODV and DSR. The node information is translated into information about the rough set. In this work, we evaluated the forwarded nodes using 1-hop and

Table 2 Properties of covering lower and upper ($C(X) and \bar{C}(X)$).

S. no	Properties	
1	$X_L \subseteq X \subseteq X_U$	**Contraction and extension**
2	$\varnothing_U = \varnothing, \varnothing_L = \varnothing$	Normality
3	$U_U = U, U_L = U$	Co-normality
4	$(X \cup Y)_U = X_U \cup Y_U$	Addition
5	$X \subseteq Y \Rightarrow X_U \subseteq Y_U$	
6	$X \subseteq Y \Rightarrow X_L \subseteq Y_L$	Monotone
7	$X_L \subseteq X_U$	Appropriateness
8	$(X \cap Y)_L = X_L \cap Y_L$	Multiplication
9	$(X_L)_L = X_L$	
10	$(X_U)_U = X_U$	Idempotency

Algorithm: Using 1-hop nodes to find equivalent nodes based on CRS
Input: *Matrix with parameters of 1-hop MNs*
Output: *Equivalent class with 1-hop MNs*
Step 1: *Apply Covering relation on 1-hop nodes.*
Step 2: *Finally, Compute the equivalence class of the 1-hop node's covering Objects (i. e $C_1, C_2, C_3, C_4, \ldots.. C_n$).*
Step 3: *Output the 1-hop equivalence class nodes.*

Fig. 14 Using 1-hop nodes to find equivalent nodes based on CRS.

Algorithm: Selection process at the source node
Input: *Source node's Source and destination address, hop count, the sequence number, and broadcasting id.*
Output: *Selected 1-hop and 2-hop nodes.*
If (source node's neighbour node is the destination)
then the route request packet will be immediately forwarded to the Destination node's address.
Else
The route request packet will be forwarded to every equivalent class of 1-hop and 2-hop nodes.

Fig. 15 1-hop and 2-hop MNs in CRS lower and upper approximations.

2-hop mobile nodes. If we consider the equivalence class to find 1-hop neighbor nodes, then it is a self-pruning strategy. At a unique instance of the ad hoc network, similar conditions are extended to all neighboring nodes. Figs. 14 and 15 explain the proposed CRS-based route selection in MANETs to minimize the route request packets. The following points explain the limitation of the rough set and the difference between the rough set and covering a rough set.

A downside of conventional rough sets is that they can accommodate only discrete databases. Therefore, before attribute reduction, continuous databases need to be discretized. The advantages of CRS are to avoid attribute discretization and to be able to deal directly with numerical data. Figs. 14 and 15 explain how CRS-based 1-hop and 2-hop nodes are generated for route selection algorithms.

The orthodox rough set theory highlights two limitations:

- In the modern world, databases are numerical such that they are not explicitly handled by conventional RST.
- Before attribute reduction, numerical data must be discrete, which eventually leads to information loss.

Why CRS?

- On equivalent relations or partitions, conventional rough sets are created.
- However, for many applications, the necessity of an equivalence relationship as the relation of indiscernibility is too restrictive.
- Various extensions of a partition or equivalence relationship have been suggested, such as coverings, to solve this problem.
- Taking the definition of partition as main, several writers have extended the concept of rough sets to include rough sets, where the universe's coverage substitutes the relationship of equivalence.

4.3 Applying CRS-based route selection to AODV

This section describes information about the application of the proposed CRS to the existing AODV routing protocol to minimize broadcasting. In RREQ, we have made improvements to the current routing algorithm, AODV. The RREQ packet is received in the proposed covering rough set-based AODV (CRS-AODV) protocol only by selected nodes. Node selection relies on two ways: first, using self-pruning, i.e., using 1-hop nodes. These 1-hop nodes are divided by constructing a covering connection in it, and another technique is based on dominant pruning, which divides the 1-hop nodes into lower and higher approximations using information from both neighbor nodes for the hop and 2-hop [14].

We made a few changes to the current AODV parameters, which are the time to leave (TTL) parameter and the number of withdrawals for route requests. TTL has been set to 20, and the path request number has been set to 10. In our implementation, we changed the RREQ packet to hold information about 1-hop and 2-hop neighbors, similar to hello packets to preserve neighbor information. Figs. 16–21 explain the steps of the selection process at the source and destination nodes. Also, the pruning algorithm has been used to control the redundant route request message.

```
Algorithm: Selection process at the intermediate node
Input: Intermediate node's Source and destination address, hop
       count, sequence number, and broadcasting id.
Output: Selected 1-hop forwarding nodes.
If (the packet of path route requests is outdated)
        then the route request packet will be removed
Else
        If (the route request packet is new and the neighbour node is the
          destination)
              the route request packet will be broadcast to the destination
              node.
        Else
              the route request packet will be forward to selected 1-hop
              nodes n such a way that one node from each equivalence
              the class covers all of the forwarding node's 1-hop nodes.
```

Fig. 16 Steps of the selection process at the source node.


```
Algorithm: Selection process at the intermediate node
Input: Intermediate node's Source and destination address, hop
       count, sequence number, and broadcasting id.
Output: Selected 1-hop forwarding nodes.
If (the packet of path route requests is outdated)
        then the route request packet will be removed
Else
        If (the route request packet is new and the neighbour node is the
          destination)
              the route request packet will be broadcast to the destination
              node.
        Else
              the route request packet will be forward to selected 1-hop
              nodes n such a way that one node from each equivalence
              the class covers all of the forwarding node's 1-hop nodes.
```


Fig. 17 Steps of the selection process at an intermediate node.

This section explains not only the AODV protocol but is also applied to the DSR protocol. Each node will decide its status as a forward or nonforward node for broadcasting based on self-pruning. The self-pruning algorithm only exploits the knowledge of neighborhood information that is directly associated. A node need not retransmit a packet if the previous transmission covered all of its neighbors. The dominant pruning algorithm uses knowledge about the 2-hop neighborhoods.

Algorithm: CRS-AODV (RREQ) using self-pruning for single-hop node
Input: *Node's source and destination address, hop count, sequence number.*
Output: *Selected single-hop self-pruning nodes.*

Step 1: *Each node collects and verifies the attribute values of single-hop neighbour nodes within a time interval.*
Step 2: *Apply covering relation (relationship of equivalence) on the 1-hop nodes. This step divides the neighbouring 1-hop nodes into equivalent classes.*
Step 3: *Source node Initializes the route request process when the path to the destination needs to be identified.*

Fig. 18 CRS-AODV (RREQ) using self-pruning for a single-hop node.

Algorithm: CRS-AODV (RREQ) using dominant pruning for 2-hop nodes
Input: *Node's source and destination address, hop count, sequence number.*
Output: *Selected lower and upper approximations of 1-hop and 2-hop dominant pruning nodes.*

Step 1: *Compute Lower and upper approximation for 1-hop and 2-hop discrete data using Equations (7 and 8)*
Step 2: *The route request packet is initiated by the source node when it needs to discover the Route to the destination address node.*

Fig. 19 CRS-AODV (RREQ) using dominant pruning for 2-hop nodes.

Algorithm: Selection process at source node 2-hop
Input: *Source node's Source and destination address, hop count, The sequence number, and broadcasting id.*
Output: *Selected 2-hop lower approximation nodes.*

If (source node's neighbour node is the destination)
The RREQ packet will then be immediately forwarded to the destination.
Else
In each equivalent class of nodes, RREQ will be forwarded to CRS lower approximation MNs which covers all the 2-hop nodes.

Fig. 20 Steps of the selection process at the source node 2-hop.

RREQ consists of the source IP address, destination IP address, number of source sequences, number of distance sequences, broadcast ID, and hop count. The source node initializes all these values. As specified in Fig. 18, the nodes are selected, they are imposed

Algorithm: Selection process at intermediate node 2-hop
Input: *Intermediate node's source and destination address, hop count, the sequence number, and broadcasting id.*
Output: *Selected 2-hop lower approximation nodes.*

If (the packet of path route requests is outdated)
    then the route request packet will be removed.
Else
    If (the route request packet is new and its neighbour node is the destination)
        then the route request packet will be broadcast to the destination node.
    Else
        If (the CRS lower approximation node covers all the 2-hop nodes)
            then the route request packet will be forwarded to CRS lower approximation 2-hop nodes.
        Else
            Change the lower approximation nodes in such a way that all 2-hop nodes are protected. Then the route request packet is broadcast to lower approximation nodes that have been updated.

Fig. 21 Steps of the selection process at the intermediate node 2-hop.

into equivalent classes, one node is selected from each class, and the RREQ is sent. This process is replicated with a given number of requests for routes.

5. Experimental results and simulation environment set-up

5.1 Simulation tool

The network simulator (NS 2.34) tool is used for this work. The NS2 is a discrete-event simulator tool for networking research, and it is a combination of the tool command language (TCL) and C++ languages. We can simulate both wired and wireless networks, and the role of TCL is to specify the scenarios and events. The role of C++ is to design protocols. Table 3 depicts the simulation parameters and their values. An initial simulation starting time is 5 s and a simulation ending time is 250 s. This tool produces two files named trace and nam. The trace file accommodates the source and destination addresses, packet size, and packet received time. Using the nam window, we obtain a graphical view of mobile nodes [79].

This section presents the performance of the different route selection algorithms. The performances of the proposed CRS-AODV and CRS-DSR algorithms are compared with various existing route selection algorithms such as AODV, DSR, R-AODV,

Table 3 CRS Simulation parameter setup.

Parameters	Simulation values
Simulation initiate time	05.0 s
Simulation ending time	250.0 s
Antenna type	Omni directional
Coverage area	1000 m * 1000 m
Channel type	Wireless
Energy	Default in NS2
Propagation model	TwoRayGround
Base routing protocols	DSR and AODV
Mobility model	RWP
Pause time	10–30 s
Number of nodes	100
Network type	Wireless
MAC type	802.11

Table 4 Performance analysis of CRS route selection under AODV.

Route selection algorithm	Number of nodes	No. of RREQ packets	Average throughput (Kb/s)	Delay (s)
AODV	25	5621	72.6	0.01752
	50	4992	70.4	0.03344
	75	4850	60.4	0.08374
	100	4552	63.3	0.01374
R-AODV	25	4689	76.5	0.01845
	50	4985	73.6	0.01523
	75	4600	75.4	0.05526
	100	4722	69.5	0.02475
CRS-AODV	25	3985	80.6	0.00452
	50	3674	76.4	0.02575
	75	3584	79.8	0.03945
	100	3900	70.8	0.02588

and R-DSR. The performance analyses of route selection (AODV and DSR) are presented in Tables 4 and 5.

5.2 Evaluation metrics

The following metrics are evaluated in the simulation.

Delay: Using Eq. (9), the average delay of packet delivery is calculated between the arrival time of data packets and the sent time of data packets, followed by the total number of connections in topology.

$$\text{Delay} = \sum PAT_i - PST_i \tag{9}$$

Table 5 Performance analysis of CRS route selection under DSR.

Route selection algorithm	Number of nodes	No. of RREQ packets	Average throughput (Kb/s)	Delay (s)
DSR	25	4651	75.6	0.01672
	50	4592	69.4	0.02474
	75	4650	62.25	0.01842
	100	4300	63.4	0.01637
R-DSR	25	4056	76.5	0.01175
	50	4481	71.6	0.01644
	75	4330	70.45	0.01654
	100	4152	70.5	0.01869
CRS-DSR	25	3607	84.5	0.00366
	50	3032	79.4	0.01514
	75	3451	78.23	0.01588
	100	3956	78.6	0.01447

where PAT is the packet arrival time and PST is the packet start time, for the ith packet [13].

Throughput: Throughput is calculated by data packets successfully delivered from one node to another node over a communication network. Normally, the below equation shows the calculation of throughput where n is the number of data packets, which usually takes bits/second.

$$\text{Throughput} = \sum\nolimits_i \frac{PD}{PAT - PST} \tag{10}$$

where PD is the packet delivery, PAT is the packet arrival time, and PST is the packet start time [13].

The number of RREQ packets [14]: A path request packet is sent to the on-demand routing protocol node, whenever it needs to be identified. RREQ packets are sent across the network to find the route to the destination. The total number of route requests sent by all nodes consumes the bandwidth in consequence [14].

The performance of the proposed methods is compared with that of other route selection algorithms in practice and is depicted in Figs. 22–27 in terms of throughput, PDR, and delay. It is noted from this line chart that CRS-AODV and CRS-DSR have shown higher throughput and fewer RREQ packets and delays than other algorithms.

It is observed that the CRS-AODV and CRS-DSR algorithms have produced the highest accuracy of 80% and 84%, respectively, for throughput. This has a higher accuracy rate compared with other route selection algorithms such as AODV, DSR, R-DSR, and R-AODV.

In AODV, the delay is lower since paths from the source to the destination are found in this short duration. The routing can be achieved quickly and so the average delay for

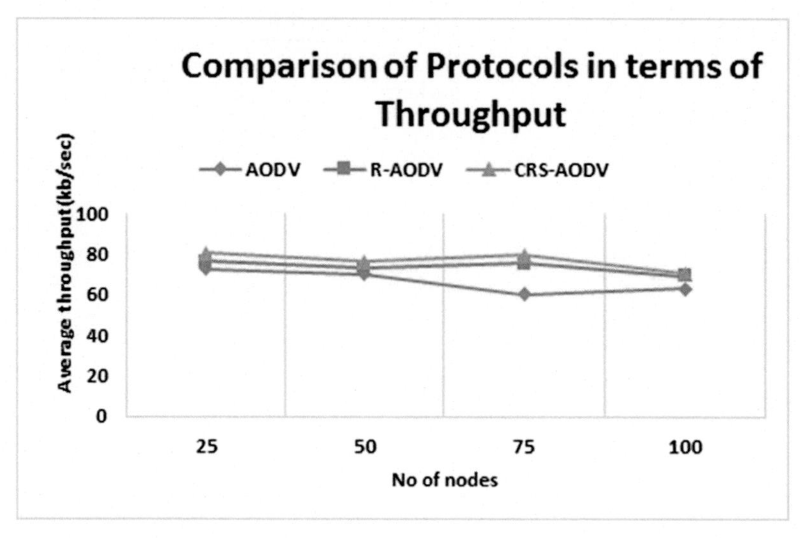

Fig. 22 Performance of AODV based on throughput.

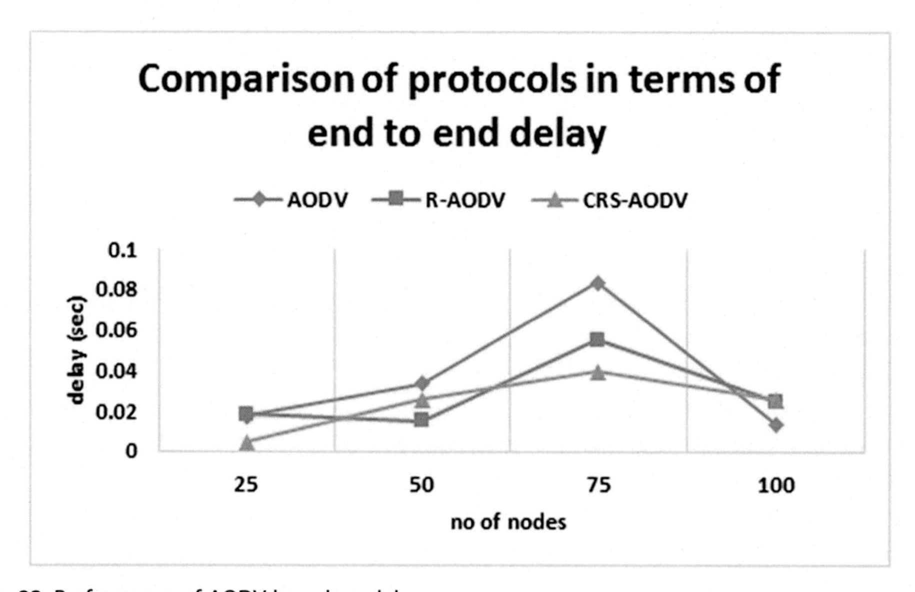

Fig. 23 Performance of AODV based on delay.

pause time is less than 20–30 s. A node's pause time increases as the node is active in managing the routing of more route lengths. In this case, CRS-AODV and CRS-DSR perform well during a node's high pause period. In this situation, long path lengths are specified.

It is noted that minimum delay and RREQ are recorded in CRS-based route selection algorithms when compared to other algorithms. Hence, it has been proved that the

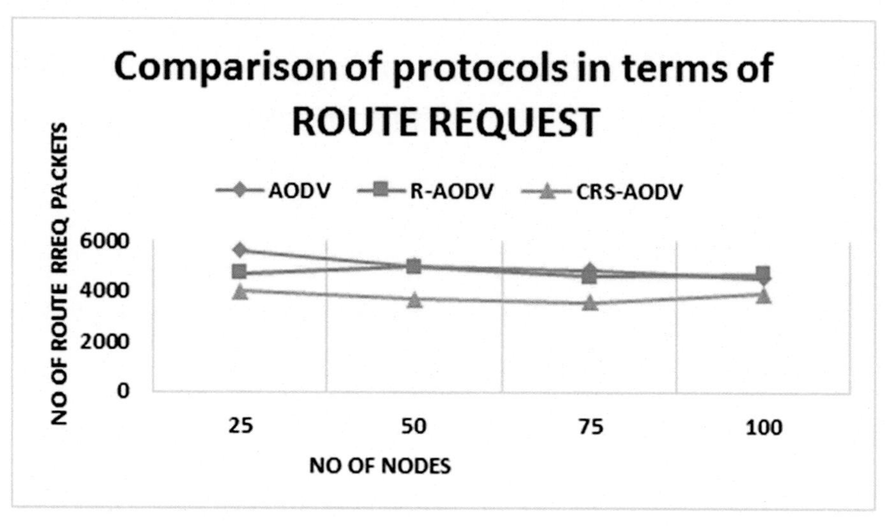

Fig. 24 Performance of AODV based on RREQ.

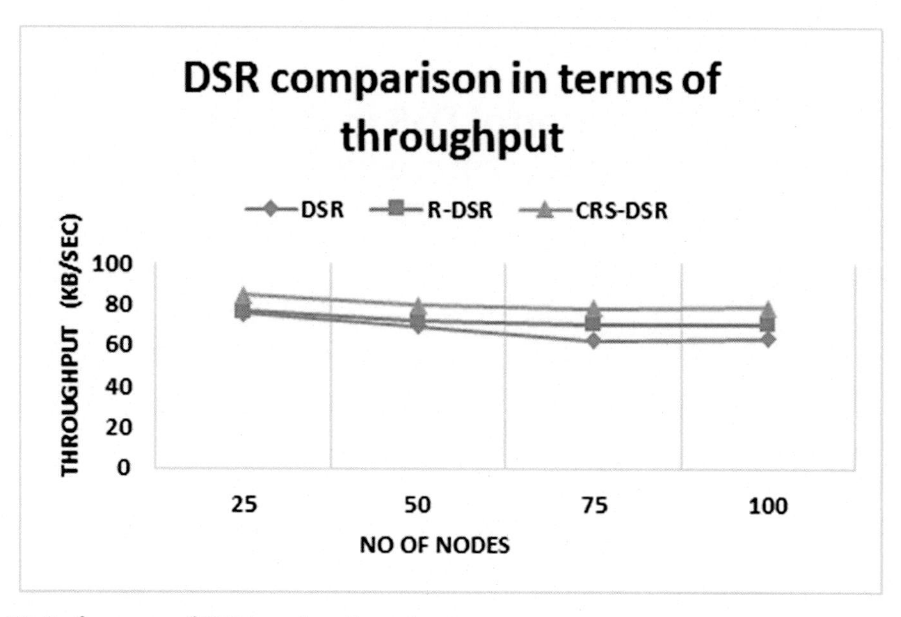

Fig. 25 Performance of DSR based on throughput.

proposed route selection algorithm is capable of selecting the route with fewer RREQ packets.

5.3 Simulation analysis

In this section, the proposed covering rough set-based routing protocols have attained higher accuracy in terms of packet delivery ratio, throughput, and reducing the number

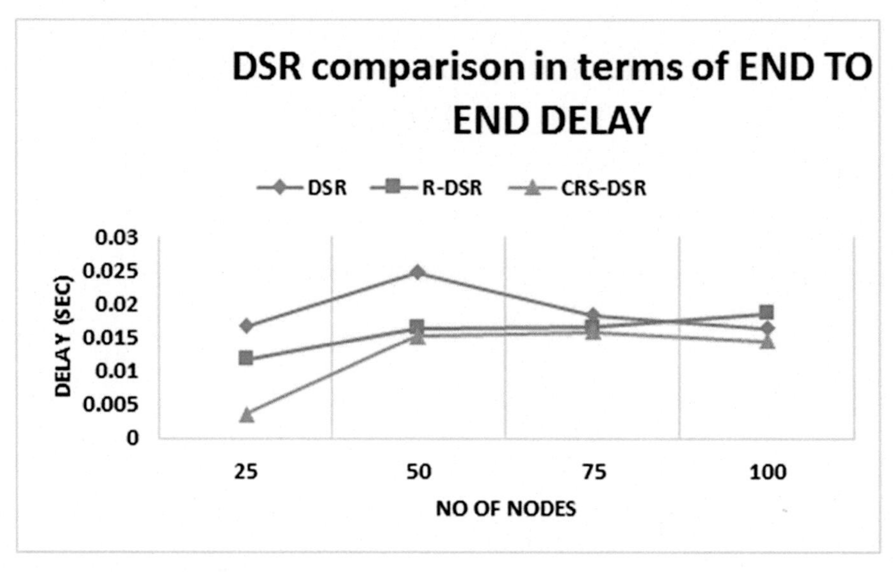

Fig. 26 Performance of DSR based on delay.

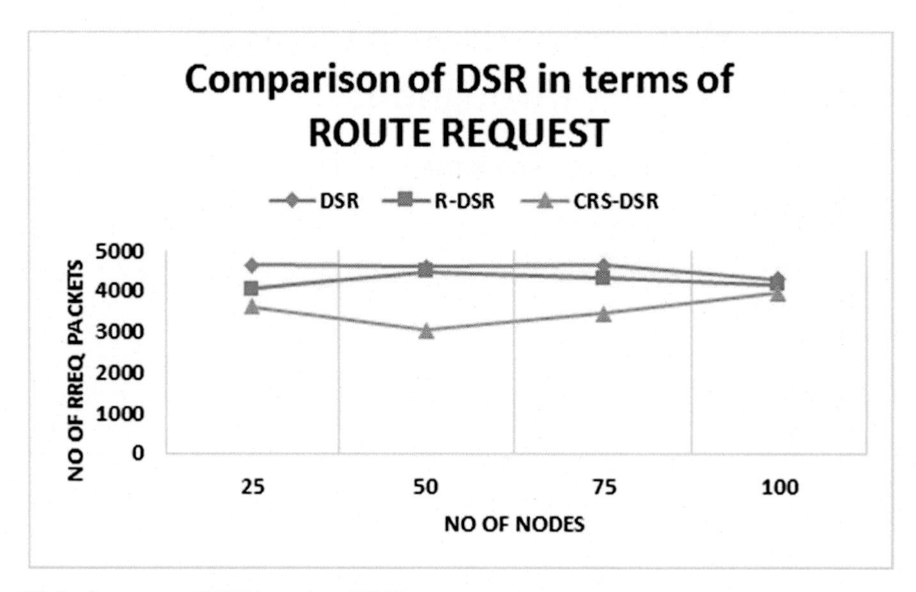

Fig. 27 Performance of DSR based on RREQ.

of route request packets. Meanwhile, pruning algorithms are used to reduce the number of route request packets. In this work, we have used a single pruning algorithm for a single-hop network.

Fig. 24 explains that the number of route request packets has been reduced when compared with the general rough set concept. The proposed covering rough set approach has been outperformed by other models such AODV-based route request and DSR-based route request. When the number of nodes is increased, the proposed approach can reduce the redundant control packets.

Furthermore, the delay of the proposed work has been reduced when compared with AODV and DSR routing protocols demonstrated in Figs. 23 and 26. When the number of nodes in a network grows, the network's performance suffers.

6. Conclusion and future direction

Ad hoc networks are stable for a limited stipulated period in certain cases, and this stability is used to gather information about the neighboring node that is retained with each node. In this chapter, two CRS-based on-demand route selection algorithms have been used to reduce the control packets in a wireless network. Different route selection algorithms, namely, AODV and DSR, were employed to assess the performance of the route selection algorithms. From the experimental results, it is noted that the proposed algorithms produce better results compared to other route selection algorithms in terms of evaluation metrics. It is also noted that the proposed route selection increases the throughput and minimizes the delay. Further, the selected routes can be used to retransmit the packets. Besides, this long-term, legitimate path reduces the number of unwanted control packets for route requests. In future, the number of nodes can be increased and other soft computing techniques such as optimization and fuzzy concepts can be applied to improve the performance of networks.

References

[1] T. Sudhakar, H.H. Inbarani, S. Kumar, Route classification scheme based on covering rough set approach in mobile ad hoc network (CRS-MANET), Int. J. Intell. Unmanned Syst. 8 (2) (2019) 85–96.
[2] A. Nayyar, B. Mahapatra, Effective classification and handling of incoming data packets in mobile Ad Hoc networks (MANETs) using random forest ensemble technique (RF/ET), in: Data Management, Analytics and Innovation, Springer, Singapore, 2020, pp. 431–444.
[3] F. Bai, A. Helmy, A survey of mobility models, in: Wireless Adhoc Networks, 206, University of Southern California, USA, 2004, pp. 1–30.
[4] M.G. Rubinstein, I.M. Moraes, M.E.M. Campista, L.H.M. Costa, O.C.M. Duarte, A survey on wireless ad hoc networks, in: IFIP International Conference on Mobile and Wireless Communication Networks, Springer, Boston, MA, 2006, August, pp. 1–33.
[5] M.N. Alslaim, H.A. Alaqel, S.S. Zaghloul, A comparative study of MANET routing protocols, in: The Third International Conference on e-Technologies and Networks for Development (ICeND2014), IEEE, 2014, April, pp. 178–182.
[6] I. Ahmad, U. Ashraf, A. Ghafoor, A comparative QoS survey of mobile ad hoc network routing protocols, J. Chin. Inst. Eng. 39 (5) (2016) 585–592.
[7] M. Khari, P. Kumar, G. Shrivastava, Test optimisation: an approach based on modified algorithm for software network, Int. J. Adv. Intell. Paradig. 17 (3/4) (2020) 208–237, https://doi.org/10.1504/IJAIP.2020.109508.

[8] C. Kumar, B. Bhushan, S. Gupta, Evaluation of MANET performance in presence of obstacles, Int. J. Ad Hoc Ubiquitous Comput. 3 (3) (2012) 37–46.
[9] S. Taneja, A. Kush, A survey of routing protocols in mobile ad hoc networks, Int. J. Innov. Technol. Manag. 1 (3) (2010) 279–285.
[10] B. Mahapatra, S. Patnaik, A. Nayyar, Effect of multiple-agent deployment in MANET, Recent Pat. Comput. Sci. 12 (3) (2019) 180–190.
[11] H.Y. Hsieh, R. Sivakumar, On using the ad-hoc network model in cellular packet data networks, in: Proceedings of the 3rd ACM International Symposium on Mobile Ad Hoc Networking & Computing, 2002, June, pp. 36–47.
[12] A. Nayyar, Simulation based evaluation of reactive routing protocol for MANET, in: 2012 Second International Conference on Advanced Computing & Communication Technologies, IEEE, 2012, January, pp. 561–568.
[13] V.V. Mandhare, R.C. Thool, Improving QoS of mobile ad-hoc network using cache update scheme in dynamic source routing protocol, Procedia Comput. Sci. 79 (2016) 692–699.
[14] A. Nagaraju, S. Ramachandram, Reduce redundant broadcasting in MANETs using rough sets, Int. J. Wirel. Mob. Comput. 7 (2) (2014) 103–122.
[15] M. Sadeghi, S. Yahya, Analysis of wormhole attack on MANETs using different MANET routing protocols, in: 2012 Fourth International Conference on Ubiquitous and Future Networks (ICUFN), IEEE, 2012, July, pp. 301–305.
[16] A. Nayyar, Cross-layer system for cluster based data access in MANET'S, 2001. Special Issue of International Journal of Computer Science & Informatics (IJCSI), ISSN (PRINT).
[17] I. Ullah, S.U. Rehman, Analysis of Black Hole Attack on MANETs Using Different MANET Routing Protocols (Thesis), School of Computing, Blekinge Institute of Technology, Sweden, 2010, pp. 1–40.
[18] R. Kaur, M.K. Rai, A novel review on routing protocols in MANETs, Undergrad. Acad. Res. J. 1 (1) (2012) 103–108.
[19] D. Dhiman, A. Nayyar, Complete scenario of routing protocols security leaks and attacks in MANETs, IJARCSSE 3 (10) (2013).
[20] S. Radha, S. Shanmugavel, Mobility models in mobile ad hoc network, IETE J. Res. 53 (1) (2007) 3–12.
[21] Z. Ye, S.V. Krishnamurthy, S.K. Tripathi, A routing framework for providing robustness to node failures in mobile ad hoc networks, Ad Hoc Netw. 2 (1) (2004) 87–107.
[22] H. Xu, X. Wu, H.R. Sadjadpour, J.J. Garcia-Luna-Aceves, A unified analysis of routing protocols in MANETs, IEEE Trans. Commun. 58 (3) (2010) 911–922.
[23] J. Broch, D.A. Maltz, D.B. Johnson, Y.C. Hu, J. Jetcheva, A performance comparison of multi-hop wireless ad hoc network routing protocols, in: Proceedings of the 4th Annual ACM/IEEE International Conference on Mobile Computing and Networking, 1998, October, pp. 85–97.
[24] T. Camp, J. Boleng, V. Davies, A survey of mobility models for ad hoc network research, Wirel. Commun. Mob. Comput. 2 (5) (2002) 483–502.
[25] Random_Walk, 2020. Retrieved December 5, 2020, from https://en.wikipedia.org/wiki/Random_walk.
[26] F. Maan, N. Mazhar, MANET routing protocols vs mobility models: a performance evaluation, in: 2011 Third International Conference on Ubiquitous and Future Networks (ICUFN), IEEE, 2011, June, pp. 179–184.
[27] R.R. Roy, Random walk mobility, in: Handbook of Mobile Ad Hoc Networks for Mobility Models, Springer, Boston, MA, 2011, pp. 35–63.
[28] E. Syukur, J. García-Villalba, A. George, A. Kumar, S. Srinivasan, A multi-hop mobility management protocol for heterogeneous wireless networks, Int. J. Pervasive Comput. Commun. 5 (2) (2009) 187–207.
[29] S.H. Bae, S.J. Lee, W. Su, M. Gerla, The design, implementation, and performance evaluation of the on-demand multicast routing protocol in multihop wireless networks, IEEE Netw. 14 (1) (2000) 70–77.
[30] E.M. Royer, C.K. Toh, A review of current routing protocols for ad hoc mobile wireless networks, IEEE Pers. Commun. 6 (2) (1999) 46–55.

[31] S.J. Lee, M. Gerla, C.K. Toh, A simulation study of table-driven and on-demand routing protocols for mobile ad hoc networks, IEEE Netw. 13 (4) (1999) 48–54.
[32] S. Radha, S. Shanmugavel, Performance evaluation of routing algorithms for mobility models in ad hoc network, IETE Tech. Rev. 21 (3) (2004) 199–210.
[33] P. Hiranvanichchakorn, S. Lertvorratham, Using regression analysis for improving multipath ad hoc network performance, Int. J. Comput. Appl. 32 (2) (2010) 206–214.
[34] C.E. Perkins, E.M. Royer, S.R. Das, M.K. Marina, Performance comparison of two on-demand routing protocols for ad hoc networks, IEEE Pers. Commun. 8 (1) (2001) 16–28.
[35] Chakravarthy, Secured Right Angled and ANT Search Multipath Routing Protocol for Mobile Ad Hoc Networks (Doctoral thesis), University of Madras, Shodhganga Thesis Publishing, 2018.
[36] R. Bai, M. Singhal, DOA: DSR over AODV routing for mobile ad hoc networks, IEEE Trans. Mob. Comput. 5 (10) (2006) 1403–1416.
[37] B. Renu, T. Pranavi, Routing protocols in mobile ad-hoc network: a review, in: International Conference on Heterogeneous Networking for Quality, Reliability, Security and Robustness, Springer, Berlin, Heidelberg, 2013, January, pp. 52–60.
[38] Z. Pawlak, Rough sets, Int. J. Comput. Inform. Sci. 11 (5) (1982) 341–356.
[39] Z. Pawlak, Vagueness and uncertainty: a rough set perspective, Comput. Intell. 11 (2) (1995) 227–232.
[40] Z. Pawlak, Rough classification, Int. J. Hum. Comput. Stud. 51 (2) (1999) 369–383.
[41] T. Yang, Q. Li, Reduction about approximation spaces of covering generalized rough sets, Int. J. Approx. Reason. 51 (3) (2010) 335–345.
[42] G. Liu, Y. Sai, A comparison of two types of rough sets induced by coverings, Int. J. Approx. Reason. 50 (3) (2009) 521–528.
[43] P. Zhu, Covering rough sets based on neighborhoods: an approach without using neighborhoods, Int. J. Approx. Reason. 52 (3) (2011) 461–472.
[44] Q. Zhang, Q. Xie, G. Wang, A survey on rough set theory and its applications, CAAI Trans. Intell. Technol. 1 (4) (2016) 323–333.
[45] H.H. Inbarani, A.T. Azar, G. Jothi, Supervised hybrid feature selection based on PSO and rough sets for medical diagnosis, Comput. Methods Prog. Biomed. 113 (1) (2014) 175–185.
[46] A.T. Azar, Fast neural network learning algorithms for medical applications, Neural Comput. & Applic. 23 (3–4) (2013) 1019–1034.
[47] A.T. Azar, S.M. El-Metwally, Decision tree classifiers for automated medical diagnosis, Neural Comput. & Applic. 23 (7–8) (2013) 2387–2403.
[48] P. Li, S. Guo, S. Yu, A.V. Vasilakos, Reliable multicast with pipelined network coding using opportunistic feeding and routing, IEEE Trans. Parallel Distrib. Syst. 25 (12) (2014) 3264–3273.
[49] A.Y. Barnawi, I.M. Keshta, Energy management in wireless sensor networks based on naive Bayes, MLP, and SVM classifications: a comparative study, J. Sens. 2016 (2016) 1–12.
[50] I. Stojmenovic, M. Seddigh, J. Zunic, Dominating sets and neighbor elimination-based broadcasting algorithms in wireless networks, IEEE Trans. Parallel Distrib. Syst. 13 (1) (2002) 14–25.
[51] M. Khabbazian, V.K. Bhargava, Reducing broadcast redundancy in wireless ad hoc networks (IEEE GLOBECOM 2007-IEEE Global Telecommunications Conference), IEEE, 2007, November, pp. 769–774.
[52] A. Nagaraju, R. Sirandas, C.R. Rao, Applying heuristic technique to ad-hoc on demand distance vector routing to reduce broadcast, in: World Congress on Engineering, 2007, pp. 1530–1533.
[53] V.M.A. Rajam, V. UmaMaheswari, A. Siromoney, Temporal information systems and their application to mobile ad hoc routing, Ubiquit. Comput. Commun. J. 3 (4) (2008) 417–424.
[54] N. Aitha, R. Srinadas, A strategy to reduce the control packet load of aodv using weighted rough set model for manet, Int. Arab J. Inf. Technol. 8 (1) (2009) 108–116.
[55] A. Nagaraju, G.C. Kumar, S. Ramachandram, Ad-hoc on demand distance vector routing algorithm using neighbor matrix method in static ad-hoc networks, in: International Conference on Computer Science and Information Technology, Springer, Berlin, Heidelberg, 2011, January, pp. 44–54.
[56] P. Seethalakshmi, M. Gomathi, G. Rajendran, Path selection in wireless mobile ad hoc network using fuzzy and rough set theory, in: 2011 2nd International Conference on Wireless Communication,

Vehicular Technology, Information Theory and Aerospace & Electronic Systems Technology (Wireless VITAE), IEEE, 2011, pp. 1–5.
[57] S.S. Kumar, P. Manimegalai, S. Karthik, A rough set calibration scheme for energy effective routing protocol in mobile ad hoc networks, Clust. Comput. 22 (6) (2019) 13957–13963.
[58] R. Kulsum, S. Anand, S. Sinha, A soft computing approach to analyse Aodv routing protocol, Int. J. Innov. Technol. Explor. Eng. 8 (2019) 1443–1446.
[59] B. Wu, S. Nazir, N. Mukhtar, Identification of attack on data packets using rough set approach to secure end to end communication, Complexity 2020 (2020).
[60] M. Ahmad, A. Hameed, F. Ullah, A. Khan, H. Alyami, M.I. Uddin, A. ALharbi, Cluster optimization in mobile ad hoc networks based on memetic algorithm: memeHoc, Complexity 2020 (2020).
[61] B.S. Kumar, Soft computing approach to enhance the performance of AODV (ad-hoc on-demand distance vector) routing protocol using active route TimeOut (ART) parameter in MANETs, Turk. J. Comput. Math. Educ. 12 (2) (2021) 3060–3068.
[62] D. Kanellopoulos, F. Cuomo, Recent Developments on Mobile Ad-Hoc Networks and Vehicular Ad-Hoc Networks, MDPI, 2021.
[63] H.H. Inbarani, M. Bagyamathi, A.T. Azar, A novel hybrid feature selection method based on rough set and improved harmony search, Neural Comput. & Applic. 26 (8) (2015) 1859–1880.
[64] D. Dubois, H. Prade, Rough fuzzy sets and fuzzy rough sets, Int. J. Gen. Syst. 17 (2–3) (1990) 191–209.
[65] P. Konar, M. Saha, J. Sil, P. Chattopadhyay, Fault diagnosis of induction motor using CWT and rough-set theory, in: Computational Intelligence in Control and Automation (CICA), 2013 IEEE Symposium on, IEEE, 2013, April, pp. 17–23.
[66] M. Kondo, On the structure of generalized rough sets, Inf. Sci. 176 (5) (2006) 589–600.
[67] D. Meng, X. Zhang, K. Qin, Soft rough fuzzy sets and soft fuzzy rough sets, Comput. Math. Appl. 62 (12) (2011) 4635–4645.
[68] T. Medhat, Missing values via covering rough sets, Int. J. Data Min. Intell. Inf. Technol. Appl. 2 (1) (2012) 10–17.
[69] L. Ma, On some types of neighborhood-related covering rough sets, Int. J. Approx. Reason. 53 (6) (2012) 901–911.
[70] S.S. Kumar, H.H. Inbarani, A.T. Azar, K. Polat, Covering-based rough set classification system, Neural Comput. & Applic. 28 (10) (2017) 2879–2888.
[71] W. Zhu, Topological approaches to covering rough sets, Inf. Sci. 177 (6) (2007) 1499–1508.
[72] K. Fall, K. Varadhan, The ns manual (formerly ns notes and documentation), VINT Project 47 (2005) 19–231.
[73] W. Zhu, Relationship among basic concepts in covering-based rough sets, Inf. Sci. 179 (14) (2009) 2478–2486.
[74] Y. Yao, B. Yao, Covering based rough set approximations, Inf. Sci. 200 (2012) 91–107.
[75] X. Ge, X. Bai, Z. Yun, Topological characterizations of covering for special covering-based upper approximation operators, Inf. Sci. 204 (2012) 70–81.
[76] C. Wang, D. Chen, B. Sun, Q. Hu, Communication between information systems with covering based rough sets, Inf. Sci. 216 (2012) 17–33.
[77] Y.S. Sandeep, P.V.S. Reddy, C. Manoj, K.A. Lakkshmanan, Identifying the vague regions by using covering based rough sets, Int. J. Adv. Res. Comput. Sci. Softw. Eng. 3 (7) (2013) 743–746.
[78] C. Degang, W. Changzhong, H. Qinghua, A new approach to attribute reduction of consistent and inconsistent covering decision systems with covering rough sets, Inf. Sci. 177 (17) (2007) 3500–3518.
[79] M.S. Hasan, C. Harding, H. Yu, A. Griffiths, Modeling delay and packet drop in networked control systems using network simulator NS2, Int. J. Autom. Comput. 2 (2) (2005) 187–194.

CHAPTER 5

Optimization of hybrid broadcast/ broadband networks for the delivery of linear services using stochastic geometry

Ahmad Shokair[a], Matthieu Crussière[a], Youssef Nasser[b], Oussama Bazzi[c], and Jean-Francois Hélard[a]
[a]INSA-Rennes, Rennes, France
[b]Huawei, Paris, France
[c]Lebanese University, Beirut, Lebanon

1. Introduction

The explosion of capabilities of communication networks over the past years is redefining the lifestyle of the human race. In particular, multimedia services have reshaped how people interact with the world. Modern consumers rely on multimedia services for entertainment, education, information, mass communication, business management, and personal relations. During this period, technology has been racing to keep up with the increasing demand. The different challenges that this type of service introduces lead to the development of a different solution, which, unsurprisingly, follows a historical trend.

Ever since people started communicating, there were two main types of communication, one with a single side addressing a large number of receiving sides, and one with two sides exchanging messages privately. One of the earliest examples is the speech versus a conversation. In a speech, a person speaks loud enough so a certain number of people surrounding him can hear and listen to the content. This method is time-efficient, since the speaker has to deliver the message once, but it is limited to the power of the speaker's voice and has poor feedback, preventing the speaker from making sure that the audience gets the message properly. On the other hand, in a conversation, the feedback is fairly easy, which ensures proper reception of the message, yet if the message is to be delivered to a large number of people, it rapidly becomes overwhelmingly consuming in terms of time resources. At each milestone of human history, this duality is held. The trade-off was between publishing a book and exchanging written messages, and with the introduction of wireless technologies, it quickly becomes between broadcasting a message over the radio and personally delivering the message to recipients over the early forms of cellular phones. With the advancement in communication technologies, this duality continued.

Comprehensive Guide to Heterogeneous Networks
https://doi.org/10.1016/B978-0-323-90527-5.00005-8

A live event can now be watched using a broadcast (BC) service like TV BC, or using a broadband (BB) network that specifically sends the information to the user in a personalized manner. Over the last couple of decades, multimedia streaming services have been growing exponentially on both sides of the duality, starting from the first digital BC platforms and standards in the late 1990s to the first cellular networks with real multimedia capabilities with third-generation partnership project (3GPP), universal mobile telecommunications system (UMTS), and high-speed packet access (HSPA).

Modern portable devices, such as smartphones and tablets, have the capability to connect to numerous network technologies to obtain services such as mobile TV and multimedia streaming from different sources. This includes cellular BB networks such as long-term evolution (LTE) and fifth-generation (5G) new radio (NR), and terrestrial BC networks like DVB-terrestrial-2 (DVB-T2), and advanced television system committee (ATSC) 3.0 [1–11]. However, as in previous technologies, each network type suffers from its own limitations and challenges. For instances, BC networks lack feedback and personalization, since it addresses a large number of users with the same stream regardless of their conditions. Moreover, the large transmitted power makes it less efficient if the service or the content is not popular. On the other hand, BB networks have limited resources, especially for demanding services like high-quality multimedia streams.

The hybrid network concept emerged to overcome such drawbacks. In a hybrid network, multiple network technologies collaborate to bear the content to the end user. The BB/BC hybridization can take several forms, like service sharing, where each network provides a part of the service [12–15], stream sharing, where each network delivers a part of the data stream to each user [16–18], and user sharing, where each network is responsible of delivering the service to part of the users set [19–24]. The later can be as an offload mechanism from the BB to the BC network. It can also be seen as an extension to the range of the BC network. In this chapter, we focus on the user-sharing hybrid BC/BB networks.

The work is divided into three main sections. The first exploits the different scenarios over which hybridization is used. Section 2 gives a novel in-depth analysis of one of the scenarios. Section 3 provides an insight into the performance of the hybrid network under each scenario.

2. Hybrid models

In this section, we discuss a set of possible hybrid network models. The focus here is on the delivery of linear services such as live stream to a number of users via a user-sharing hybrid BC/BB network. Yet, such networks have the ability to work under different conditions based on the choice of the following factors:

- The BB cellular network operation mode: The BB cellular network is equipped to work in either unicast (UC) or multicast (MC) modes.

- The deployment of the BC network: Several deployments of the BC can be used, two of which are the single terrestrial broadcast (TBC) transmitter at the center deployment, and the multiple TBCs distributed over a wider zone deployment.
- User-sharing scheme: Assigning users to either cellular or BC network is accomplished through predefined criteria that the network later uses. Two of those schemes are signal-quality-based, or more precisely signal to interference and noise ratio (SINR)-based allocation, and location-based or zone-based allocation.

Different combinations of these factors produce different network models. The choice depends on the use cases, the availability of resources, and the expected outcome. In this section, these factors, their use cases, and their modeling are discussed. In the following sections, the analysis of their combinations' performance is detailed.

2.1 The modeling of the broadband network

It is well known that the cellular network is one of the most popular communication networks since the late 1980s of the 20th century.

Even though the second-generation cellular networks such as global system for mobile (GSM) communication were capable to send data, but it was limited to rates as low as 400 kbps with enhanced data rates for GSM evolution (EDGE), which is certainly not enough to support multimedia streams. With the introduction of the third-generation (3G) networks such as UMTS achieved up to 2 Mbps of data rate, which was later enhanced to 14.4 Mbps with HSPA, and up to 23 Mbps with HSPA+ with the introduction of multiple input multiple output (MIMO) among other modifications [25–28].

Later, 3GPP released LTE as the fourth-generation (4G) network standard. It aimed at delivering data rates up to 100 Mbps with the introduction of orthogonal frequency division multiplexing (OFDM). With 3GPP release 10, LTE-advanced (LTE-A) emerged with the introduction of 4×4 MIMO schemes [29–31]. More recently, the 5G networks are getting deployed, after major changes in the network architecture as well as the radio access technologies. The new modifications such as the introduction of the new mmWave bands and the network slicing and virtualization allow bit rates up to 20 Gbps [32–35].

BB cellular networks normally operate in UC. However, since 3G, MC mode was introduced as an optional operation mode [36–39]. Next, both modes and their corresponding modeling characteristics are discussed.

2.1.1 Modeling the common features for both modes

Conventionally, the cellular networks used to be modeled with the grid model. This simplistic model made the planning easier especially for frequency reuse, yet it is far from being accurate. More recently, stochastic geometry emerged as a tool for the modeling and analysis of wireless networks. In stochastic geometry, the nodes in a network are modeled by a random point process. It provides a set of mathematical tools for studying

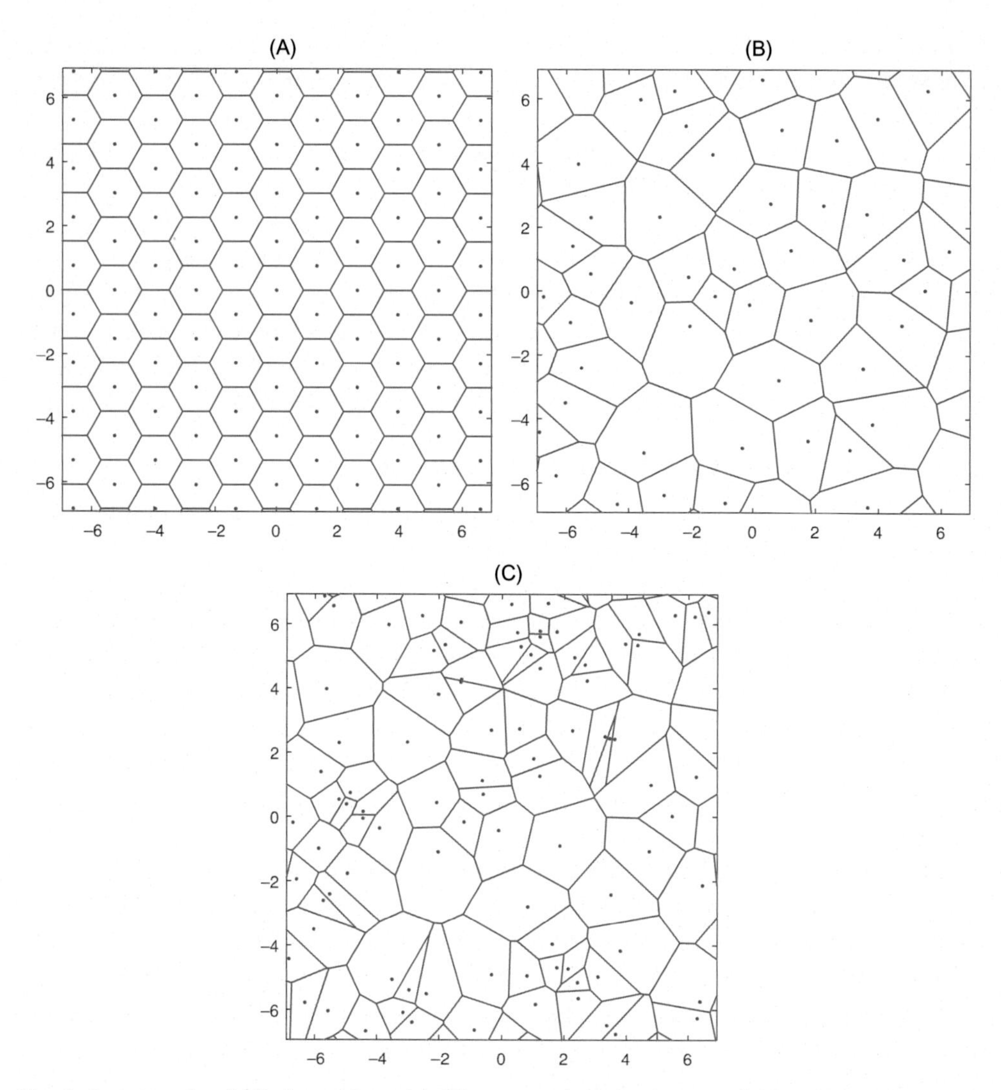

Fig. 1 An example of (A) the grid model, (B) an actual deployment, and (C) a PPP with Voronoi tessellation.

the performance of the network. One of the well-known processes used is the homogeneous Poisson point process (PPP) in $\mathbb{R}^2$ with a fixed density.[a] A typical user in the network is normally served by the closest base station (BS). This creates what is called a Voronoi tessellation. Fig. 1 shows how the PPP with Voronoi tessellation can be more accurate in modeling an actual network deployment compared to the grid model.

[a] For more information on stochastic geometry or PPPs and their applications in wireless networks, the reader is invited to see Chapter 2 in Ref. [40], or Refs. [41–45].

Normally, two important random processes control the performance of the BB cellular network, the interference, and the quality of the useful signal. Any received signal, useful on interfering, it is a function of the distance to the serving BS and the fading channel, which are also a random variable. The power received by a user device from each BS is calculated based to the following definition:

$$\mathcal{P}_R = \mathcal{P}_T h r^{-\alpha} \tag{1}$$

Although, the received power being counted as a useful power or an interference power depends on the mode of operation of the network. Here, $\mathcal{P}_R$ and $\mathcal{P}_T$, respectively, represent the received and the transmitted powers. In addition, the random variable h denotes the Rayleigh fading channel due to multipath, r represents the link distance in-between the transmitter and the receiving device. Moreover, α denotes the pathloss exponent. The probability distribution of the channel coefficient h is critical in the analysis.

In a cellular network, the BS allocates some time-frequency resources to the users. In OFDM-based systems such as LTE and 5G-NR, resource blocks (RBs) are allocated to the users. An RB is the minimum amount of temporal-spectral resources that can be assigned to the user. An RB includes some set of subcarriers for one-time frame. For instance, an RB in LTE contains 12 subcarriers, with 15 kHz spacing, resulting in a sum of 180 kHz of bandwidth (BW) for one-time slot of 0.5 ms. 5G-NR brings more flexibility by allowing several subcarrier spacing values (15, 30, 60, 120, and 240 kHz). To simplify the analysis, one can assume that the assigned RBs per user has flat spectrum.

2.1.2 Unicast: The default mode

UC is the default mode of operation for BB cellular networks. In UC, each BS assign a number of unique RBs to each user being served by it. This allows personalized transmission (technically and content wise) for each user, yet it limits the total number of served devices.

Since the band allocated to the BB cellular services is limited, it is important to manage the spectral resources among the different cells in the network. One option is to set the frequency reuse factor to one, this means that all the BSs use the same frequency band for transmission. This option, while it magnifies the interference, it allows each BS to use all the available band, providing flexibility for the system and enhances its capacity.

The other factor controlling the signal quality is the link distance between a user and the BS. The probability distribution of the random link distances between a point and a node in a PPP has been evaluated in multiple references. In particular, the probability density function (PDF) of the distances r from a point in the space to its ith nearest neighbor in a homogeneous PPP with a constant density λ in the $\mathcal{D}$-dimensional space is calculated as follows:

$$f_{R_i}(r) = \exp\left(-\lambda c_{\mathcal{D}} r^{\mathcal{D}}\right) \frac{\mathcal{D}\left(\lambda c_{\mathcal{D}} r^{\mathcal{D}}\right)^i}{r\Gamma(i)} \tag{2}$$

where $c_{\mathcal{D}} r^{\mathcal{D}}$ is the volume of the $\mathcal{D}$-dimensional sphere of radius r and $\Gamma(\cdot)$ denotes the gamma function, see Ref. [46] for the complete proof.

By limiting the space to $\mathbb{R}^2$, and the neighbor index to 1 in Eq. (2) (taking the closet BS as the serving one), the previous expression can be reduced to the following:

$$f_R(r) = 2\pi\lambda r \exp\left(-\lambda\pi r^2\right) \tag{3}$$

2.1.3 Multicast mode and the SFN deployment

The main challenge for the UC mode is the limited BW, especially when serving high data rate services such as live streaming, which can cause an overload to the network. To solve this problem, MC was introduced as an optional mode to avoid blocking users or downgrading the quality of service. In MC, a set of time-frequency resources are allocated to a set of users requesting the same service. The main difference between MC and BC is that MC is controlled by the BS, which decides to whom the joined resources are allocated, in contrary to BC, where a user decides to tune in to the transmission.

MC was first introduced in 3GPP's UMTS release 4, but it was limited to low-rate services like weather and traffic reports. It was not until release 6 that multimedia broadcast multicast service (MBMS) was introduced to broadcast Internet protocol (IP) packets to users. An enhanced version of MBMS was included in 3GPP release 9 for LTE under the name enhanced-MBMS (eMBMS) to support spectral efficiency up to 3 bit/s per Hz, and 30 fps, 720p high definition (HD) resolution [47].

To reduce intercell interference (ICI), the concept of single-frequency network (SFN) is often used with MC mode. Multiple neighboring cells use the same BW for the MC services. This means that the receiver will get different versions of the same signal, while it can enhance the received signal, it can introduce synchronization issues, and sometimes interference.[b] In OFDM, the guard intervals are usually helpful to ensure that multiple received duplicates of the signal do not interfere with one another. Assuming the receiving device is taking the first received signal (normally from the nearest BS) as a reference for synchronization, the other copies are received with a time delay that falls into a case of the following three:

- The arrival delay is lower than the symbol's guard interval: The power of the signal is fully considered as useful power.
- The arrival delay is between the guard interval and the symbol duration: Part of the power is considered as useful power; the other part is considered to be interference.

[b] This is not to be confused with the multipath effect when different versions of the signal sent by the same transmitter, arrive with certain delay from each other due to the following different paths. In the case here, different transmitters are sending the same signal.

- The arrival delay is higher than the total symbol duration: The received signal is fully out of phase, and all the received power is considered as interference.

Fig. 2 shows an example of these cases.

2.2 Broadcast network deployments

A BC can take mainly one of two forms. The first is when a single isolated BC high power high tower (HPHT) is used to cover a certain area. The second is when a set of BC HPHT operates together as a network to cover a much wider zone.

2.2.1 Single BC transmitter

A HPHT positioned at the center of the targeted zone broadcasts the signals with enough power at a preallocated frequency. Interested users will then tune in to that frequency to get the service. Normally, a device is considered to be covered by the BC if the received SINR exceeds a certain threshold, which allows the device to properly decode the transmission.

In the analysis of such networks, it can also be useful to use stochastic geometry. A PPP can be used here to model the positions of the users. Moreover, the same path loss model as in Eq. (1) can be used to estimate the received power. Due to the absence of external interference, the transmission becomes noise limited, contrary to the interference-limited BB network.

2.2.2 Multiple BC transmitters

The case of multiple BC network can be seen as a generalization of the previous case. Normally, it is used to cover a larger service area, using an SFN scheme to enhance the signal-to-noise ratio (SNR). This is illustrated in Fig. 3.

Since the BC network normally has low density, it is rarely modeled by a PPP, but rather have some deterministic positions. However, since the topography (which can be quite random) usually controls the locations of the broadcast transmitter (BCT), and since stochastic geometry helps to make the analysis tractable, some variations of PPP can be used, especially with the network is denser.

As in the case of MC, BC is often used in an SFN deployment. The receiver will get multiple versions of the signal, and the usefulness of these signals depends on the time delay of the arrival.

2.3 Hybrid broadcast/broadband model

Besides the settings of the two members of the hybrid network, the third important factor is sharing mechanism. As we are focusing on user-sharing hybrid setup, that is, the users are allocated to either network, it is crucial to carefully define the criteria with which the decision of joining any of the two networks is taken. This results in some user allocation schemes.

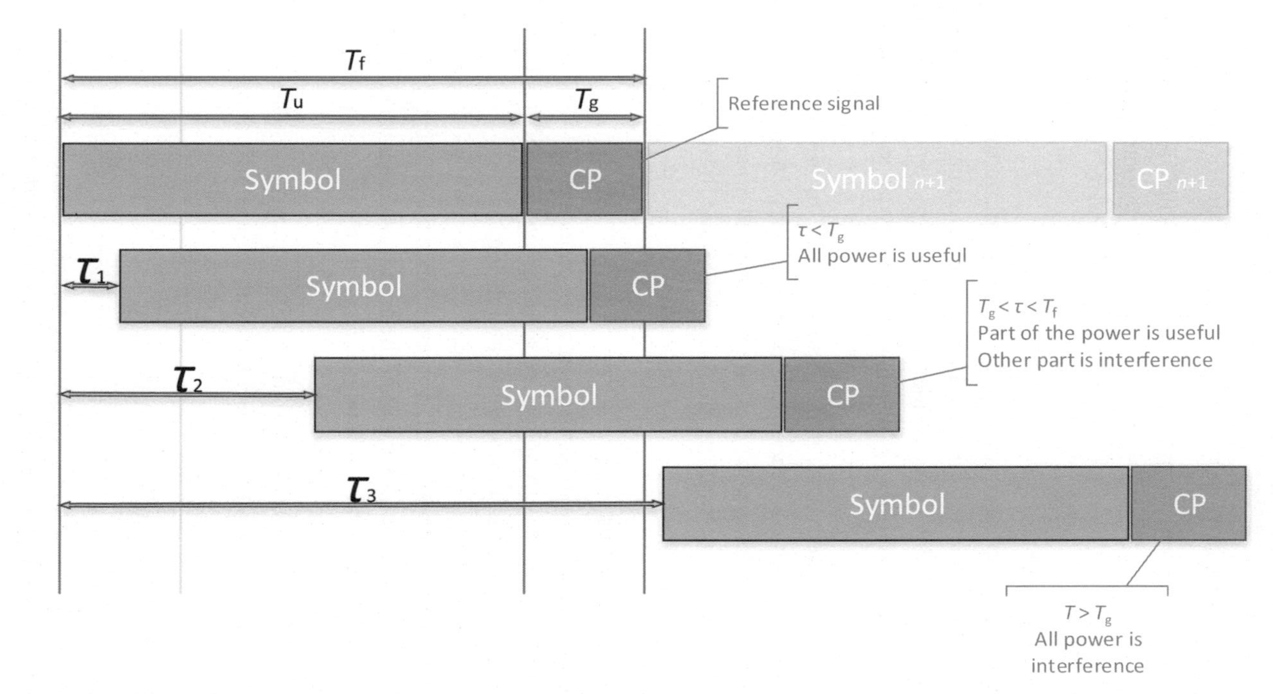

Fig. 2 An illustration of the reception of four duplicates from different BSs. The one on the top is the reference, and each of the other three corresponds, respectively, to a case of the mentioned three: All useful, partially useful, and total interference.

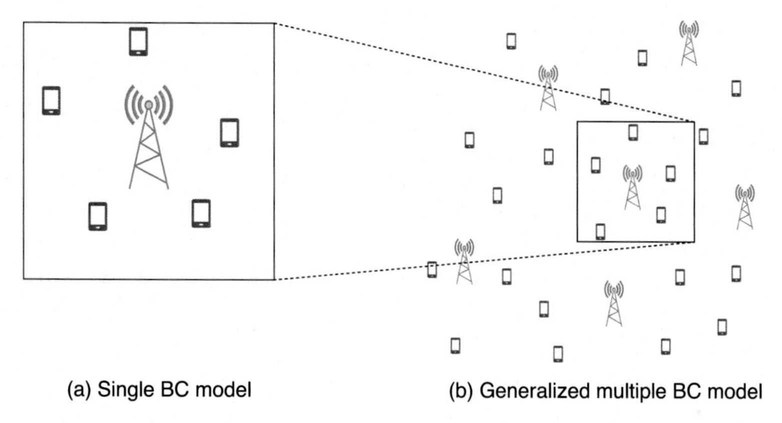

Fig. 3 An illustration of the two deployments: (a) single BC and (b) multiple BC.

Here, we discuss two of the possibilities, which are user allocation based on location, and user allocation based on signal quality. As it will be used in the later analysis (which is probabilistic in nature), and even if it seems trivial, we set some assumptions: a user device is capable of connecting to either network, and the service of both networks are available at all time.

2.3.1 Location-based user association

In location-based schemes, the users' relative position from the network deployment is the criterion for the allocation. For example, a user within a certain distance from the BCT is assigned to the BC network. This creates a disk-shaped area, within which the users are served by BC, and the BB BSs are not delivering the shared service.

This scheme is network centric. The user sharing is controlled by the network management units. This gives the advantage of preventing an overload of the BB network. However, it comes with some drawbacks. For instance, it adds to the overhead of the network, and it requires a collaboration between the BC and the cellular branches to control which cells are offering the service, and which are not.

Location-based schemes can be used in scenarios, where certain areas are expected to be overpopulated, like in the cases of an event gathering (sports games, concerts, etc.), or the city centers and main hubs. It can also be used when the frequency band coexistence between the BC and BB networks is needed. Spatial separation into zones can reduce the mutual interference of the two networks.

The optimization of this deployment is done by controlling the radius of the BC zone. Increasing this radius expands the zone, and consequently adds more users to the BC subset, which reduces the load on the BB network. On the other hand, this can affect the BC edge users, since the BCT has an upper limit to the available power.

2.3.2 Signal-quality-based user association

As the name suggests, in signal-quality-based schemes, a user joins the network that provides the best quality. Usually, SINR is the metric used to judge this signal quality. To relief the BB cellular network, a priority can be given to the BC side. So if the BC SINR is good enough for acceptable reception of the service, the receiving device uses BC network, else, the user joins the cellular network.

From the described mechanism, it can be easily seen that this scheme is user centric. The allocation is controlled by the users' devices, not by the network. This brings the advantage of reducing the overhead for the network. However, at the same time, it limits the ability of the hybrid network control to manage the users and can lead to an overload on the BB network in some cases.

To optimize the hybrid network while using a signal-quality-based scheme, the network controller can tune the BC transmission power. Rising the BC power increases the probability of users being linked to the BC network, but at the same time reduces the power efficiency of the transmission, especially if the number of users is not large enough.

2.4 Summary of the hybrid combinations

Fig. 4 summarizes the options for a hybrid network deployment presented in this section. Three main factors are to consider in a user-sharing hybrid BC network. Each factor has a couple of options to select from. This creates a total of eight possible combinations. In each combination, on BC deployment option is selected, one BB operation mode is set, and one allocation scheme is used. So one red, one blue, and one green ball are thrown into the mix whenever a network is to be setup.

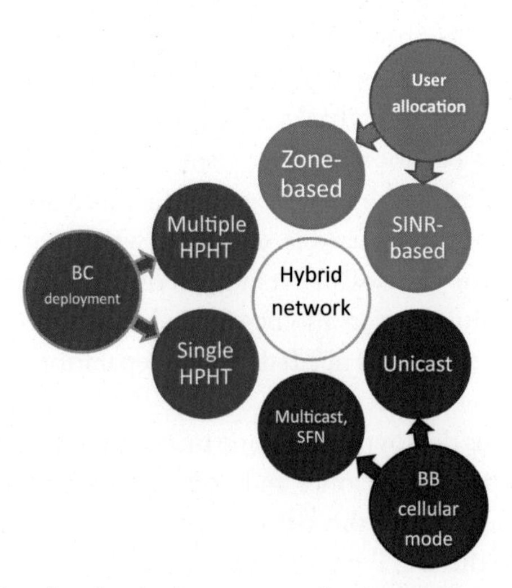

Fig. 4 The considered options for the deployment, mode of operation, and user allocation.

For the BC deployment options, single BCT ensures the coverage of a small area, without interference. Generalizing into an SFN deployment increases the coverage area, but introduces interference due to delay.

As for the BB network operation mode, UC being the most popular mode assures proper personalization for each user, yet it is vulnerable to overload if the number of users exceeds the BS's capacity. On the other hand, MC immunes the system against overload, yet it inherits the limitations of the BC network.

Moreover, assigning users to either network based on location gives control to the network and it can be suitable to cover occasionally crowded zones. This scheme also allows a level of coexistence in frequency because of the geographical separation. Yet, this can reduce the quality of service (QoS) for edge users, and create additional overhead for the network control. On the other side, a signal-quality-based user assignment reduces the overhead for the network, yet prevents the possibility for coexistence.

In the next section, we take one of the combinations and go through an extensive analysis of the performance.

3. Analytical analysis of the hybrid model

In this section, we use stochastic geometry to analyze the performance of a signal-quality-based, UC BB operating, single BCT hybrid network model. Fig. 5 shows the different choices made for different factors of the system design.

3.1 Model settings and definitions

The network consists of two subnetworks: a BC subnetwork, made out of one BCT at the center, and a BB subnetwork, which consists of a set of BS spread over the area of

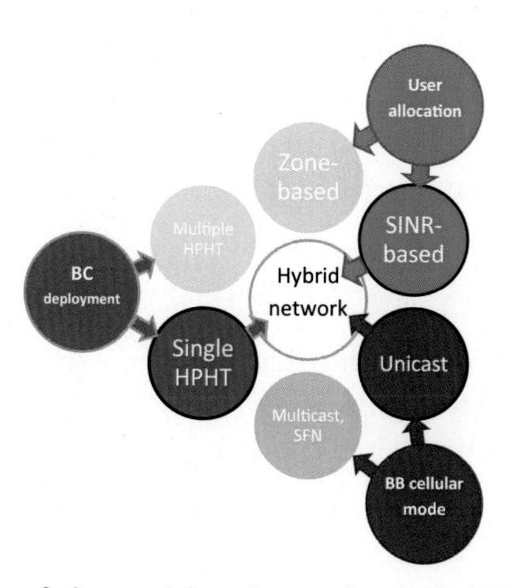

Fig. 5 The selected options of the model under study: single BC deployment, quality-based, and UC mode.

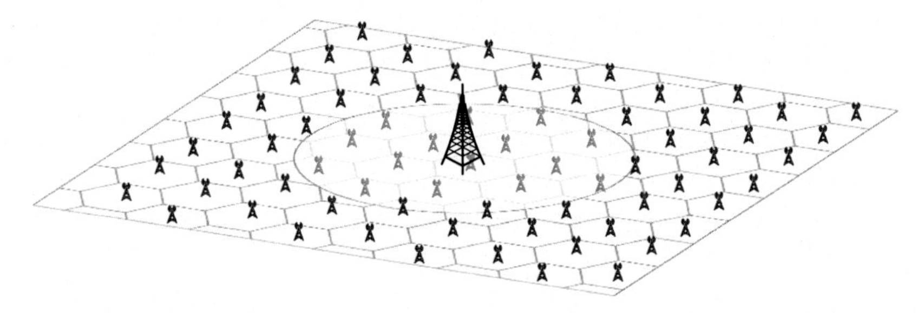

Fig. 6 The overlay deployment.

service. The BB network operates in UC mode. A homogeneous PPP Φ having λ_{BS} as density is used to model the network deployment. The set of users is likewise modeled by an independent homogeneous PPP Ψ with density λ_U. These settings create what is called an overlay setup, as shown in Fig. 6.

The BCT transmits over the frequency f_B, with power equal to $\mathcal{P}_B$. The BB BSs use a different frequency f_L and transmits a power equal to $\mathcal{P}_U$, where $\mathcal{P}_B > \mathcal{P}_U$. As it is assumed that the service is always available, the duty cycle is set to one. Moreover, to maximize the capacity of the system, a unity frequency reuse factor is selected, that is all the BSs utilize the same frequency band. Connection to the BC has a priority, so if the BC SINR is adequate, the device is connected to the BCT, else, it is connected to the nearest BS.

The first step of the analysis is to define the SINR for both networks. For BB, the received SINR is as follows:

$$S_{UC} = \frac{\mathcal{P}_U h r_l^{-\alpha}}{P_{N,UC} + I} \tag{4}$$

where r_l denotes the link distance in-between the user device and its serving BS, α denotes the pathloss exponent for the BB network environment, and $P_{N,\ UC}$ denotes the noise at the BB receiver. Moreover, h represents the fading channel's power. Rayleigh channel is assumed, so h will have an exponential distribution with rate μ. $I_{U/U}$ denotes the interference from other BSs in the network, and it is given by

$$I = \sum_{i \in \Phi'} \mathcal{P}_U h_i q_i^{-\alpha} \tag{5}$$

where Φ' is the set of all nodes with the exception of the serving node, and q_i denotes the link distance to the ith interfering BS.

As for the BC transmission, there is no source of interference, so the received SNR will be as follows:

$$S_{BC} = \frac{\mathcal{P}_B g r_v^{-\beta}}{\mathcal{P}_{N_{BC}}} \tag{6}$$

where g represents the fading channel power. The channel is also considered to be a Rayleigh fading channel, so g has an exponentially distribution with rate υ. In addition, r_v denotes the link distance between the user and the HPHT, β is the pathloss exponent in the BC environment, and $\mathcal{P}_{N_{BC}}$ represents the noise at the BC receiver.

In Eqs. (4), (6), the distances are random since the users and the BSs are allocated randomly. So, before deriving the performance metrics of the network, it is important to find the distributions of these distances. For BC, the distribution of r_v is as follows:

$$f_{r_v}(r_v) = \frac{2}{R_m^2} r_v, \quad r_v \subset [0, R_m] \tag{7}$$

where R_m denotes the maximum available distance. On the other hand, for the BB network, the PDF of the link distance from a user to the serving BS is similar to Eq. (3):

$$f_{r_l}(r_l) = 2\pi\lambda_{BS} r_l \exp\left(-\lambda_{BS}\pi r_l^2\right) \tag{8}$$

3.2 The coverage probability

For a network, one of the most important evaluation figures that are used to measure the performance is the coverage probability. It is the probability that a user in the service area receives a signal with enough quality for the device to be able to properly decode it. It can also translate to the ratio of the properly served user to the total number of users. It can be defined as follows:

$$P_c = \mathbb{P}[\text{SINR} > \mathcal{T}] \tag{9}$$

where $\mathcal{T}$ is the SINR threshold needed for proper reception.

In the context of the hybrid network model, the coverage probability is the probability of being connected to either network. It is in fact equal to the complementary of the blockage probability, that is, not being served by any network. We start by calculating the probability coverage for each network separately, then combine them at the end.

3.2.1 Coverage probability for a BC user

Based on the definition in Eq. (9), the coverage probability in the BC at a distance r_v from the HPHT is as follows:

$$\begin{aligned} P_{c_{BC}|r_v} &= \mathbb{P}[S_{BC} > T_{BC}|r_v] \\ &= \mathbb{P}\left[\frac{\mathcal{P}_B g r_v^{-\beta}}{\mathcal{P}_{N,BC}} > \mathcal{T}_{BC} \middle| r_v\right] \\ &= \mathbb{P}\left[g > \frac{\mathcal{T}_{BC}\mathcal{P}_{N,BC} r_v^{\beta}}{\mathcal{P}_B} \middle| r_v\right] \end{aligned} \tag{10}$$

Now obtain the average probability of coverage in BC regardless of the position:

$$
\begin{aligned}
P_{c_{BC}} &= \mathbb{E}_{r_v}\left[\exp\left(-\frac{\upsilon \mathcal{T}_{BC}\mathcal{P}_{N,BC}r_v^{\beta}}{\mathcal{P}_B}\right)\right] \\
&= \int_0^{R_m} f_{r_v}(r_v)\exp\left(-\frac{\upsilon \mathcal{T}_{BC}\mathcal{P}_{N,BC}r_v^{\beta}}{\mathcal{P}_B}\right)dr_v \\
&= \frac{2}{R_m^2}\int_0^{R_m} r_v \exp\left(-\frac{\upsilon \mathcal{T}_{BC}\mathcal{P}_{N,BC}r_v^{\beta}}{\mathcal{P}_B}\right)dr_v
\end{aligned}
\tag{11}
$$

where $\mathbb{E}(\cdot)$ denotes the average operator, and the first step follows the exponential distribution of g. By a simple change of variable, letting $x = \upsilon \mathcal{T}_{BC}\frac{\mathcal{P}_{N,BC}}{\mathcal{P}_B}r_v^{\beta}$, the above integral can be reduced into

$$
P_{c_{BC}} = \frac{2\mathcal{P}_B\gamma\left(\frac{2}{\beta}, \frac{\upsilon \mathcal{T}_{BC}\mathcal{P}_{N,BC}R_m^{\beta}}{\mathcal{P}_B}\right)}{\upsilon \mathcal{T}_{BC}\mathcal{P}_{N,BC}R_m^2\beta}
\tag{12}
$$

where $\gamma(z, s)$ denotes the lower incomplete gamma function.

3.2.2 Coverage probability for a UC user

The coverage probability of the BB network is slightly more complicated. We again evaluate it at a certain distance and then generalize it. At a distance r_l from its serving BS, the probability of coverage for a user is given by

$$
\begin{aligned}
P_{c_{UC}|r_l} &= \mathbb{P}[S_{UC} > \mathcal{T}_{UC}|r_l] \\
&= \mathbb{P}\left[\frac{\mathcal{P}_U h r_l^{-\alpha}}{\mathcal{P}_{N,UC} + I_{U/U}} > \mathcal{T}_{UC}|r_l\right] \\
&= \mathbb{P}\left[h > \frac{\mathcal{T}_{UC}(\mathcal{P}_{N,UC} + I_{U/U})r_l^{\alpha}}{\mathcal{P}_U}\middle| r_l\right]
\end{aligned}
\tag{13}
$$

Here, average over r_l to obtain the universal probability of coverage, and knowing the distribution of h to be exponential, we get

$$
\begin{aligned}
\mathcal{P}_{c_{UC}} &= \mathbb{E}_{r_l}\left[\exp\left(-\frac{\mu \mathcal{T}_{UC}(\mathcal{P}_{N,UC} + I_{U/U})r_l^{\alpha}}{\mathcal{P}_U}\right)\right] \\
&= \int_0^{\infty} f_{r_l}(r_l)\mathbb{E}_{I_{U/U}}\left[\exp\left(-\frac{\mu \mathcal{T}_{UC}(\mathcal{P}_{N,UC} + I_{U/U})r_l^{\alpha}}{\mathcal{P}_U}\right)\right]dr_l \\
&= 2\pi\lambda_{BS}\int_0^{\infty} r_l \exp(-\lambda_{BS}\pi r_l^2)\exp\left(-\frac{\mu \mathcal{T}_{UC}\mathcal{P}_{N,UC}r_l^{\alpha}}{\mathcal{P}_U}\right)\mathcal{L}\left(\frac{\mu \mathcal{T}_{UC}r_l^{\alpha}}{\mathcal{P}_U}\right)dr_l
\end{aligned}
\tag{14}
$$

where $\mathcal{L}(\cdot)$ denotes the Laplace transform (LT) of the interference that will be evaluated shortly. Substituting $f_{r_l}(r_l)$ by its equivalent from Eq. (8) produces the final expression:

$$\mathcal{P}_{c_{UC}} = 2\pi\lambda_{BS}\int_0^{\infty} r_l \exp\left(-\lambda_{BS}\pi r_l^2\right)\exp\left(-\frac{\mu T_{UC}\mathcal{P}_{N,UC}r_l^{\alpha}}{\mathcal{P}_U}\right)\mathcal{L}\left(\frac{\mu \mathbb{T}_{UC}r_l^{\alpha}}{\mathcal{P}_U}\right)dr_l \tag{15}$$

Now, we turn our attention on evaluating the LT. let $s = \frac{\mu T_{UC}r_l^{\alpha}}{\mathcal{P}_U}$, by definition of the LT we have

$$\begin{aligned}
\mathcal{L}_{I_{U/U}}(s) &= \mathbb{E}_{I_{U/U}}\left[\exp\left(-sI_{U/U}\right)\right] \\
&= \mathbb{E}_{\Psi,h}\left[\exp\left(-s\sum_{i\in\Psi}\mathcal{P}_U h_i q_i^{-\alpha}\right)\right] \\
&\overset{(a)}{=} \mathbb{E}_{\Psi}\left[\prod_{i\in\Psi}\mathbb{E}_h\left[\exp\left(-s\mathcal{P}_U h_i q_i^{-\alpha}\right)\right]\right] \\
&= \mathbb{E}_{\Psi}\left[\prod_{i\in\Psi}\frac{1}{1+\frac{s\mathcal{P}_U}{\mu q^{\alpha}}}\right] \\
&\overset{(b)}{=} \exp\left(-\lambda_{BS}\int_{\mathbb{R}^2}\frac{1}{1+\frac{\mu q^{\alpha}}{s\mathcal{P}_U}}\right) \\
&\overset{(c)}{=} \exp\left(\frac{-2\pi\lambda_{BS}}{\alpha}\int_{r^{\alpha}}^{\infty}\frac{x^{\frac{2}{\alpha}-1}}{1+\frac{\mu}{s\mathcal{P}_U}x}dx\right)
\end{aligned} \tag{16}$$

where (a) follows the independence of the PPP and the random channel, and (b) follows Campbell's theorem of the product over a PPP. The integral in (b) is applied on the $\mathbb{R}^2$ plane starting at a distance r. In (c), we switch to polar coordinates and then perform a simple change of variables by setting $x = q^{\alpha}$. Now to solve the integral, we use Eq. (3.194) from Ref. [48] that indicates that

$$\int_w^{\infty}\frac{x^{u-1}}{(1+\beta x)^{v}}dx = \frac{w^{u-v}}{\beta^{v}(v-u)}\,{}_2F_1\left(v, v-u; v-u+1; -\frac{1}{\beta w}\right) \tag{17}$$

where ${}_2F_1(\cdot)$ is the Gaussian hypergeometric function. This results in the following expression:

$$\mathcal{L}_{I_{U/U}}(s) = \exp\left(\frac{-2\pi\lambda_{BS}r_l^{2-\alpha}s\mathcal{P}_U}{\mu(\alpha-2)}\,{}_2F_1\left(1, 1-\frac{2}{\alpha}; 2-\frac{2}{\alpha}; \frac{-s\mathcal{P}_U}{\mu r_l^{\alpha}}\right)\right) \tag{18}$$

Replacing s by its equivalent results in this final form:

$$\mathcal{L}_{I_{U/U}}\left(\frac{\mu \mathbb{T}_{UC} r_I^{\alpha}}{\mathcal{P}_U}\right) = \exp\left(\frac{-2\pi\lambda_{BS}\mathbb{T}_{UC} r_I^2}{\alpha-2} {}_2F_1\left(1, 1-\frac{2}{\alpha}; 2-\frac{2}{\alpha}; -\mathbb{T}_{UC}\right)\right) \quad (19)$$

3.2.3 Coverage probability for any user

The general coverage probability is the complementary to the outage probability. A user is considered to be out of coverage if neither network covers it, so the outage probability is the product of the outage probability of BC and UC subnetworks, following the independence of the two processes. Hence, the probability of coverage becomes

$$P_c = 1 - P_{o_{BC}} P_{o_{UC}} \quad (20)$$

where P_o denotes the outage probability. The system's probability of coverage is, therefore,

$$P_c = 1 - (1 - P_{c_{BC}})(1 - P_{c_{UC}}) \quad (21)$$

3.2.4 Verifying the formulas

The derived expressions are set side by side against the results of a Monte-Carlo simulation for different requirements and different environments. The settings used for the simulation are summarized in Table 1. The results are shown in Fig. 7.

The results show the accuracy of the derived expressions in estimating the probability of coverage for different path loss values and under a range of SINR threshold requirements. Moreover, as expected, it is shown that increasing the SINR requirements reduces the coverage probability, and so does increasing the path loss.

3.3 The power efficiency optimization

As helpful as it can be, the coverage probability alone cannot be used to optimize the network. For instance, increasing the BC power massively means that all the users become BC users, and all would be covered. However, if the user density does not justify

Table 1 The parameters of the simulation.

Parameter	Value
R_m	30 km
$\mathcal{P}_B$, $\mathcal{P}_U$	17 kW, 1.3 kW
BW_{BC}, BW_{UC}	8 MHz, 20 MHz
μ, υ	1, 1
$\mathcal{P}_N$	−141 dBm/Hz
λ_{BS}	0.5 BS/km^2
λ_U	1.5 user/km^2

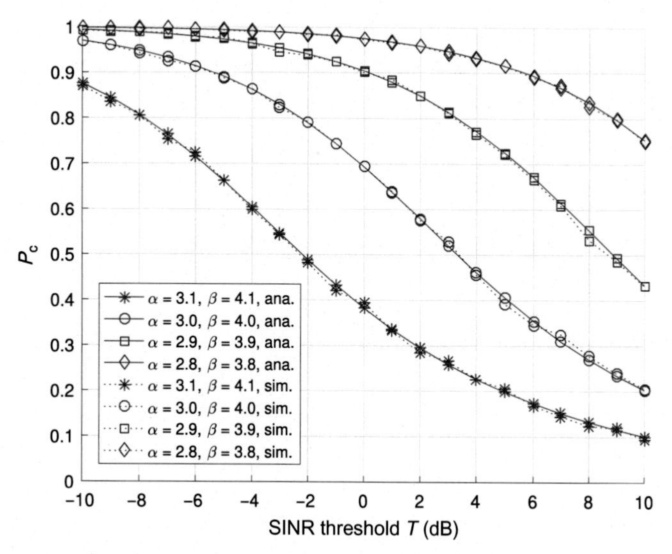

Fig. 7 The coverage probability as a function of the threshold SINR.

that increase, the network will be greatly inefficient. To solve this problem, we introduce another metric: the power utilization factor. This factor can be defined by

$$\mathcal{U} = \frac{\mathbb{E}\left[\sum_j \mathcal{P}_{T_j}\right]}{\mathbb{E}\left[\sum_i \zeta_i\right]} \tag{22}$$

where $\mathcal{P}_{T_j}$ denotes the transmitted power by the jth, and transmitter in the network ζ_i denotes the binary variable, which is equal to 1 if the ith user is served and equal to 0 otherwise. The numerator represents the average total transmitted power in a network, and the denominator represents the average number of the served users. Therefore, $\mathcal{U}$ is measured in Watts per served user.

3.3.1 Power utilization factor

Based on Eq. (22), one can write the power utilization factor as follows:

$$\mathcal{U} = \frac{\mathbb{E}\left[\mathcal{P}_B + \sum_{i \in \Psi} \mathcal{P}_{U_i}\right]}{\mathbb{E}\left[\sum_{i \in \Phi} \zeta_i\right]} \tag{23}$$

This can be reduced to the following:

$$\mathcal{U} = \frac{\mathcal{P}_B + A\lambda_{BS}(1 - P_{c_{BC}})\mathcal{P}_U}{P_{c,UC} A \lambda_U} \tag{24}$$

The second term in the numerator in Eq. (24) follows from the fact the total average total power is equal to the individual power multiplied by the average number of BSs. The latter is equal to the product of the BS density and the service area A. The term $1 - P_{c_{BC}}$ represents the probability that a BS is out of the BC coverage, and therefore actually transmitting the service. The denominator follows from the fact that the average total number of covered users is the mean of the number of users (the product of user density and service area A), multiplied by the coverage probability.

3.3.2 Optimization of power utilization

Optimizing a user-sharing hybrid network is executed by finding the best ratio of users that are served by each network. Since the signal-quality-based allocation is used, the network does not directly control the users' allocation. Alternatively, the network can control the transmission power of the BCT, and consequently indirectly control the segmentation of the user set. The derived expressions can be used to estimate the coverage probabilities, and therefore the power utilization factor. These expressions are fairly generic and include numerous parameters that reflect the environment, the service requirements, the network settings, and the users' behavior.

Common observations

Since the analytical optimization of the derived expressions is intractable without taking further assumptions regarding the network, numerical optimization can be conducted, by evaluating the metrics over a range of BC power. This is done here, taking into account the effect of different parameters like the BS density, required data rates, path loss exponents, size of the service area, and user density. Figs. 8–12, respectively, illustrate the results.

The first common, and most important observation that can be made from all the obtained results is that an optimal point exists under different conditions. The range of BC power values is taken. BC power of zeros indicates a pure BB network, while a very high value indicates a full BC network. Having the optimal point in between shows that the hybrid approach can provide gain to any of the two networks.

At low BC power, as most of the users are BB users, the coverage probability is normally less than one due to interference, and therefore, $\mathcal{U}$ is high. As $\mathcal{P}_B$ increases, more users join the BC network, where the reception is guaranteed[c] and, therefore, $\mathcal{U}$ drops. Beyond a certain value, additional BC power will no longer result in enhancing the service, and therefore the efficiency starts to drop, which is reflected in an increase in the power utilization factor.

[c] This is a condition for joining the BC network from model definition.

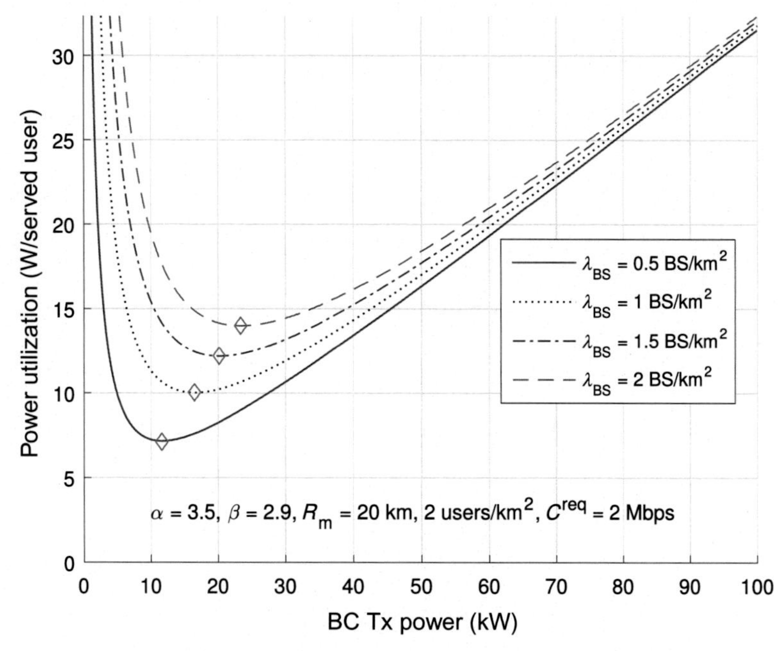

Fig. 8 Optimization of the power efficiency while considering the network deployment density.

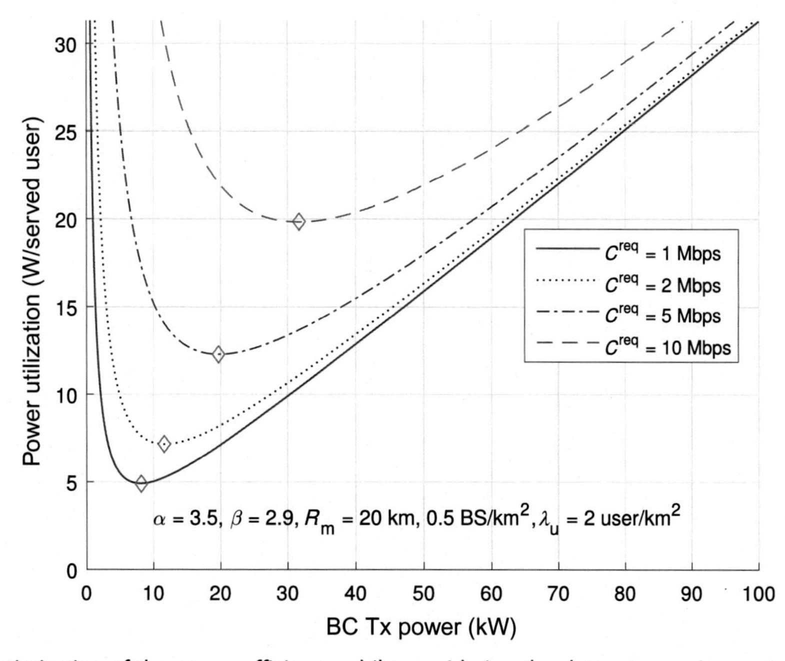

Fig. 9 Optimization of the power efficiency while considering the data rate requirements.

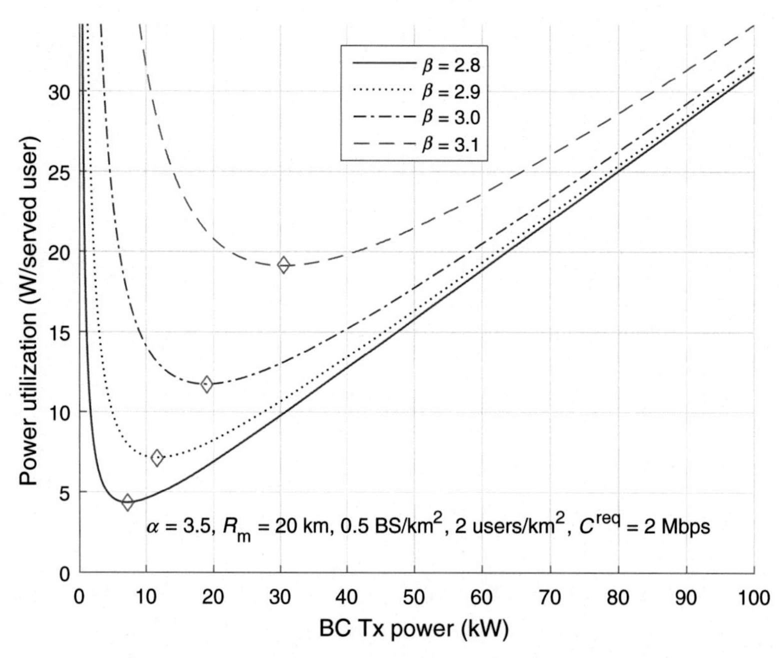

Fig. 10 Optimization of the power efficiency while considering the environment pathloss.

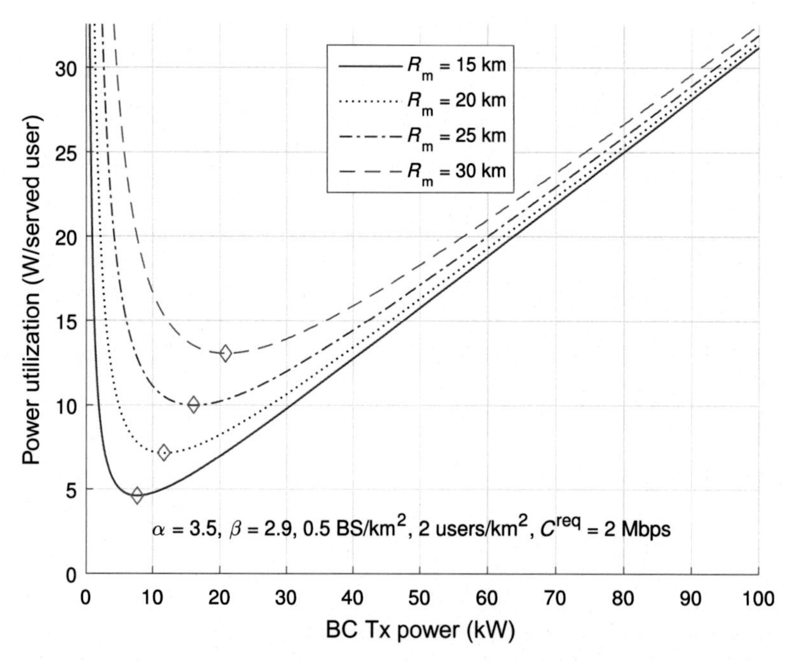

Fig. 11 Optimization of the power efficiency while considering the size of the service area.

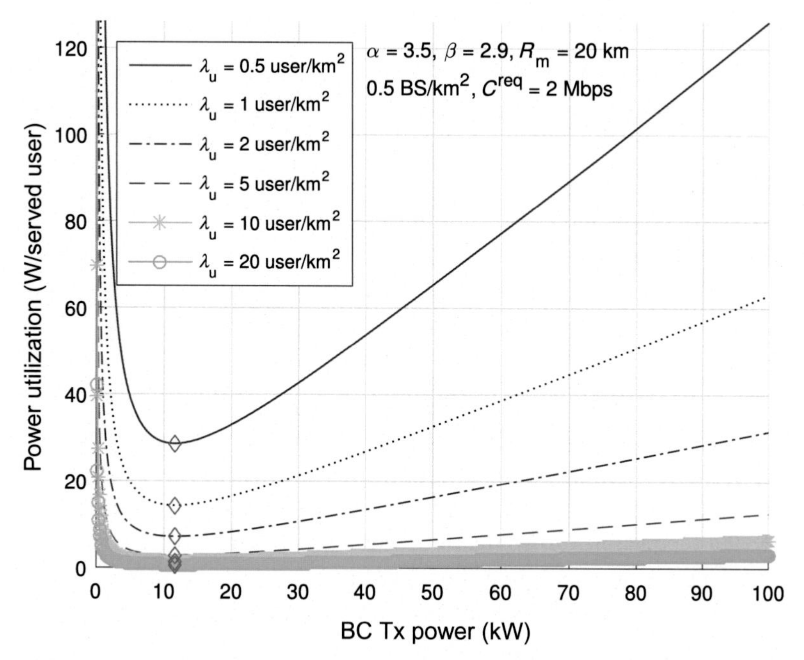

Fig. 12 Optimization of the power efficiency while considering the data rate requirements user density.

BB network deployment

The variation of power utilization factor with the BC power for several values of BS density λ_{BS} is illustrated in Fig. 8. Even though adding more BSs to the network reduces the average link distance to the serving BS, it has two main drawbacks: (1) magnifies the levels of interference, and (2) adds on the power cost of the network. Those two effects lead to the degradation of power efficiency (from around 7 W per user to 14 W per user) as λ_{BS} increased (from 0.5 to 2BS/km^2), and the shift of the optimal point toward the BC side (from 11 to 22 kW for the same range).

Service requirements

The nature of the service under consideration here requires special attention to the data rates and signal quality. In this context, Fig. 9 shows the optimization of $\mathcal{U}$ for different required capacity values C^{req}. C^{req} is defined as the needed link capacity for proper reception of the service. Although the term C^{req} is not a variable in the derived expressions, it affects other variables such as $\mathcal{T}$. For some value of W, $\mathcal{T}$ and C^{req} are linked as follows:

$$C^{req} = W \log 2(1 + \mathcal{T}) \tag{25}$$

The results show that having higher data rate regiments raise the power requirements from the network. This appears in the result when the $\mathcal{U}$ curves are elevated when higher C^{req} values are set. Moreover, higher capacity requirements call for extended BC involvement. This can be seen from the shift in the optimal point toward the right as we increase C^{req}.

Transmission environment

Two of the main factors that dictate the choice of the used model are the nature and size of the service area. The nature of the service area, being urban, suburban, or rural environments, changes the path loss exponent in the propagation model. In addition, the size of the targeted area alters the power requirements of the BC component.

Fig. 10 studies the effect of the path loss exponent of the BC component on $\mathcal{U}$. For higher path loss, indicating a more urban environment, higher power is required from the BC component to serve a user. This also pushes the optimal point to the right, since more power is needed to obtain the optimal ratio between BC and BB users.

A similar effect can be seen when the size of the service is studied. Fig. 11 shows the optimization of the power utilization factor for different service area sizes. Wider areas require more power to cover, and more power to reach the optimal number of users, which push the optimal point to the right as well.

Users' density

The user density affects the power efficiency in two main levels. The first one is fairly obvious, as it appears in the denominator of $\mathcal{U}$. This level is related to the justification of the power usage, and therefore the power efficiency. But user density appears on a deeper level. It is related to the BB network capacity. As the spectral resources of the BB network are limited, when the number of users in a cell increase, the BS has to increase the SINR requirements to maintain the needed data rate.

To simplify the analysis, it is assumed that the available spectrum at each cell, denoted by W^{BS}, is evenly allocated to the users in that cell. On average, the bandwidth allocated to user is defined as follows:

$$W = \frac{\lambda_{BS}}{\lambda_U} W^{BS} \tag{26}$$

substituting Eq. (26) into Eq. (25), we can now find the relation between the SINR threshold T and the user density.

The optimization of the power utilization for different user densities is shown in Fig. 12. The results indicate that as more users utilize the network, the more efficient it gets from a power perspective. It can also be noticed that the optimal power is independent of the user density. This can go back to the fact that for any setting, there is a point where all the users are covered with BC, beyond that point, any power boost will hardly translate to a boost in coverage, and therefore the efficiency drops.

4. Discussion and recommendations

As discussed earlier, each of the two networks that are conventionally used for live multimedia delivery services, that is, BC and BB networks suffer from some limitations. Cellular BB networks have a limited spectrum to be distributed over numerous users. On the other hand, BC networks lose their efficiency as the number of users drops for a certain transmission setting. Different scenarios of a user-sharing hybrid BC/BB network are introduced as a solution.

The designer of such a network has to make choices regarding three major factors: (1) the user allocation scheme, (2) the BC deployment, and (3) the BB network operation mode. Each choice brings its own improvements but limits the network in other aspects. In the previous section, we presented an example of the analysis of one combination of these factors. Other combinations can be used under different circumstances.

In the cases of limited spectral resources, location-based schemes can be useful, since it allows a geographical separation between networks, and therefore it permits a spectral coexistence. This eliminates the need for two bands. This can affect the service of edge users, but it is a price to be paid for the reduction of the spectrum cost.

When direct cooperation between the two networks is not possible, or when the overhead of the network is costly, the signal-quality-based scenarios relax such constraints, since the user device makes the allocation decision. This can be the case when the networks are run by different providers.

Moreover, in the scenarios where a small area is to be covered, like an event venue, a sports game, or a musical concert, a single BC unit is preferred. This deployment limits the budget expenses and reduces the need for extra infrastructure. On contrary, covering a larger service might need multiple BCTs.

In addition, MC mode is preferred when the user density is high. By sharing the spectral and temporal resources, the load on the BB network is reduced. Furthermore, the SFN mode helps to reduce the ICI to enhance the capacity. For lower user densities, it might be very efficient to reserve resources for a service that can be delivered to few individuals. UC can also be the preferable mode as MC is not yet popular neither among service providers nor among device manufacturers.

Table 2 summarizes some of the combinations of the user-sharing hybrid network, their advantages, disadvantages, and use cases.

Table 2 Pros, cons, and use cases for different user-sharing hybrid models.

User allocation	BC deployment	BB mode	Pros	Cons	Use cases	Refs.
Location-based	Single	UC	Possibility of coexistence, full control of operator	Network overhead, edge users conditions	Limited area with limited BW	[23, 49]
Location-based	Multiple	UC	Possibility of coexistence cooperative BC	Network overhead, edge users conditions	Wide area with limited BW	[24]
SINR-based	Single	UC	Reduced overhead	Coexistence not possible	Limited area with overhead/cooperation constraints	Here
SINR-based	Multiple	MC	Independent of user density, high SINR capacity	Coexistence not possible	User dense limited area	[40, 50]

5. Conclusions and future scope

The demand for high-quality streaming services is driving technological advancements for both BC and BB cellular networks. The collaboration of these networks in the form of a hybrid network can cover both networks' limitations in order to optimize the service for the end user, especially in the context of the live multimedia stream. Different forms of user-sharing hybrid BC/BB networks can be used for different scenarios. These forms include different combinations of the network's operation modes, deployments, and collaboration schemes.

In this chapter, the idea of the user-sharing hybrid network is discussed. To have an insight on the gains that this approach can provide compared to the conventional methods, a thorough analysis was conducted to one of the possible scenarios. This was preceded by a clear definition for the different network models and the hybrid model as well. Stochastic geometry was used to conduct the analysis, which resulted in expressions that estimate the coverage probability and the power efficiency of the studied model.

Numerical analysis using the developed expressions shows first that the hybrid approach provides gain compared to a full BB or a full BC network. The optimal point here is in terms of the share of the users' set, which is controlled by the BC transmission power. The analysis also shows that both the optimal point and the gain achieved by the hybridization depend on different factors. The different results show a general trend: the BC is more efficient in higher loads. This was illustrated, for example, in the effect of the user density or the service data rate requirements. Other factors such as the environment, the size and nature of the service area, and the network deployment were also studied.

The discussion included four of eight possible combinations of the deployment, operation mode, and sharing criteria factors. One of these combinations was analyzed thoroughly here, and the others were analyzed in other publications for the authors. The other four are yet to be analyzed. The analysis here can represent general guidelines to analyze the other scenarios and deployments. Each one of these deployments can operate under different conditions to provide optimal service.

Moreover, machine learning can provide a suitable tool to achieve an optimal operation. Machine learning algorithms can be exploited especially in the aspect of dynamic control of the sharing mechanism, to ensure that the network is flexible to changes in the nature of the network both on small and large scales.

References

[1] M. El-Hajjar, L. Hanzo, A survey of digital television broadcast transmission techniques, IEEE Commun. Surv. Tutorials 15 (4) (2013) 1924–1949.

[2] D. Gómez-Barquero, P. Angueira, Y. Wu, Next-generation mobile multimedia broadcasting, in: Next Generation Mobile Broadcasting, CRC Press, 2013, pp. 44–71.

[3] J. Calabuig, J.F. Monserrat, D. Gomez-Barquero, 5th generation mobile networks: a new opportunity for the convergence of mobile broadband and broadcast services, IEEE Commun. Mag. 53 (2) (2015) 198–205.
[4] J. Zhou, Z. Ou, M. Rautiainen, T. Koskela, M. Ylianttila, Digital television for mobile devices, IEEE Multimedia 16 (1) (2009) 60–71.
[5] L. Polak, O. Kaller, T. Kratochvil, SISO/MISO performances in DVB-T2 and fixed TV channels, in: 2015 38th International Conference on Telecommunications and Signal Processing (TSP), IEEE, 2015, pp. 768–771.
[6] S. Jeon, J. Kim, Y.-S. Shin, S. Kim, S. Hahm, Y.-W. Suh, S.-I. Park, J.-Y. Lee, J.-S. Seo, Preliminary field trial results for DVB-T2 indoor reception in Seoul: a single transmitter case, in: 2017 IEEE International Symposium on Broadband Multimedia Systems and Broadcasting (BMSB), IEEE, 2017, pp. 1–5.
[7] N. Cornillet, M. Crussière, J.-F. Hélard, Performance of the DVB-T2 system in a single frequency network: analysis of the distributed Alamouti scheme, in: 2011 IEEE International Symposium on Broadband Multimedia Systems and Broadcasting (BMSB), IEEE, 2011, pp. 1–4.
[8] D. Gómez-Barquero, D. Vargas, M. Fuentes, P. Klenner, S. Moon, J.-Y. Choi, D. Schneider, K. Murayama, MIMO for ATSC 3.0, IEEE Trans. Broadcast. 62 (1) (2016) 298–305.
[9] S. Park, J. Lee, S. Kwon, B. Lim, S. Ahn, H.M. Kim, S. Jeon, J. Lee, M. Simon, M. Aitken, K. Gage, Y. Wu, L. Zhang, W. Li, J. Kim, ATSC 3.0 physical layer modulation and coding performance analysis, in: 2018 IEEE International Symposium on Broadband Multimedia Systems and Broadcasting (BMSB), 2018, pp. 1–3.
[10] S. Ahn, S. Park, J. Lee, S. Kwon, B. Liml, H.M. Kim, N. Hur, Y. Wu, L. Zhang, W. Li, H. Kim, J. Kim, Performance evaluation of ATSC 3.0 mobile service with LDM/TDM under TU-6 channel, in: 2018 IEEE International Symposium on Broadband Multimedia Systems and Broadcasting (BMSB), 2018, pp. 1–9.
[11] R. Chernock, J.C. Whitaker, Y. Wu, ATSC 3.0—the next step in the evolution of digital television, IEEE Trans. Broadcast. 63 (1) (2017) 166–169.
[12] A.A. Razzac, S.E. Elayoubi, T. Chahed, B. El Hassan, Practical implementation of mobile TV delivery in cooperative LTE/DVB networks, in: 2014 12th International Symposium on Modeling and Optimization in Mobile, Ad Hoc, and Wireless Networks (WiOpt), IEEE, 2014, pp. 662–666.
[13] A.A. Razzac, S.E. Elayoubi, T. Chahed, B. El Hassan, Planning of mobile TV service in standalone and cooperative DVB-NGH and LTE networks, in: 2013 11th International Symposium on Modeling & Optimization in Mobile, Ad Hoc & Wireless Networks (WiOpt), IEEE, 2013, pp. 609–614.
[14] N. Cornillet, M. Crussière, J.-F. Hélard, On the hybrid use of unicast/broadcast networks under energy criterion, in: 2012 IEEE 23rd International Symposium on Personal Indoor and Mobile Radio Communications (PIMRC), IEEE, 2012, pp. 1256–1261.
[15] N. Cornillet, M. Crussière, J.-F. Hélard, Optimization of the energy efficiency of a hybrid broadcast/unicast network, in: 2013 IEEE Wireless Communications and Networking Conference Workshops (WCNCW), IEEE, 2013, pp. 39–44.
[16] D. Gómez-Barquero, N. Cardona, A. Bria, J. Zander, Affordable mobile TV services in hybrid cellular and DVB-H systems, IEEE Netw. 21 (2) (2007) 34–40.
[17] H.-Y. Seo, B. Bae, J.-D. Kim, A transmission method to improve the quality of multimedia in hybrid broadcast/mobile networks, in: Information Science and Applications, Springer, 2015, pp. 151–155.
[18] M.M. Anis, X. Lagrange, R. Pyndiah, A simple model for DVB and LTE cooperation, in: 2013 IEEE International Symposium on Broadband Multimedia Systems and Broadcasting (BMSB), IEEE, 2013, pp. 1–6.
[19] A. Abdel-Razzac, S. Elayoubi, T. Chahed, B. El Hassan, Impact of LTE and DVB-NGH cooperation on QoS of mobile TV users, in: International Conference on Communications 2013 (ICC 2013), 2013, pp. 2265–2270.
[20] H. Bawab, P. Mary, J.-F. Hélard, Y. Nasser, O. Bazzi, Spectral overlap optimization for DVB-T2 and LTE coexistence, IEEE Trans. Broadcast. 64 (1) (2018) 70–84.
[21] D. Rother, S. Ilsen, F. Juretzek, A software defined radio based implementation of the "tower overlay over LTE-A+" system, in: 2014 IEEE International Symposium on Broadband Multimedia Systems and Broadcasting (BMSB), IEEE, 2014, pp. 1–6.

[22] P.A. Fam, S. Paquelet, M. Crussière, J.-F. Hélard, P. Brétillon, Analytical derivation and optimization of a hybrid unicast-broadcast network for linear services, IEEE Trans. Broadcast. 62 (4) (2016) 890–902.
[23] A. Shokair, Y. Nasser, O. Bazzi, J.-F. Hélard, M. Crussìere, On the coexistence of broadcast and unicast networks for the transmission of video services using stochastic geometry, EURASIP JWCN (2018) (submitted).
[24] A. Shokair, M. Crussière, Y. Nasser, J.-F. Hélard, O. Bazzi, Analysis and optimization of the hybrid broadcast/unicast network with multiple broadcasting stations, IEEE Access (2018) (submitted).
[25] A.U. Gawas, An overview on evolution of mobile wireless communication networks: 1G-6G, Int. J. Recent Innov. Trends Comput. Commun. 3 (5) (2015) 3130–3133.
[26] Z.U. Rahman, E. Ali, S. Shah, K. Ali Shah, et al., Overview of smart communication in light of LTE, Int. J. Adv. Comput. Tech. Appl. 4 (2) (2017) 247–251.
[27] M.R. Bhalla, A.V. Bhalla, Generations of mobile wireless technology: a survey, Int. J. Comput. Appl. 5 (4) (2010) 26–32.
[28] H. Holma, A. Toskala, P. Tapia, HSPA+ Evolution to Release 12: Performance and Optimization, John Wiley & Sons, 2014.
[29] I.F. Akyildiz, D.M. Gutierrez-Estevez, E.C. Reyes, The evolution to 4G cellular systems: LTE-advanced, Phys. Commun. 3 (4) (2010) 217–244.
[30] E. Dahlman, S. Parkvall, J. Skold, 4G, LTE-Advanced Pro and the Road to 5G, Academic Press, 2016.
[31] S.K. Routray, K.P. Sharmila, 4.5 G: a milestone along the road to 5G, in: 2016 International Conference on Information Communication and Embedded Systems (ICICES), IEEE, 2016, pp. 1–6.
[32] M. Vaezi, Z. Ding, H.V. Poor, Multiple Access Techniques for 5G Wireless Networks and Beyond, Springer, 2019.
[33] Y. Yang, J. Xu, G. Shi, C.-X. Wang, 5G Wireless Systems, Springer, 2018.
[34] S. Parkvall, E. Dahlman, A. Furuskar, M. Frenne, NR: The new 5G radio access technology, IEEE Commun. Stand. Mag. 1 (4) (2017) 24–30.
[35] A. Mämmelä, Energy efficiency in 5G networks, in: IFIP Networking 2015, Toulouse, France, 2015.
[36] G.K. Walker, J. Wang, C. Lo, X. Zhang, G. Bao, Relationship between LTE broadcast/eMBMS and next generation broadcast television, IEEE Trans. Broadcast. 60 (2) (2014) 185–192.
[37] A.A. Razzac, S.E. Elayoubi, T. Chahed, B. El-Hassan, Comparison of LTE eMBMS and DVB-NGH mobile TV solutions from an energy consumption perspective, in: 2013 IEEE 24th International Symposium on Personal, Indoor and Mobile Radio Communications (PIMRC Workshops), IEEE, 2013, pp. 16–20.
[38] G. Xylomenos, V. Vogkas, G. Thanos, The multimedia broadcast/multicast service, Wirel. Commun. Mob. Comput. 8 (2) (2008) 255–265.
[39] F. Hartung, U. Horn, J. Huschke, M. Kampmann, T. Lohmar, MBMS–IP multicast/broadcast in 3G networks, Int. J. Digit. Multimed. Broadcast. 2009 (2009) 1–25.
[40] A. Shokair, Optimization of Hybrid Broadcast/Broadband Networks for the Delivery of Linear Services Using Stochastic Geometry (Ph.D. thesis), INSA Rennes, 2019.
[41] M. Haenggi, Stochastic Geometry for Wireless Networks, Cambridge University Press, 2012.
[42] S.P. Weber, X. Yang, J.G. Andrews, G. De Veciana, Transmission capacity of wireless ad hoc networks with outage constraints, IEEE Trans. Inf. Theory 51 (12) (2005) 4091–4102.
[43] H.Q. Nguyen, F. Baccelli, D. Kofman, A stochastic geometry analysis of dense IEEE 802.11 networks, in: IEEE INFOCOM 2007—26th IEEE International Conference on Computer Communications, IEEE, 2007, pp. 1199–1207.
[44] J.G. Andrews, F. Baccelli, R.K. Ganti, A tractable approach to coverage and rate in cellular networks, IEEE Trans. Commun. 59 (11) (2011) 3122–3134.
[45] M. Di Renzo, A. Guidotti, G.E. Corazza, Average rate of downlink heterogeneous cellular networks over generalized fading channels: a stochastic geometry approach, IEEE Trans. Commun. 61 (7) (2013) 3050–3071.
[46] M. Haenggi, On distances in uniformly random networks, IEEE Trans. Inf. Theory 51 (10) (2005) 3584–3586.
[47] D. Lecompte, F. Gabin, Evolved multimedia broadcast/multicast service (eMBMS) in LTE-advanced: overview and Rel-11 enhancements, IEEE Commun. Mag. 50 (11) (2012) 68–74.

[48] I.M. Ryzhik, I.S. Gradshtein, Table of Integrals, Series, and Products, Academic Press, New York, NY, 1965.
[49] A. Shokair, M. Crussière, J.-F. Hélard, O. Bazzi, Y. Nasser, Power efficiency of the hybrid broadcast unicast network with suitable resource allocation, in: IEEE International Symposium on Broadband Multimedia Systems and Broadcasting (BMSB 2018), 2018.
[50] A. Shokair, J.-F. Helard, O. Bazzi, M. Crussiere, Y. Nasser, Analysis of the coverage probability of cellular multicast single frequency networks, in: 2019 International Conference on Wireless and Mobile Computing, Networking and Communications (WiMob), IEEE, 2019, pp. 1–6.

CHAPTER 6

A comprehensive survey on heterogeneous cognitive radio networks

Indu Bala[a], Kiran Ahuja[b], Komal Arora[c], and Danvir Mandal[a]
[a]Lovely Professional University, Phagwara, Punjab, India
[b]DAV Institute of Engineering and Technology, Jalandhar, Punjab, India
[c]CGC Group of Colleges, Mohali, India

1. Introduction

Over the past few years, we all have witnessed the explosive growth of Internet traffic. To overcome the spectrum scarcity problem and to enhance utilization of the TV spectrum, various governing bodies worldwide have started developing new communication paradigms, such as a cognitive radio, that allow unlicensed users to access these licensed bands with no or minimum interference to the licensed user. Thus, to accommodate more Internet of Things (IoT) devices in the future, one can utilize these unoccupied TV channels, known as TV whitespace (TVWS), opportunistically. Therefore, many unlicensed networks/users with different transmission power, bandwidth, device types, and system architectures may be trying to exploit these TV bands sporadically.

To overcome the existing spectrum scarcity problem by adopting and implementing the new communication paradigms that could access the spectrum more efficiently through dynamic spectrum access (DSA)-based cognitive radio (CR) technology can be implement in TV whitespace [1]. Moreover, with digital TV (DTV) transmission in the United States in 2009, many opportunities have been opened up in TV white space to deploy and enhance new and existing wireless applications. However, the availability of the TV bands is very much noncontiguous [2]. These TV bands are available sporadically over the radio spectrum: 54–72, 76–88, 174–216, and 470–806 MHz. This dynamic appearance of TVWS with strict average or peak power constraints imposed by licensed users/networks poses new and subtle challenges to the coexistence of the both types of users [3]. In this chapter, we discuss some key challenges for the coexistence of heterogeneous cognitive wireless networks (HetCWNs) in whitespace in Section 2. In Section 3, RA to the HetCWNs is discussed in detail. An overview of medium access control (MAC) strategies for HetCWN is presented in Section 4, followed by a detailed discussion on security concerns in the heterogeneous environment in Section 5. In Section 6, progress on the global standardization

Comprehensive Guide to Heterogeneous Networks
https://doi.org/10.1016/B978-0-323-90527-5.00010-1

activities is highlighted. Finally, conclusions and future recommendations are given to make HetCWN a reality.

2. Coexistence challenges for HetCWNs in TVWS

To utilize the available TVWS by multiple unlicensed users simultaneously requires a regulatory framework. These networks are heterogeneous in nature as they are comprised of variety of different network architecture, different type of handsets/devices, varied transmission power levels and bandwidth requirements, distinct modulation/demodulation types, encryption/decryption methods, and MAC protocols, etc. [4]. Thus, new standards need to be developed to coexist with these networks known as HetCWNs. Fig. 1 shows the main challenges in HetCWNs, which that can be categorized as:

1. Detection of available TV channels.
2. Spectrum sharing.
3. Interference mitigation.

2.1 Detection of available TV channels

2.1.1 Detection of licensed networks in a TV channel

This requires identifying available TV channels and utilizing them without exceeding the interference limit imposed by licensed users. At the same very time, it is equally important to detect the presence of coexisting unlicensed networks struggling to access the available TV spectrum for opportunistic use. Thus, the spectrum detection process includes:

a. *Noncentralized detection of available TVWS:* To detect the available TV channel, HetCWNs follow many practices that have been adopted to determine the occupancy level of the TV channel by creating a central database and using sensors such as energy detectors.

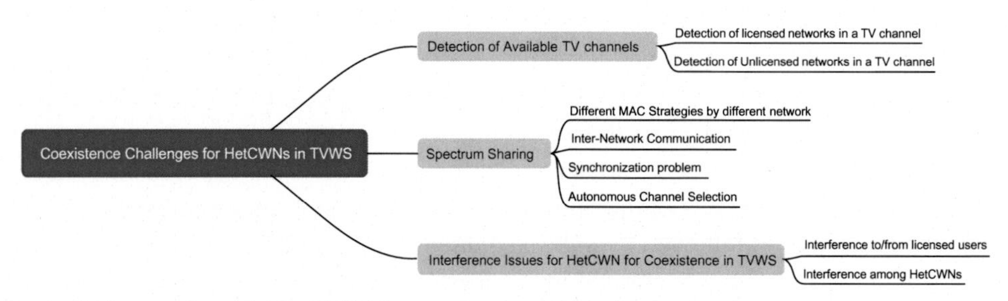

Fig. 1 Challenges faced by HetCWNs for coexistence in TV white space.

b. *Centralized detection of TV channels through location coordinates and database:* In this method, a central repository called TV white space database (TVWSD) is managed by a secure and reliable authority. This authority continuously monitors and updates the database about licensed user operations such as location coordinates, transmission power levels, channel occupancy duration, frequency of channel usage, etc. [5]. Any unlicensed user looking for transmission opportunities can access this database by sending a query message to the TVWSD after sharing its location coordinates. In response to the query message, the TVWSD will share the list of available TV channels in that region with permissible transmission power levels on those channels [6].

2.1.2 Detection of unlicensed networks in a TV channel

a. *Spectrum sensing*: It is a process that monitors the available radio spectrum to detect the presence of licensed user signals. Generally, energy detectors perform this activity by setting certain sensing thresholds depending upon the noise in a channel. The performance of the energy detector is measured in terms of: (a) *probability of false alarms*, and (b) *probability of missed detections* [4,7]. The other popular spectrum sensing techniques are [8]: *waveform-based sensing, matched filtering,* and *cyclostationary feature detection*. Each technique has its pros and cons. But, due to the ease of implementation and the lack of a prior information requirement, the energy detection scheme is the most popular.

b. *Unlicensed user detection*: It is possible for a typical HetCWN that the detected TV channel is already being shared by other unlicensed communication systems. Under such circumstances, the identification of available TV channels become more critical as it involves the detection of different air interfaces [9].

The issue of coexistence of heterogeneous unlicensed networks, that is, the networks that operate on similar technologies and protocols, has been addressed by many standards, such as 802.22 [10]. To implement any self-coexistence mechanism successfully, the network requires the ability to detect neighboring unlicensed networks. If not, this could lead to the following complications:

a. The network performance may deteriorate due to excessive interference.

b. The asynchronous sensing and therefore transmission among unlicensed networks may cause high collisions.

c. The hidden terminal problem may result in high data loss.

The issues highlighted above justify the importance of sensing similar unlicensed networks over TV channels. Some of the challenges in identifying unlicensed networks include:

a. *Increase in network discovery overhead*: In most existing standards, beacon transmission is used to facilitate network discovery. For example, in 802.22, base station (BS)regularly transmits a super frame that contains information about the cell [6]. In a heterogeneous cognitive radio environment, the detection of such beacon signals is difficult as the list of available TV channels is highly dynamic in nature, and thus frequent channel scanning is required to protect licensed users from interference [11].

b. *Coordination complexities and in-band signalling overheads: The* idea of using a dedicated control channel for network discovery is already in use for heterogeneous networks [12]. However, the same idea is applicable for upcoming standards being developed for TVWS for coordinated spectrum access HetCWNs. For such networks, in-band signalling approaches are required for licensed user detection, which significantly increases the signalling overhead and sensing duration [13].

Another challenge posed in HetCWNs is the detection of networks with noncompatible technologies and protocols. Some of the prominent challenges include:

a. *Variation in channel bandwidth for coexisting networks*: For HetCWNs, the channel bandwidths may vary from network to network. For example, 802.22 has an operating bandwidth of 6 MHz whereas 802.11a/b/g has a 20 MHz bandwidth. Similarly, 802.11af uses 5 MHz or groups of multiple channels such that the consolidated channel bandwidth is up to 20 MHz [14].

b. *Variation in transmission power among various operating standards*: For heterogeneous networks, the users may have different transmission power requirements. For example, in the 802.22 network, the users can transmit up to 4 W EIRP, whereas personal/portable devices are restricted to transmitting up to 100 mW EIRP. In such scenarios, the detection of low-power users poses challenge [2].

c. *Variety of signal characteristics among heterogeneous networks:* For a typical broadcast DTV standard, pilot signals and/or preambles are an inherent characteristic that could be used to check spectrum availability. However, in typical HetCWNs, the characteristics may vary from one communication standard to another. Therefore, to detect other unlicensed networks effectively and efficiently, the signal characteristics must be known before deciding the sensing technique based on the signal characteristics [15].

d. *Provision to detect licensed and unlicensed users in upcoming HetCWN standards:* To improve the detection reliability in heterogeneous scenarios, new ideas must be explored to coordinate sensing. For example, in 802.22 and ECMA 392, the concept of quiet periods is used for sensing during which no transmission is allowed. Thus, the coordination and synchronization of quiet times can be explored as a viable option in HetCWNs to detect licensed and unlicensed users over the TVWS [16].

Based on the transmission characteristics of other HetCWNs, the spectrum sensing technique can also be exploited to reduce the false alarm probability. Moreover, it also imposes the requirement of standardizing sensing thresholds while reducing overhead to meet the regulatory guidelines [17].

Some recommendations that could be adopted in upcoming standards for efficient and reliable sensing of unlicensed HetCWNs in TVWS are:

a. *Efficient band sensing management*: Future HetCWN standards should incorporate efficient out-of-band sensing mechanisms when a network is idle by exploiting cooperative sensing mechanisms. Moreover, the centralized approach could be used with reliable reporting mechanisms to send spectrum utility updates to other opportunity-seeking licensed users to coordinate TV channel access.

b. *Unique preamble for HetCWN*: In packet switched networks, the data packet is made up of three fields: the preamble, the header, and the data payload. By designing a unique preamble for the HetCWN, the detection process could be simplified for a heterogeneous environment.

c. *Unlicensed network database*: To identify unlicensed networks quickly, the database approach can also be used. It could be beneficial for the detection of networks such as fixed 802.22; however, for mobile networks such as 802.11af or ECMA392, it is difficult to identify [18].

2.2 Spectrum sharing

On detecting the available TV spectrum, it is important to share the spectrum among all the competitive users using cooperative or noncooperative schemes [19]. In a noncooperative scheme, every unlicensed user has a selfish approach and detects the spectrum autonomously using the listen-before-talk approach such as carrier sense multiple access (CSMA) in 802.11 networks. However, in cooperative schemes, the sensing results of every unlicensed user are shared with a central controller if using a centralized approach or among other secondary networks if using a noncentralized approach. The spectrum sharing is easy to implement in a cooperative or noncooperative manner if all networks sharing the spectrum are using the same physical and medum access control (PHY/MAC) protocols. On the contrary, in heterogeneous HetCWNs, all unlicensed networks use different protocols, so spectrum sharing is a bit challenging in that case. Some challenges in heterogeneous environments are listed below:

a. *Different MAC strategies by different networks*: In HetCWNs, the networks may be operating on different MAC techniques. The variety of MAC protocols within the network limits the effectiveness of the noncooperative schemes for fair existence in TVWS.

b. *Internetwork communication*: In the present scenario, due to noncompatible MAC/PHY standards, different secondary users are not able to communicate over the channel and therefore spectrum sharing is not possible.

c. *Synchronization problem:* In the future, if it became feasible to share the spectrum by heterogeneous networks by using some cooperative mechanism, there would be a stringent requirement of time synchronization among secondary networks. Thus, to communicate over a heterogeneous environment, a number of personal/portable networks with different protocols must have universal reference clocks to share the spectrum.

d. *Autonomous channel selection:* In a heterogeneous environment, autonomous channel selection is very challenging as many unlicensed users may try to access detected TV channels with cooperation. To overcome this issue, a dynamic list of available unlicensed channels can be shared among users for spectrum sharing. Although it will increase the overhead of the network, it is still a cost-effective solution.

2.3 Interference issues for HetCWN for coexistence in TVWS

Most heterogeneous networks operate on the unlicensed frequency band of 2.4 GHz. Therefore, interference is a challenging issue for their coexistence, especially when the channel availability is limited [8]. It is anticipated that similar issues could exist as most of these technologies are migrating into the TVWS where low-power portable mobile devices such as 802.11af and high-power immobile systems such as 802.22 operate on the same TV channel. Moreover, due to good propagation characteristics, the TVWS is the first choice for many unlicensed networks to operate heterogeneously, and this may contribute to increased interference. The interference issues in TVWS can be broadly classified as:

a. Interference to/from incumbent licensed users [3,20].

b. Interference among HetCWNs.

2.3.1 Interference to/from licensed users

For unlicensed networks to coexist in TV channels, it is mandatory to control the out band emissions to reduce interference among devices/networks. To do so, it is important to assess the location of the interferer, the channel gain between the TV station and the unlicensed interferer, and the operating frequency deviation between them [21]. The interference mitigation techniques used in HetCWNs can be broadly classified as:

a. Interference avoidance.

b. Interference control.

c. Interference mitigation.

In the first type of interference avoidance technique, both licensed and unlicensed users share TVWS in time division multiple access (TDMA) or frequency division multiple access (FDMA) mode. The interference control scheme allows both licensed and unlicensed users to coexist on the same TV channel, provided the interference from the unlicensed user does not exceed the predefined limit by the licensed receiver. For example, by using a multiple input multiple output (MIMO) scheme as proposed in ref. [22], the interference from unlicensed users can be avoided by using null space for signal transmission to licensed users. In addition, the unlicensed users can adapt their transmission power levels to ensure that minimum or no interference is experienced by incumbent users.

The interference mitigation techniques are further divided into two categories:

a. Opportunistic interference cancellation (OIC).

b. Asymmetric cooperation.

In OIC, the unlicensed users have access to the licensed users' codebooks to decode their transmissions. Thus, the unlicensed channel transmission rates can be increased by subtracting them from their received signals. In asymmetric cooperation [23], the side information is used by the unlicensed users to mitigate interference while cooperating with licensed users to boost their signal with the help of codebooks at the receiver side.

2.3.2 Interference among HetCWNs

In a heterogeneous environment, due to uncoordinated spectrum access, many HetCWNs may access the same TV channel due to limited TV channel availability. Under such circumstance, the chances of data collision in the channel are very high. The main aspects that contribute to interference in TVWS scenarios are:

a. *Different transmission powers*: In a typical heterogeneous environment, the interference among unlicensed users can be controlled by transmission power control under received power constraints and average power constraints, as discussed in ref. [3]. This means that the intended receiver may have better reception and nonintended receivers may have minimum or no interference because of this transmission.

b. *Variable channel bandwidth*: In a heterogeneous network, different secondary networks operate on different bandwidths. As discussed earlier, 802.22 WRANs operate on a 6 MHz bandwidth whereas 802.11af may have signal bandwidths of 5, 10, and 20 MHz [10]. Thus, the amount of interference can be controlled by controlling the transmission bandwidth.

c. *Transmission time and data packet size*: The transmission time required to transmit data is directly proportional to the size of the data packet for a given data rate. Moreover, the duration when the interference will affect the network is also related to the transmission time. Thus, by shrinking the data packet size, the amount of interference among the heterogeneous networks can be controlled.

d. *Low ratio of signal to interference plus noise*: The ratio of signal to interference plus noise (SINR) is used to measure the performance of the network in terms of the packet error rate (PER). Thus, the chances of error are greater in heterogeneous scenarios under low SINR, which may lead to incorrect estimates of PER.

To overcome interference issues in TVWS, some considerations are made in this section for upcoming TV standards:

a. *Cooperative schemes can* be used to synchronize the quiet periods and exchange spectrum sensing information among networks.

b. *Spatial diversity* techniques can also be exploited for MIMO systems using smart antennas to avoid interference.

Receiver threshold: By strictly controlling the receiver sensitivity threshold, the interference and false alarm situations can be easily dealt with in a typical heterogeneous environment.

3. RA in HetCWNs

The tremendous growth in HetCWNs in recent years makes efficient RA (RA) necessary [8]. The increasing traffic demands can be fulfilled by the use of heterogeneous networks. The main issue in these networks is the use of excessive power. Specially for mobile operators, this issue is critical [24,25]. Device-to-device (D2D) communication along with the use of relays is a hopeful answer to achieve capacity and energy efficiency in these networks [26]. For ultradense networks with heterogeneous properties, a novel scheme for resource allocation based on radio is presented in ref. [27]. This novel scheme utilized subband allocation, which is relatively faster than existing techniques.

In ref. [8], RA is done by considering the spatial and temporal dimensions for BSs and user equipment (UE), respectively. An investigation of the performance was done with stochastic geometry along with the queueing theory. In order to maximize the energy efficiency for RA, optimization techniques were used by researchers. The use of the mixed integer nonlinear optimization (MINO) problem with two-phase optimization was illustrated in ref. [28] to enhance the energy efficiency of RA for multihomed user equipment in heterogeneous wireless networks.

RA based on the barrier method and the Lagrangian optimization algorithm for macro and femtocells in heterogeneous networks, a cross-layer-based security aware algorithm for RA, and mobility-based RA are presented in refs. [29–31], respectively. In ref. [32], RA along with cell selection for long-term evolution (LTE) advanced networks, which are heterogeneous in nature, is presented. A technique based on a quantum coral reef optimization algorithm (QCROA) for RA in heterogeneous networks is illustrated in ref. [33].

To meet the growing demand of bandwidth to satisfy the fairness issue among users of heterogeneous networks, the concept of cognitive small cell networks is introduced in ref. [34] that integrate the concept of cognitive radio with small cells to propose a fairness-based distributed resource allocation algorithm to maximize the total throughput of the network by jointly considering interference management, fairness-based resource allocation, average outage probability, and channel reuse radius. In ref. [35], a general heterogeneous cellular network model is investigated in context to the spectrum allocation problem by proposing an equivalent orthogonal network model. An approximate signal-to-interference ratio analysis based on the Poisson point process (ASAPPP) is used to provide accurate approximations to the spectral and energy efficiency in the network to the spectrum allocation process, while considering the future 5G networks with limited on board battery backup, the resource allocation problem is investigated for energy harvesting D2D communication system in ref. [36].

Many techniques using joint content caching and radio RA, joint power control and user association along with RA, node placement and RA, and RA based on a learning approach, which is further based on deep reinforcement, are illustrated in refs. [25,37–39], respectively. In ref. [40], schemes for RA based on a fair energy efficient method and a technique for networks with imperfect spectra are proposed and implemented. An approach for RA for cognitive radio networks is addressed in ref. [41].

3.1 RA schemes

The current scenario-based RA schemes for various kinds of heterogeneous wireless networks are classified as per the following categories:

a. RA for D2D communication in HetCWNs.
b. RA for ultradense HetCWNs.
c. RA for HetCWNs with MINO problem.
d. Security aware RA for HetCWNs.
e. Fairness-based RA for HetCWNs.
f. RA for LTE-based HetCWNs.

a. *RA for D2D Communication in HetCWNs* The goals of fifth-generation networks can be fulfilled using relays and D2D communication in heterogeneous networks. To meet the requirements of quality of service in HetCWNs, power allocation, the RA of the spectrum, and cell selection are the key areas that need improvement and optimization. To enhance the efficiency of the spectrum and the rate of data transmission, D2D communication is a possible solution [33].

In ref. [26], to efficiently perform RA, the programming problem for optimization, which was a nonlinear fractional problem, was converted into a concave optimization problem using the Charnes-Cooper transformation. After this conversion, to solve the problem, the authors proposed and implemented an outer approximation algorithm (OAA). The QCROA was also used for RA to enhance the total throughput of all the links in the D2D communication to maximum values [26]. The heterogenous network (Hetnet) design utilized in ref. [26] is illustrated in Fig. 2.

In order to maximize the throughput for RA in D2D HetCWNs, a greedy strategy along with a convex optimization problem is another promising technique to solve the formulated problem. In this context, a novel algorithm known as the mode selection and RA algorithm (MSRA) is shown in ref. [36].

b. *RA for ultradense HetCWNs* In ultradense HetCWNs, computational complexity becomes very high due to the dynamic traffic patterns. In addition, communication overhead will also increase when outdated schemes are used for RA. RA based on radio is another promising method for HetCWNs [27]. In this scheme, the computational overhead along with the rate of handoff for the subband was reduced to 30%. In ref. [27], a novel estimation method based on interference was proposed and implemented for RA.

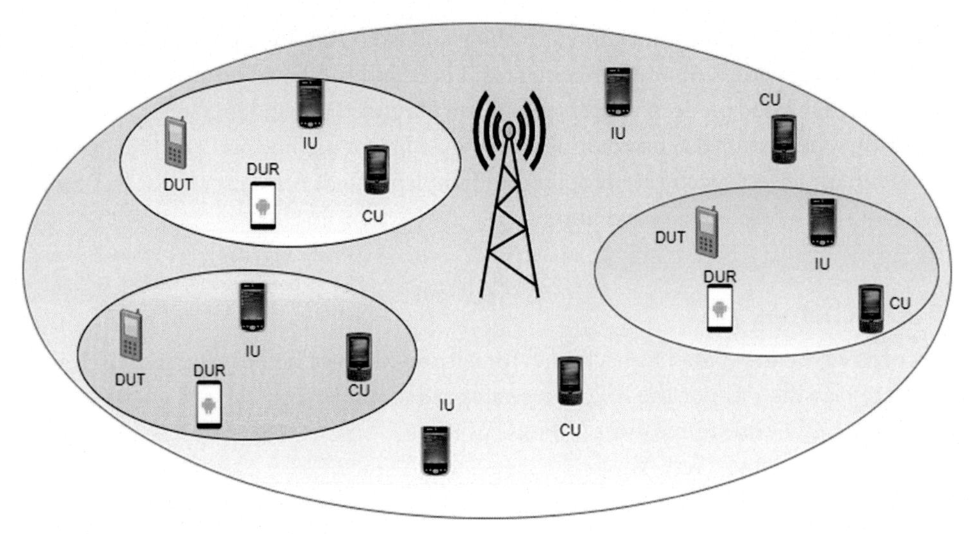

Fig. 2 D2D HetCWN used in ref. [26].

It is a well-known fact that for 4G and 5G networks, an ultradense HetCWN is a scheme that can easily increase their efficiency in a notable manner. The main challenge is the interference and management of mobility. A new approach for RA, based on a hybrid cell, is illustrated and implemented in ref. [31]. The proposed method in ref. [31] is developed for a practical scenario, as the presented scheme is adjustable to the changes in the parameters of the dynamic HetCWNs.

In this technique, the main problem of RA was divided further into two subproblems: a hybrid cell RA and a hybrid partial cell RA [31].

c. *RA for HetCWNs with MINO problem:* For multihomed user equipment in HetCWNs, RA using a MINO problem was investigated, formulated, and presented in ref. [28]. The RA was considered a nonconcave problem for the maximization of energy efficiency. This problem was converted into the MINO problem using the fractional theory for programming. The HetNet employed is presented in Fig. 3. Further, the problem was solved by employing the Lagrange dual and continuity relaxation method. Finally, an optimization method based on two phases was used to achieve RA, which was energy efficient and had a minimum rate guaranteed [28].

d. *Security aware RA for HetCWNs:* The formulation of RA with a security aware feature, as a maximization problem for energy efficiency, is proposed and illustrated in ref. [30]. First, at the physical layer, packet delay and packet dropping probability with average values are converted into a minimum secrecy rate constraint for each mobile terminal. Then, the maximization problem for secrecy energy efficiency, which is nonconvex in nature, is approximated by the representation of an epigraph, which is convex in nature. The implemented scheme in ref. [30] shows that the secrecy

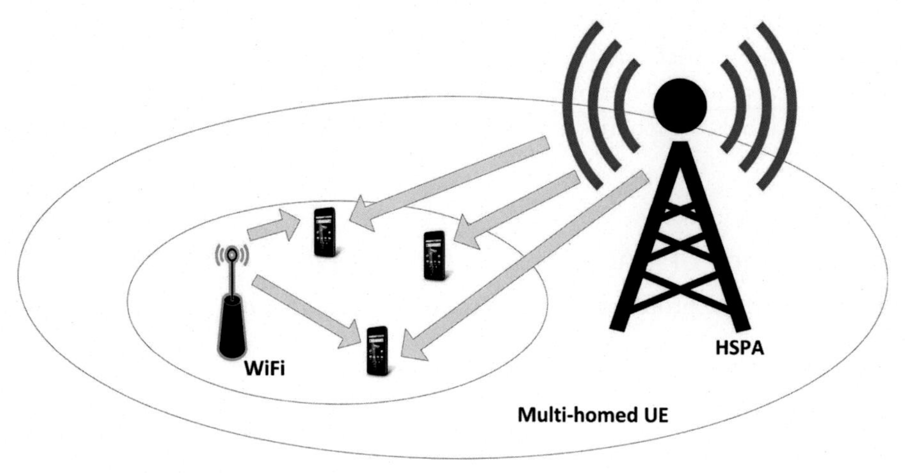

Fig. 3 HetCWN utilized in ref. [28].

energy efficiency and throughput were improved along with the quality of service at the packet level.

e. *Fairness-based RA for HetCWNs:* RA, distributed in nature and fairness based, is proposed and implemented in ref. [34]. Further, on the basis of traffic load, which is time varying, and various geographical situations, an improved algorithm is also illustrated in ref. [34]. Using small cells with cognitive radio, the scheme introduced small cell networks with cognitive features. The actual problem was divided into three subproblems as the original problem was inflexible. The proposed and implemented algorithms offered better fairness compared to the results available in the literature for other traditional schemes.

f. *RA for LTE-based HetCWNs:* In HetCWNs with a multiinput multioutput (MIMO) feature, the RA is an even more difficult and complex problem [25]. The energy efficiency is optimized in addition to spectrum efficiency by allocating the resource blocks dynamically. To achieve lower complexity, an algorithm, which is distributed in nature, was operated on two levels. The results of the numerical experiments were also satisfactory and better than the existing greedy-based algorithm.

4. MAC protocols for HetCWN

MAC plays a very vital role in HetCWN to perform functions such as TVWS detection, spectrum access and management. and spectrum mobility [3]. For the opportunistic access of TV channels to unlicensed users to meet the end user quality of service (QoS) requirements without exceeding the interference to the licensed primary user, sophisticated MAC protocols are required. Multichannel MAC protocols set the foundation for developing a MAC strategy for HetCWN. However, the main difference compared to the homogenous environment is that the number of channels per user remains fixed in a multichannel network, whereas the number of channels varies sporadically in a heterogeneous environment. Moreover, HetCWNs also need periodic scanning of TV channels, which makes them unique compared to other wireless ad hoc radio networks.

4.1 Classification of MAC protocols for HetCWNs

Generally, the MAC protocols can be classified on the basis of:
(1) Protocol architecture.
(2) Complexity.
(3) Cooperation level.
(4) Signalling and data transfer management.

Fig. 4 shows the main classification of MAC protocols for HetCWNs. These are:
a. Direct access based (DAB).
b. Dynamic spectrum allocation (DSA).

In DAB protocols, every sender and receiver pair works toward its own optimization problem and thus does not requires global network optimization. In DSA protocols, networks work on a global optimization problem in an adaptive manner. Both types of MAC protocols are applicable to the centralized as well as decentralized network architecture. The networks with distributed protocols are more robust as they are capable of making independent decisions about resource access without involving the central coordinator. However, the networks with centralized architecture can efficiently utilize resources by exploiting the global information [42].

a. *Direct access-based MAC protocols*

The DBA protocols are further divided into two categories:
a. *Contention-based protocols*: In these protocols, unlicensed transmitters and receivers exchange their sensing results while exchanging handshake signals called a channel

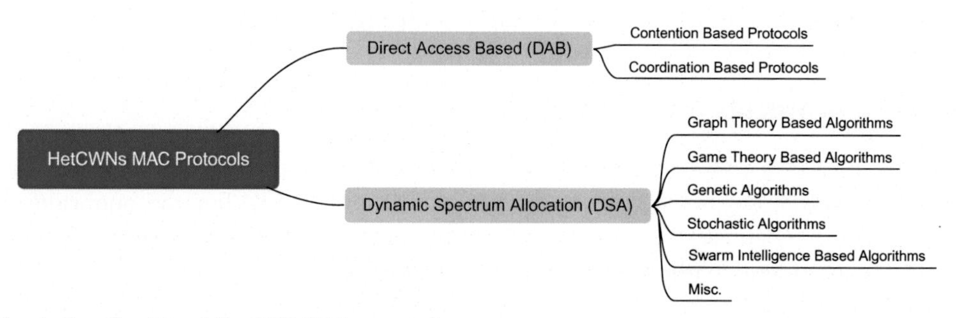

Fig. 4 Classification of HetCWN MAC protocols.

filtering sender receiver (CFSR) handshake. During handshaking, they compare available resources and negotiate about the channel for future communication.

b. *Coordination-based protocols:* These protocols are more efficient and reliable as every node shares its sensing information with its neighbor.

Table 1 lists some of the direct access-based MAC protocols, and their details can be found in the references therein.

c. *MAC protocols for dynamic spectrum allocation*

The advanced optimization algorithms are used in DSA-driven MAC protocols for a fair spectrum to the users. Every opportunity-seeking unlicensed user adapts its transmission parameters to avail the available resource efficiently. DSA algorithms have better performance compared to DAB algorithms; however, due to the low scalability, the negotiation delay and complexity of the system increase. The most popular approaches in DSA protocols are:

a. Graph coloring theory based.
b. Game theory based.
c. Stochastic theory based.
d. Genetic algorithms.
e. Swarm intelligence algorithms.

Table 2 lists some DSA methods, and the details can be found in the literature therein.

(a) *Graph theory algorithms:* In such algorithms, the HetCWN modeling is done with a graph G=(V,E), where V and E indicate the vertex vs. the edge sets. They are further classified as a node contention graph (NCG) and a link contention graph (LCG). In NCG, the primary unlicensed users are represented by nodes, whereas the edges represent nodes that are in the interference range of each other. Similarly, in LCG,

Table 1 Direct access-based MAC protocols.

S. no	Cooperation	Protocol	Reference
1.	Coordination based	IEEE 802.22	[10]
2.	Coordination based	SYN-MAC	[43]
3.	Contention based	HC-MAC	[44]
4.	Contention based	COMAC	[45]
5.	Contention based	DOSS	[46]
6.	Coordination based	C-MAC	[47]
7.	Coordination based	HD-MAC	[48]
8.	Coordination based	OS-MAC	[49]
9.	Coordination based	MMAC-CR	[50]
10.	Coordination based	SU08	[51]
11.	Coordination based	OS-MAC	[49]
12.	Coordination based	GHaboosr08	[50]

Table 2 DSA-based MAC protocols.

S. no.	Protocol	Algorithm type	Reference
1.	BIOSS	Swarm intelligence	[52]
2.	DC-MAC	Stochastic model	[53]
3.	DH	Graph coloring	[54]
4.	G-MAC	Game theory	[55]
5.	Nainary08	Genetic algorithm	[56]
6.	Zheng05	Graph coloring	[54]
7.	Zou08	Game theory	[57]

the vertex set represents active flows, whereas the edges represent a contention between different flows.

(b) *Stochastic algorithms:* The channel availability can also be represented by a stochastic process. One of the most popular stochastic approaches is Markov chain formulation. Based on the statistics of local spectrum sensing and its historical access experience, every node estimates the channel usage to maximize the utility function.

(c) *Game theoretic algorithms:* The communication between unlicensed cognitive radios is represented as a game. This scheme efficiently models the dynamics of a cognitive network. In a game, each player may have a unique utility function to achieve desired objectives.

(d) *Genetic algorithms:* These algorithms are based on the evolutionary ideas of natural selection. The randomly generated solutions called population, start the iteration process and utility function are evaluated to select the best solution. With crossover and/or mutation, the optimum solution is obtained as from the offspring. These algorithms work well in RA problems due to their fast convergence rate.

(e) *Swarm intelligence algorithms:* The swarm intelligence (SI) algorithms model network users as a population of simple agents interacting with the surrounding environment. In this type of algorithm, the group of users uses its intelligence to achieve the desired objective.

5. Security issues/challenges for heterogeneous cognitive radio networks

A cognitive radio network resolves the bandwidth scarcity issue by providing dynamic spectrum access to unlicensed users. A cognitive radio network (CRN) is practically a kind of heterogeneous network as it has the profuse user's demands based on transmission rate, voice or video call requirements, and delay-time tolerance [58]. In addition, heterogeneous networks are needed for next-generation wireless communications. Because of their diverse requirements, these networks incorporate various challenges including stringent spectrum management, bandwidth sharing, and spectrum security issues. In order to efficiently use the spectrum, a CRN utilizes methodologies of spectrum reusing and spectrum sharing. Spectrum reusing allows the unlicensed users to use the vacant spectrum not utilized by licensed users while spectrum sharing allows simultaneous data transmission for licensed and unlicensed users. These CRN features have created enormous security hazards, thus making CRNs more susceptible to security attacks compared to conventional wireless systems [23]. Nowadays, CRNs are finding applications in various emerging fields, including medical, military, and mobile communications. Thus, transferring information has to be done in such a manner that it is not decoded by intruders. Hence, security issues are prime concerns for CRNs. In this chapter, CRN security issues are described in detail. Table 3 lists the parameters to characterize CRN security.

5.1 Security intrusions

This section describes different types of security intrusions/attacks faced by CRNs. Security intrusions are generally divided into active and inactive intrusions [61]. In active intrusions, the intruder attempts to amend the message and change the information to be transmitted, thus sending the wrong information to the user. Inactive intrusions are attacks where the intruder attempts to quote the data from the continuing conversation without the consent of the authorized user. Security intrusions can also be classified according to the affected layers, as shown in Fig. 5. This section provides a detailed description of service denial intrusions (SDI) corresponding to the affected layer.

5.1.1 Service denial intrusions at physical layer

(a) *PUEA* (primary user emulation attack): CRNs are highly susceptible to intrusions, mainly while sensing the spectrum. In addition, the radio-frequency waves are also susceptible to intrusions as these can be easily tuned if the sender also transmits with a similar frequency and more power. These intruders interfere with the network

Table 3 Parameters to characterize CRN security [59,60].

Parameters	Description
Secrecy rate	Rate at which information can be transmitted securely from the transmitter to the receiver
Secrecy capacity	Upper limit on secrecy rate defines the secrecy capacity
Leakage probability	Probability that an intruder deciphers the information with an error probability less than the target bit error rate
Security gap	Security gap describes the signal-to-noise ratio corresponding to a lower-target bit error rate attained by the designed user and the signal-to-noise ratio corresponding to a higher-target bit error rate attained by the intruder

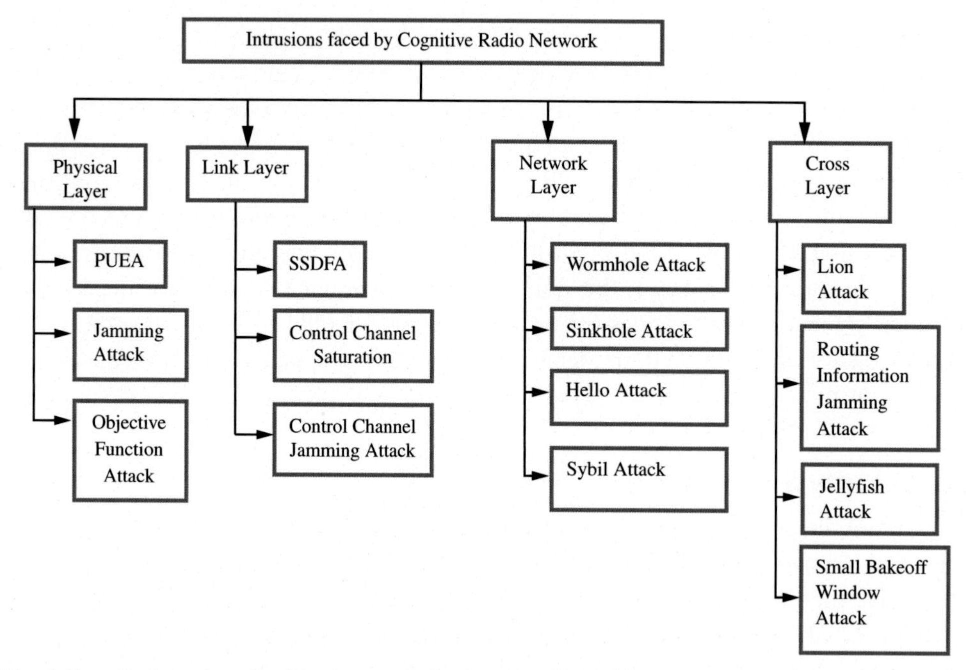

Fig. 5 Security intrusion classification according to the affected layer.

uncontrollably, thus endangering the privacy of licensed users [62,63]. The conduct of the intruders can be classified as specified in Table 4.

The intruder that performs in the above listed ways may mimic a licensed user and this type of attack is called *PUEA*. An example of this attack is illustrated in Fig. 6. This figure illustrates three nodes corresponding to a licensed user, an unlicensed user and a primary user emulator (PUE). The PUE detects that the licensed user is not using its bandwidth and thereby sends a fake signal to the unlicensed user to misguide them about the spectrum holes present in the system [64,65].

(b) *Jamming attack* (JA): A jammer deliberately transmits intrusive signals to obstruct licensed users from utilizing the spectrum holes for transmission or reception of data, thus causing a service-denial attack. Also, other situations where the network is considered congested can be because of huge traffic demand. To tackle this intrusion, unlicensed users must keep a record of licensed users' positions to verify whether the signal is coming from a licensed user or an intruder [64–67].

(c) *Objective function (OF) attack:* An OF is a function that includes multiple variables to maximize the learning ability of CRNs. These variables incorporate bandwidth, code rate, modulation schemes, encoding, and power. SUs utilize these variables to compute OF and modify them to attain maximum data rate and minimum power consumption. An OF attack targets these OF parameters and can manipulate results to entertain intruder's interests [68].

Table 4 Conduct of the intruders.

Conduct	Definition
Misconduct	This type of conduct exists where the intruder doesn't follow the network regulations
Self-centered	In this conduct, the intruder/attacker retains the bandwidth and doesn't take care of the other network users
Misleading	This type of conduct doesn't provide the actual information for the available network sources
Malignant	This user deliberately attacks the system so as to deteriorate other user's performance, thus decreasing the whole network capacity

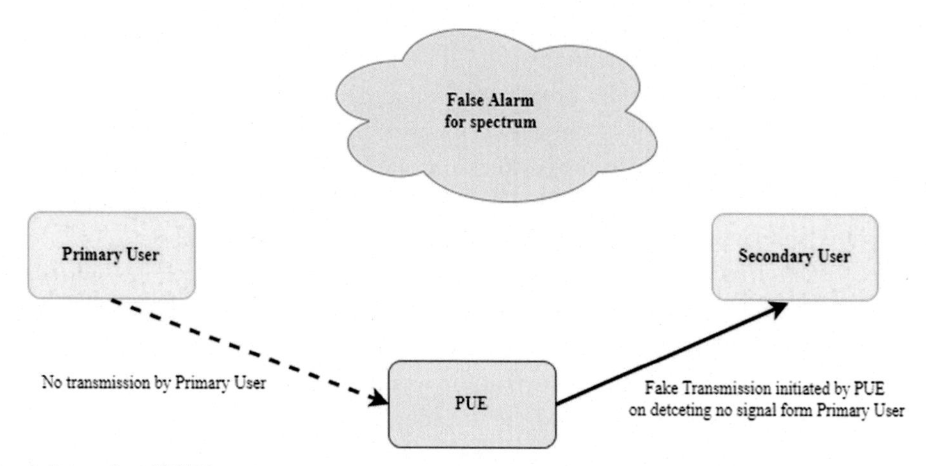

Fig. 6 Example of PUEA.

5.1.2 Service denial intrusions at link layer

(a) A spectrum-sensing data falsification attack *(SSDFA*) is an attack that occurs in cooperative spectrum sensing. In this type of attack, intruders deliberately send false sensing reports to the decision center. In this way, the decision center gets false information about the presence/absence of licensed users, which can cause unlicensed users not to use the vacant spectrum or to create an intrusion with licensed users [66].

(b) Common control channel jamming *(CCCJ*): A control channel aids in coordination among different cognitive users, which makes this method very efficacious for an intruder in demolishing the whole system. This type of jamming prohibits the receiver from getting actual control signals, leading to service denial.

(c) *Control channel saturation*: This type of attack relies upon the law that when a cognitive radio user can't accomplish successful conversations for certain defined periods, the user discontinues the transmission in the succeeding transmission. Such a situation exists in case of huge traffic demands by different users. An intruder sends several messages in order to saturate the control channel [69].

5.1.3 Service denial intrusions at network layer

(a) *Wormhole attack:* This type of attack allows two intruders to take positions judiciously inside the network. These intruders continue to hear the network and track all the information. Fig. 7 shows an example of this attack.

The intruders cunningly create a minimal distance path, as depicted in Fig. 7, and broadcast this path among all nodes in the network. To keep track of the present conversation at a specific network location and to transfer information to some other part of the network, wormhole intruders build a tunnel [70]. Different variants of this attack have been discussed in the literature,including encapsulation-based, tunnel-based, and protocol distortion-based wormhole intrusions [71,72].

(b) *Sink hole attack:* An intruder broadcasts itself as having the best path to the destination. Thus, all the nodes of the network start routing their packets to this intruding node. The intruder collects the packets, and afterward amends them before transmitting them to the actual node [73]. The intruder can also drop the whole information instead of relaying; such an attack is called a black hole attack [74].

(c) *Hello attack:* In a wireless network, an initial hello message is broadcast to the neighboring nodes to initiate communication and exchange location information among various nodes. This attack allows the intruder to broadcast hello messages with very high power to ensure other users that the intruder is a neighboring node. Thus, all nodes route their packets toward the intruder, leading to a loss of information [73].

(d) *Sybil attack:* A Sybil intruder affects the system by creating various fake identifications. The intruder attacks the licensed user to change the choice made by the licensed user, thus restricting the licensed user from getting access to the vacant frequency band. A CRN provides valid identification for every user so that they can use the channel for conversation. A Sybil attack uses this aspect by making various

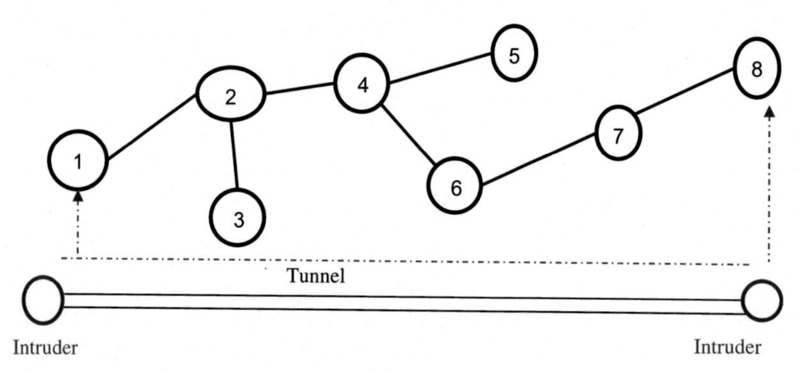

Fig. 7 Example of a wormhole attack.

identifications and pretending as if there are many users located apart, thereby creating a scenario where multiple users are contending to get access to the channel.

5.1.4 Service denial intrusions at cross layer

(a) *Lion attack:* This attack occurs at the physical layer and targets the cross layer, thereby reducing the efficiency of the transmission control protocol (TCP) and impelling frequency handoffs in the network. This intrusion is a form of PUEA. In this attack, the intruder mimics a licensed user, which leads to emptying the channel. But the TCP doesn't know about this transition. Thus, the TCP keeps making connections and transmitting packets without getting acknowledgments, which makes the TCP think that the packet has been lost. Hence, it again sends the packet, thereby decreasing the length of the congestion window. It causes a loss of data, thus reducing the efficiency of the network [51].

(b) *Routing information jamming attack:* This attack comes when there is no common control channel. It is based on the observation that there is latency while performing handovers. Because of this latency, there is crowding of the route information among various nodes. This results in the utilization of spoiled routes and false packet routing [75].

(c) *Jelly fish attack:* This attack along with the lion attack targets the TCP. In the jelly fish attack, the efficiency of the network is reduced because of incorrect ordered or retarded packets. This attack occurs at the network layer and affects the transport layer. In this type of intrusion, the intruder deliberately pierces the system by rearranging the received packets and sending them to the next node in an incorrect order. This leads to the retransmission of the packets, thereby decreasing the efficiency of the network [75].

(d) *Small back off window attack:* This type of intrusion is called a "back off manipulation attack." The intrusion modifies the TCP specifications to maintain early access to the channel. Intruders select a small contention window, which results in frequently approaching the channel. This attack is possible while utilizing the carrier sense multiple access collision avoidance protocol [11].

5.1.5 Detection methods

Several types of detection techniques are discussed in Table 5.

5.1.6 Strategies to prevent intrusions

Numerous methods have been devised so far in the literature to combat intrusions. Some of them are listed in Table 6.

Table 5 Detection methods corresponding to the affected layer.

Affected layer	Attack	Detection methods	Merits
Physical layer	PUEA	Energy detection method [76]	Easy implementation, lesser intricacy, initial knowledge of energy level of licensed user is not required
		Matched filter method [77]	Effective in noisy environment
		Intrusion detection-based method [62]	Centralized IDS system is not necessary, training of different types of network behavior is allowed
	Jamming	Frequency hopping scheme [78]	Intricacy level is less, effective for static or dynamic allocation of spectrum
		Spatial retreat method	Effective for static or dynamic allocation of spectrum
	Objective function attack	Intrusion Detection System [79]	Centralized IDS system is not necessary
Link layer	SSDFA	Bayesian Method	Centralized cooperative scheme supported.
	Control channel saturation	K-proximity method [77]	Data mining
	Control channel jamming	Dynamic silence period scheduling	No extra overhead
Network layer	Wormhole attack	Hybrid techniques [80]	Protection against all categories of wormhole attack
	Sink hole attack	Behavior tracking [81]	Routing protocols can be used
	Hello attack	Measuring signal strength	Lesser computational intricacy
	Sybil attack	Locating using received signal strength [28]	Efficiency is very high
Cross layer	Lion attack	Lion optimization method [82]	Better performance
	Routing information jamming attack	Belief propagation [73]	Highly accurate
	Jelly fish	APD-JFAD [83]	Highly efficient
	Small back off window attack	Machine learning method [84]	Higher throughput can be achieved

Table 6 Countermeasures to prevent intrusions.

Affected layer	Attack	Countermeasures
Physical layer	PUEA	Cryptographic methods, fingerprint, hybrid methods
	Jamming	Dog fight, stochastic gaming method
	Objective function attack [64]	Security- and threshold based-methods
Link layer	SSDFA	Fusion center and reputation based
	Control channel saturation	Frequency hopping method
	Control channel jamming	Machine learning methods
Network layer	Wormhole attack	Neighbor authentication
	Sink hole attack	Routing protocols
	Hello attack	Duplexity of link is verified
	Sybil attack [11]	Identity validation
Cross layer	Lion attack [51,85]	Markov decision process
	Routing information jamming attack	Authenticated routing protocols are devised
	Jelly fish [11]	Cluster and supercluster methods
	Small back off window attack	Byzantine defense methods

6. Global standardization activities on HetCWNs

With the popularity of cognitive radio technology, numerous associations have been approached for the standardization of cognitive radio organizations. In this section, a concise outline of accessible cognitive radio standards is presented in Figs. 8 and 9 according to developments over the years and the coexistence of diverse standardization authorities.

6.1 IEEE standards coordinating committee 41 (SCC-41)

A huge commitment has been made by the IEEE Standards Coordinating Committee 41 (SCC-41) to normalize cognitive radio. The IEEE started the 1900 Standards Committee to create norms for new innovations for cutting edge radios and progressed range the executives [86]. The standard board was set up by the IEEE Communications Society (ComSoc) and the IEEE Electromagnetic Compatibility (EMC) Society in 2005 [6]. The fundamental SCC-41 working gatherings and their functionalities are shown in Table 7.

6.2 IEEE 802.22 standard

The IEEE proposed a standard on cognitive radio that is, wireless regional area network (WRAN), to provide cognitive air interface for fixed, point to multipoint WRANs that work on unused channels in the VHF/UHF TV groups in the range of 54 and 862 MHz [10]. The specialized details of this standard are summed up in Fig. 10. Here, DS stands for downstream channels whereas US stands for upstream channels.

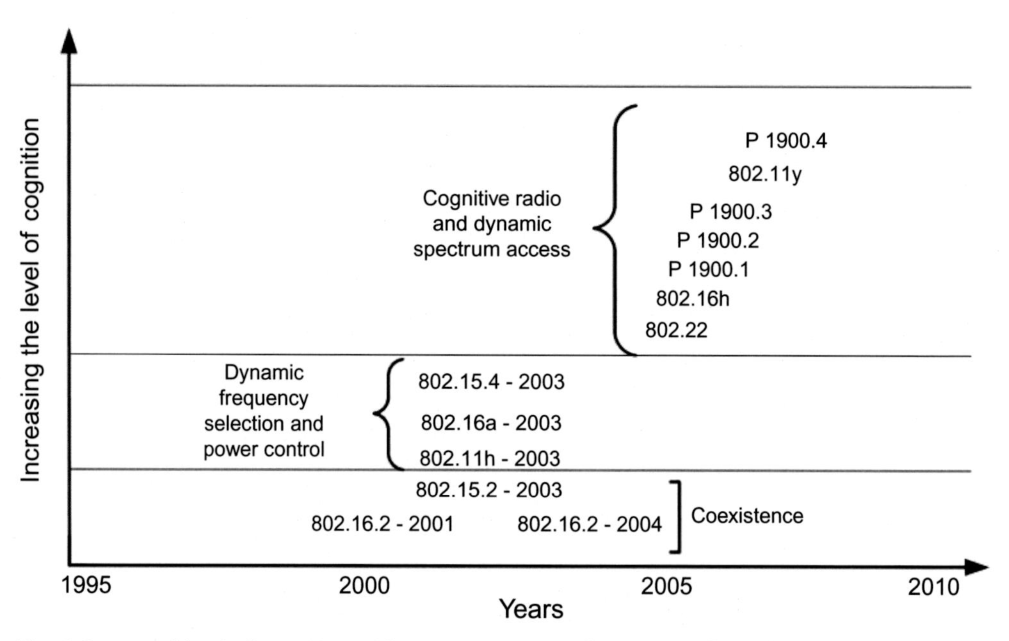

Fig. 8 Increased level of cognition with respect to years of invention of standards.

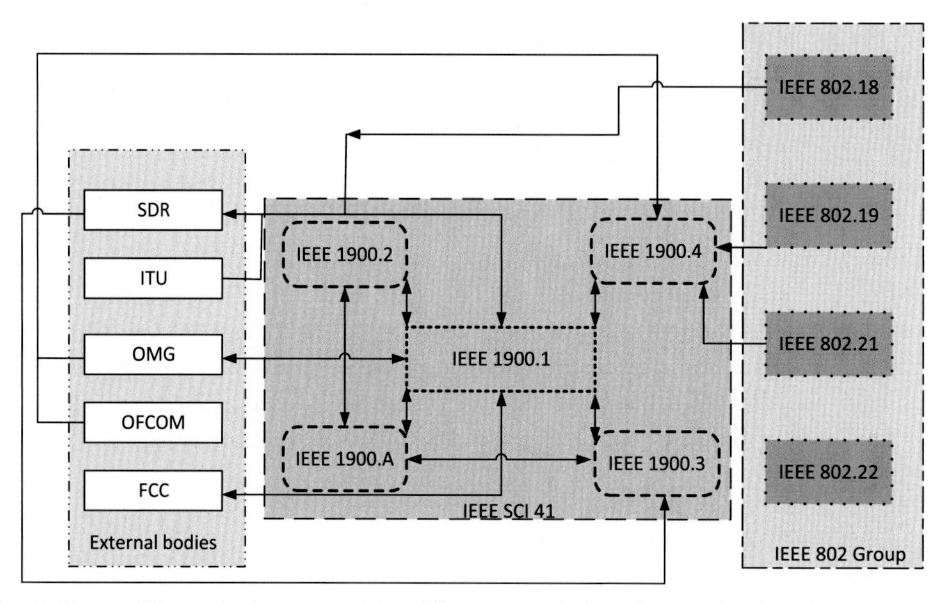

Fig. 9 Diverse CR standards proposed by different associations for working in a heterogeneous environment.

Table 7 Working of standards devised by the SCC-41 group.

S. no.	Working groups	Key function
1.	IEEE 1900.1	This standard provides technically precise definitions and key concepts related to policy defined radio, software defined radio, adaptive radio, spectrum management, and related technologies. It develops basic terminologies, concepts for next generation radio communication systems and spectrum management
2.	IEEE 1900.2	It recommends practices for interference, coexistence analysis and interference between various radio services
3.	IEEE 1900.3	It develops a standard on technical guidelines for analyzing SDR software modules to ensure compliance with regulatory and operational requirements
4.	IEEE 1900.4	Architectural building blocks enabling network-device distributed decision making for optimized radio resource usage in heterogeneous wireless access networks
5.	IEEE 1900.A	Evaluation of regulatory compliance for radio systems with dynamic spectrum access

The IEEE 802.22 working group built physical (PHY) and MAC layer particulars for WRAN activity in TVWS. The essential application is intended for fixed broadband access. The 802.22 standard embraced an orthogonal frequency division multiple access (OFDMA) PHY and an incorporated, connection-oriented MAC, where a BS controls the resource allotments inside its cell. The MAC layer is required to give client information at a data rate of 1.5 Mb/s in the downlink and 384 kb/s in the uplink per the 6 MHz TV channel [88]. The 802.22 PHY and MAC layers incorporate new CR features to secure officeholders and accomplish effective range usage, for example, reliable user identification, hybrid spectrum detection, geolocation, frequency agility, self-coexistence systems, and databases.

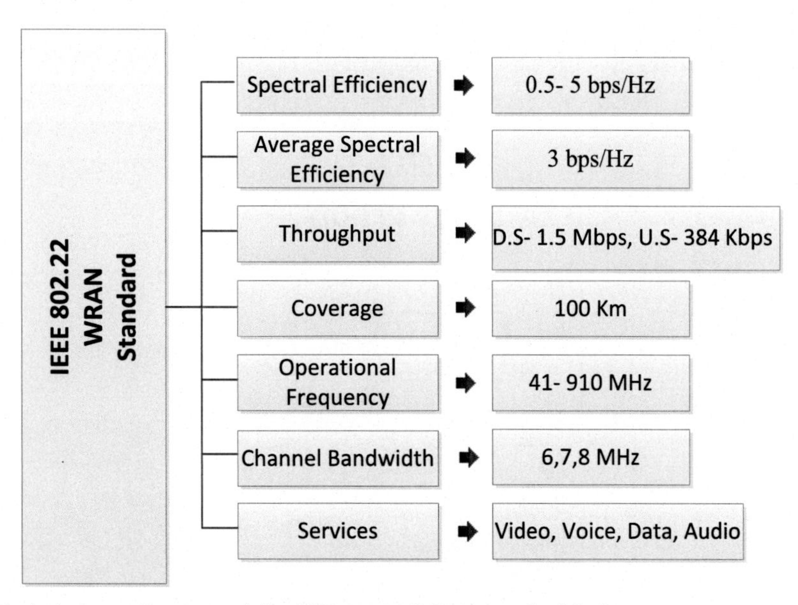

Fig. 10 Technical specifications of the IEEE 802.22 WRAN standard [87].

Various activities have additionally been proposed inside the IEEE working groups 802.11, 802.22, and 802.19 focusing on close to home/convenient gadget use cases that are viewed as future market drivers. The Task Group 802.11af (TGaf), affirmed in December 2009, is relied upon to characterize other PHY and related MAC layer alterations for TVWS activity. The 802.22 standard has additionally extended its degree to permit versatile CPEs to interface with the BSs when they are near the BS. In any case, the 802.22 standard doesn't cover completely portable CPEs at vehicular rates. In contrast to 802.22 and 802.11, the 802.19 TG1 won't build up another air-interface determination, yet will zero in on proposals for coexistence protocols and arrangements across stages to accomplish efficient spectrum usage.

Moreover, the ECMA 392 standard [6] determines the PHY and MAC layers for activity in TVWS focused on multimedia dissemination and web access for individual/versatile gadgets. Other standardization endeavors identified with CR innovation are continuous in the IEEE SCC 41 set [88]. Other than ECMA 392 and IEEE 802.22, the 802.11af and 802.19 TG1 groups are in the beginning phases, and explicit specialized arrangements have not yet been concluded. Still, a portion of the fundamental CR ideas such as spectrum detecting and geolocation components, user information base access, and dynamic frequency choice will no doubt be adjusted to the particular prerequisites of every standard.

6.3 ITU standards

The International Telecommunication Union (ITU) is additionally directing standardization exercises identified with cognitive radio networks. At present, standardization exercises are more identified with SDR, and future exercises will be more centered on cognitive radio [15]. The standardization activities are primarily completed in the testing groups, and their capacities are depicted in Fig. 11.

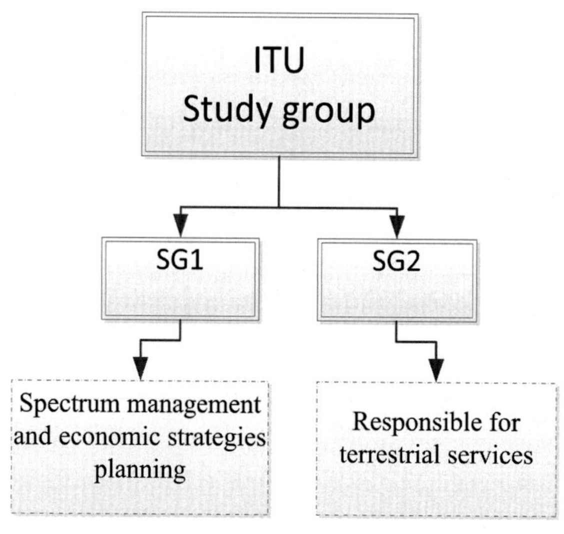

Fig. 11 ITU study groups and their functions [89].

7. Conclusion and future scope

To overcome the spectrum scarcity problem, various communication paradigms are being proposed worldwide. As the root cause of this problem is known, the dynamic spectrum access approach has emerged as a potential solution. With digital television transmission, a significant spectrum portion became available for dynamic spectrum access by heterogeneous CWNs, including licensed, typically primary users, and unlicensed users. In this chapter, we have discussed in detail the coexistence challenges faced by heterogeneous CWNs while sharing TVWS. The RA schemes or heterogeneous CWNs are elaborated upon for optimum spectrum sharing. Many DSA algorithms have been reviewed for heterogeneous coexistence in the TVWS. Various MAC protocols have been discussed in detail for the heterogeneous coexistence of different type of users/networks in TVWS. One complete section of this chapter focuses on the security threats to HetCWNs while sharing the TVWS. Moreover, the chapter also provides insights to standardization activities happening all around the world for the successful deployment of HetCWNs for efficient spectrum utilization. Based on the literature reviewed in this chapter, the following are the recommendations to deploy them successfully in the future:

(i) Energy efficient network/route discovery protocols must be designed for a heterogeneous environment.
(ii) Fast, accurate, and reliable spectrum sensing schemes must be designed to detect licensed as well as unlicensed users/networks in the TVWS.
(iii) An efficient resource sharing algorithm must be designed to choose the best channel according to the quality of service requirements of the end user.
(iv) Considering the heterogeneity of the network, security is a critical issue. Thus, new efficient and robust security protocols must be designed for the coexistence of heterogeneous users/networks simultaneously over the TVWS.

References

[1] R. Rubeena, I. Bala, Throughput enhancement of cognitive radio networks through improved frame structure, Int. J. Comput. Appl. 109 (14) (2015) 40–43, https://doi.org/10.5120/19259-1016.
[2] I. Bala, M.S. Bhamrah, G. Singh, Rate and power optimization under received-power constraints for opportunistic spectrum-sharing communication, Wirel. Pers. Commun. 96 (4) (2017) 5667–5685, https://doi.org/10.1007/s11277-017-4440-8.
[3] C. Ghosh, S. Roy, D. Cavalcanti, Coexistence challenges for heterogeneous cognitive wireless networks in TV white spaces, IEEE Wirel. Commun. 18 (4) (2011) 22–31, https://doi.org/10.1109/MWC.2011.5999761.
[4] I. Bala, Analytical modeling of ad hoc cognitive radio environment for optimum power control, Int. J. Comput. Appl. 92 (7) (2014) 19–22.
[5] H. Maloku, Z.L. Fazliu, M. Ibrani, A survey on coexistence in heterogeneous wireless networks in TV white spaces, Wirel. Commun. Mob. Comput. 2018 (2018), https://doi.org/10.1155/2018/7256835.
[6] J. Wang, et al., First cognitive radio networking standard for personal/portable devices in TV white spaces, in: 2010 IEEE Symp. New Front. Dyn. Spectrum, DySPAN 2010, 2010, https://doi.org/10.1109/DYSPAN.2010.5457855.

[7] I. Bala, M.S. Bhamrah, G. Singh, Investigation on outage capacity of spectrum sharing system using CSI and SSI under received power constraints, Wirel. Netw. 25 (3) (2019), https://doi.org/10.1007/s11276-018-1666-7.
[8] V. Rana, N. Jain, I. Bala, Resource allocation models for cognitive radio networks: a study, Int. J. Comput. Appl. 91 (12) (2014) 51–55.
[9] M.N. Hindia, F. Qamar, H. Ojukwu, K. Dimyati, A.M. Al-Samman, I.S. Amiri, On platform to enable the cognitive radio over 5G networks, Wirel. Pers. Commun. 113 (2) (2020) 1241–1262, https://doi.org/10.1007/s11277-020-07277-3.
[10] C. Cordeiro, K. Challapali, D. Birru, N. Sai Shankar, IEEE 802.22: the first worldwide wireless standard based on cognitive radios, in: 2005 1st IEEE Int. Symp. New Front. Dyn. Spectr. Access Networks, DySPAN 2005, 2005, pp. 328–337, https://doi.org/10.1109/DYSPAN.2005.1542649.
[11] N. Mahesh Kumar, G.K. Siddesh, Comprehensive survey on network and cross layers of cognitive radio networks, Int. J. Sci. Technol. Res. 8 (9) (2019) 230–235.
[12] I. Bala, K. Ahuja, Energy-Efficient Framework for Throughput Enhancement of Cognitive Radio Network by Exploiting Transmission Mode Diversity, 2021, https://doi.org/10.1007/s12652-021-03428-x.
[13] D. Gurney, G. Buchwald, L. Ecklund, S. Kuffner, J. Grosspietsch, Geo-location database techniques for incumbent protection in the TV white space, in: 2008 IEEE Symp. New Front. Dyn. Spectr. Access Networks, DySPAN 2008, 2008, pp. 232–240, https://doi.org/10.1109/DYSPAN.2008.31.
[14] I. Bala, M.S. Bhamrah, G. Singh, Capacity in fading environment based on soft sensing information under spectrum sharing constraints, Wirel. Netw. 23 (2) (2017), https://doi.org/10.1007/s11276-015-1172-0.
[15] B. Fette, Fourteen years of cognitive radio development, in: Proc.—IEEE Mil. Commun. Conf. MILCOM, 2013, pp. 1166–1175, https://doi.org/10.1109/MILCOM.2013.200.
[16] I. Bala, K. Ahuja, A. Nayyar, Hybrid spectrum access strategy for throughput enhancement of cognitive radio network, in: D.K. Sharma, L.H. Son, R. Sharma, K. Cengiz (Eds.), Micro-Electronics and Telecommunication Engineering. Lecture Notes in Networks and Systems, vol 179, Springer, Singapore, 2021, https://doi.org/10.1007/978-981-33-4687-1_11.
[17] I.F. Akyildiz, W. Lee, M.C. Vuran, S. Mohanty, Cr_Spect08, IEEE Commun. Mag. (April) (2008) 40–48.
[18] FCC, Federal Communications Commissions,47 CFR Parts 0 and 15 Unlicensed Operation in the TV Broadcast Bands; Final Rule, 2015, pp. 1–31. [Online]. Available https://www.gpo.gov/fdsys/pkg/FR-2010-12-06/pdf/2010-30184.pdf#page=23.
[19] I. Bala, M.S. Bhamrah, V. Rana, N. Jain, G. Singh, Adaptive power control scheme for the cognitive radio system based on receiver sensitivity, Lect. Notes Electr. Eng. 335 (2015) 69–79, https://doi.org/10.1007/978-81-322-2274-3_9.
[20] T. Baykas, et al., Developing a standard for TV white space coexistence: technical challenges and solution approaches, IEEE Wirel. Commun. 19 (1) (2012) 10–22, https://doi.org/10.1109/MWC.2012.6155872.
[21] I. Bala, K. Ahuja, Energy-efficient framework for throughput enhancement of cognitive radio network, Int. J. Commun. Syst. 34 (13) (2021).
[22] R. Sethi, I. Bala, Performance evaluation of energy detector for cognitive radio network, J. Electron. Commun. Eng. 8 (5) (2013) 46–51.
[23] Pratibha, S. Thangjam, N. Kumar, S. Kumar, A survey on prevention of the falsification attacks on cognitive radio networks, IOP Conf. Ser. Mater. Sci. Eng. 1033 (1) (2021), https://doi.org/10.1088/1757-899X/1033/1/012021.
[24] B. Wu, J. Shen, H. Xiang, Resource allocation with minimum transmit power in multicast OFDM systems, J. Syst. Eng. Electron. 21 (3) (2010) 355–360, https://doi.org/10.3969/j.issn.1004-4132.2010.03.002.
[25] J.S. Liu, C.H.R. Lin, Y.C. Hu, Joint resource allocation, user association, and power control for 5G LTE-based heterogeneous networks, IEEE Access 8 (2020) 122654–122672, https://doi.org/10.1109/ACCESS.2020.3007193.
[26] M. Ali, S. Qaisar, M. Naeem, S. Mumtaz, Energy efficient resource allocation in D2D-assisted heterogeneous networks with relays, IEEE Access 4 (c) (2016) 4902–4911, https://doi.org/10.1109/ACCESS.2016.2598736.

[27] C. Niu, Y. Li, R.Q. Hu, F. Ye, Fast and efficient radio resource allocation in dynamic ultra-dense heterogeneous networks, IEEE Access 5 (c) (2017) 1911–1924, https://doi.org/10.1109/ACCESS.2017.2653798.
[28] R. Liu, M. Sheng, W. Wu, Energy-efficient resource allocation for heterogeneous wireless network with multi-homed user equipments, IEEE Access 6, no. c (2018) 14591–14601, https://doi.org/10.1109/ACCESS.2018.2810216.
[29] F. Zeng, Q. Li, Z. Xiao, V. Havyarimana, J. Bai, A price-based optimization strategy of power control and resource allocation in full-duplex heterogeneous macrocell-femtocell networks, IEEE Access 6 (XX) (2018) 42004–42013, https://doi.org/10.1109/ACCESS.2018.2856627.
[30] L. Xu, H. Xing, A. Nallanathan, Y. Yang, T. Chai, Security-aware cross-layer resource allocation for heterogeneous wireless networks, IEEE Trans. Commun. 67 (2) (2019) 1388–1399, https://doi.org/10.1109/TCOMM.2018.2878767.
[31] M. Farokhi, A. Zolghadrasli, N.M. Yamchi, Mobility-based cell and resource allocation for heterogeneous ultra-dense cellular networks, IEEE Access 6 (2018) 66940–66953, https://doi.org/10.1109/ACCESS.2018.2877695.
[32] W.K. Lai, J.K. Liu, Cell selection and resource allocation in LTE-advanced heterogeneous networks, IEEE Access 6 (2018) 72978–72991, https://doi.org/10.1109/ACCESS.2018.2881093.
[33] H. Gao, S. Zhang, Y. Su, M. Diao, Joint resource allocation and power control algorithm for cooperative D2D heterogeneous networks, IEEE Access 7 (2019) 20632–20643, https://doi.org/10.1109/ACCESS.2019.2895975.
[34] X. Huang, D. Zhang, S. Tang, Q. Chen, J. Zhang, Fairness-based distributed resource allocation in two-tier heterogeneous networks, IEEE Access 7 (c) (2019) 40000–40012, https://doi.org/10.1109/ACCESS.2019.2905038.
[35] H. Wei, N. Deng, M. Haenggi, An ASAPPP approach to the spectrum allocation in general heterogeneous cellular networks, IEEE Access 7 (2019) 89141–89151, https://doi.org/10.1109/ACCESS.2019.2926398.
[36] J. Yan, Z. Kuang, F. Yang, X. Deng, Mode selection and resource allocation algorithm in energy-harvesting D2D heterogeneous network, IEEE Access 7 (2019) 179929–179941, https://doi.org/10.1109/ACCESS.2019.2956111.
[37] Y.K. Tun, A. Ndikumana, S.R. Pandey, Z. Han, C.S. Hong, Joint radio resource allocation and content caching in heterogeneous virtualized wireless networks, IEEE Access 8 (2020) 36764–36775, https://doi.org/10.1109/ACCESS.2020.2974287.
[38] J.Y. Lai, W.H. Wu, Y.T. Su, Resource allocation and node placement in multi-hop heterogeneous integrated-access-and-backhaul networks, IEEE Access 8 (2020) 122937–122958, https://doi.org/10.1109/ACCESS.2020.3007501.
[39] H. Wang, H. Ke, G. Liu, W. Sun, Computation migration and resource allocation in heterogeneous vehicular networks: a deep reinforcement learning approach, IEEE Access 8 (2020) 171140–171153, https://doi.org/10.1109/access.2020.3024683.
[40] Z.J. Ali, N.K. Noordin, A. Sali, F. Hashim, Fair energy-efficient resource allocation for downlink NOMA heterogeneous networks, IEEE Access 8 (2020) 200129–200145, https://doi.org/10.1109/access.2020.3035212.
[41] B.S. Awoyemi, B.T. Maharaj, A.S. Alfa, Resource allocation for heterogeneous cognitive radio networks, in: 2015 IEEE Wirel. Commun. Netw. Conf. WCNC 2015, 2015, pp. 1759–1763, https://doi.org/10.1109/WCNC.2015.7127734.
[42] A. De Domenico, E. Calvanese Strinati, M.G. Di Benedetto, A survey on MAC strategies for cognitive radio networks, IEEE Commun. Surv. Tutorials 14 (1) (2012) 21–44, https://doi.org/10.1109/SURV.2011.111510.00108.
[43] Y.R. Kondareddy, P. Agrawal, Synchronized MAC protocol for multi-hop cognitive radio networks, in: IEEE Int. Conf. Commun., no. Ccc, 2008, pp. 3198–3202, https://doi.org/10.1109/ICC.2008.602.
[44] J. Jia, Q. Zhang, X. Shen, HC-MAC: a hardware-constrained cognitive MAC for efficient spectrum management, IEEE J. Sel. Areas Commun. 26 (1) (2008) 106–117, https://doi.org/10.1109/JSAC.2008.080110.

[45] H.A. Bany Salameh, M.M. Krunz, O. Younis, MAC protocol for opportunistic cognitive radio networks with soft guarantees, IEEE Trans. Mob. Comput. 8 (10) (2009) 1339–1352, https://doi.org/10.1109/TMC.2009.19.

[46] L. Ma, X. Han, C.C. Shen, Dynamic open spectrum sharing MAC protocol for wireless ad hoc networks, in: 2005 1st IEEE Int. Symp. New Front. Dyn. Spectr. Access Networks, DySPAN 2005, 2005, pp. 203–213, https://doi.org/10.1109/DYSPAN.2005.1542636.

[47] S.M. Kamruzzaman, CR-MAC: a multichannel mac protocol for cognitive radio ad hoc networks, Int. J. Comput. Networks Commun. 2 (5) (2010) 1–14, https://doi.org/10.5121/ijcnc.2010.2501.

[48] J. Zhao, H. Zheng, G.H. Yang, Distributed coordination in dynamic spectrum allocation networks, in: 2005 1st IEEE Int. Symp. New Front. Dyn. Spectr. Access Networks, DySPAN 2005, 2005, pp. 259–268, https://doi.org/10.1109/DYSPAN.2005.1542642.

[49] B. Hamdaoui, K.G. Shin, OS-MAC: an efficient MAC protocol for spectrum-agile wireless networks, IEEE Trans. Mob. Comput. 7 (8) (2008) 915–930, https://doi.org/10.1109/TMC.2007.70758.

[50] M. Timmers, S. Pollin, A. Dejonghe, L. Van Der Perre, F. Catthoor, A distributed multichannel MAC protocol for multihop cognitive radio networks, IEEE Trans. Veh. Technol. 59 (1) (2010) 446–459, https://doi.org/10.1109/TVT.2009.2029552.

[51] K. Afhamisisi, H.S. Shahhoseini, E. Meamari, Defense against lion attack in cognitive radio systems using the Markov decision process approach, Frequenz 68 (3–4) (2014) 191–201, https://doi.org/10.1515/freq-2013-0048.

[52] A. Bariş, Ö.B. Akan, Biologically-inspired spectrum sharing in cognitive radio networks, in: IEEE Wirel. Commun. Netw. Conf. WCNC, 2007, pp. 43–48, https://doi.org/10.1109/WCNC.2007.14.

[53] Q. Zhao, L. Tong, A. Swami, Decentralized cognitive MAC for dynamic spectrum access, in: 2005 1st IEEE Int. Symp. New Front. Dyn. Spectr. Access Networks, DySPAN 2005, 2005, pp. 224–232, https://doi.org/10.1109/DYSPAN.2005.1542638.

[54] D. Willkomm, M. Bohge, D. Hollos, J. Gross, A. Wolisz, Double hopping: a new approach for dynamic frequency hopping in cognitive radio networks, in: IEEE Int. Symp. Pers. Indoor Mob. Radio Commun. PIMRC, no. January, 2008, https://doi.org/10.1109/PIMRC.2008.4699626.

[55] F. Wang, O. Younis, M. Krunz, GMAC: a game-theoretic MAC protocol for mobile Ad Hoc networks, in: 2006 4th Int. Symp. Model. Optim. Mobile, Ad Hoc Wirel. Networks, WiOpt 2006, 2006, https://doi.org/10.1109/WIOPT.2006.1666458.

[56] M.Y. ElNainay, D.H. Friend, A.B. MacKenzie, Channel allocation & power control for dynamic spectrum cognitive networks using a localized island genetic algorithm, in: 2008 IEEE Symp. New Front. Dyn. Spectr. Access Networks, DySPAN 2008, 2008, pp. 715–719, https://doi.org/10.1109/DYSPAN.2008.80.

[57] C. Zou, C. Chigan, On game theoretic DSA-driven MAC for cognitive radio networks, Comput. Commun. 32 (18) (2009) 1944–1954, https://doi.org/10.1016/j.comcom.2009.07.007.

[58] B. Awoyemi, B. Maharaj, A. Alfa, Optimal resource allocation solutions for heterogeneous cognitive radio networks, Digit. Commun. Netw. 3 (2) (2017) 129–139, https://doi.org/10.1016/j.dcan.2016.11.003.

[59] S.K. Leung-Yan-Cheong, M.E. Hellman, The Gaussian wire-tap channel, IEEE Trans. Inf. Theory 24 (4) (1978) 451–456, https://doi.org/10.1109/TIT.1978.1055917.

[60] H. Khodakarami, F. Lahouti, Link adaptation for physical layer security over wireless fading channels, IET Commun. 6 (3) (2012) 353–362, https://doi.org/10.1049/iet-com.2011.0319.

[61] K. Elangovan, S. Subashini, A survey of security issues in cognitive radio network, ARPN J. Eng. Appl. Sci. 11 (17) (2016) 10496–10500, https://doi.org/10.15680/ijircce.2014.0212013.

[62] R.K. Sharma, D.B. Rawat, Advances on security threats and countermeasures for cognitive radio networks: a survey, IEEE Commun. Surv. Tutorials 17 (2) (2015) 1023–1043, https://doi.org/10.1109/COMST.2014.2380998.

[63] S. Manjunath, B.R. Bhargava, V.S. Pradeep, Wireless physical layer security in cognitive radio, Int. Res. J. Eng. Technol. 7 (6) (2020) 3007–3011.

[64] F. Salahdine, N. Kaabouch, Security threats, detection, and countermeasures for physical layer in cognitive radio networks: a survey, Phys. Commun. 39 (2020), https://doi.org/10.1016/j.phycom.2020.101001, 101001.

[65] B. Wang, Y. Wu, K.J.R. Liu, T.C. Clancy, An anti-jamming stochastic game for cognitive radio networks, IEEE J. Sel. Areas Commun. 29 (4) (2011) 877–889, https://doi.org/10.1109/JSAC.2011.110418.

[66] D.L. Chaitanya, K. Manjunatha Chari, Performance analysis of PUEA and SSDF attacks in cognitive radio networks, Lect. Notes Networks Syst. 5 (2017) 219–225, https://doi.org/10.1007/978-981-10-3226-4_21.

[67] S. Sodagari, T.C. Clancy, An anti-jamming strategy for channel access in cognitive radio networks, Lect. Notes Comput. Sci 7037 (2011) 34–43, https://doi.org/10.1007/978-3-642-25280-8_5. (including Subser. Lect. Notes Artif. Intell. Lect. Notes Bioinformatics). LNCS.

[68] L. Sibomana, H. Tran, H.J. Zepernick, On physical layer security for cognitive radio networks with primary user interference, in: Proc.—IEEE Mil. Commun. Conf. MILCOM, vol. 2015-Decem, 2015, pp. 281–286, https://doi.org/10.1109/MILCOM.2015.7357456.

[69] Z. Shu, Y. Qian, S. Ci, On physical layer security for cognitive radio networks, IEEE Netw. 27 (3) (2013) 28–33, https://doi.org/10.1109/MNET.2013.6523805.

[70] D. Hlavacek, J.M. Chang, A layered approach to cognitive radio network security: a survey, Comput. Netw. 75 (Part A) (2014) 414–436, https://doi.org/10.1016/j.comnet.2014.10.001.

[71] S. Jagadeesan, V. Parthasarathy, Design and implement a cross layer verification framework (CLVF) for detecting and preventing blackhole and wormhole attack in wireless ad-hoc networks for cloud environment, Clust. Comput. 22 (2019) 299–310, https://doi.org/10.1007/s10586-018-1825-8.

[72] M. Meghdadi, S. Ozdemir, I. Güler, A survey of wormhole-based attacks and their countermeasures in wireless sensor networks, IETE Tech. Rev 28 (2) (2011) 89–102, https://doi.org/10.4103/0256-4602.78089 (Institution Electron. Telecommun. Eng. India).

[73] M. Bouabdellah, N. Kaabouch, F. El Bouanani, H. Ben-Azza, Network layer attacks and countermeasures in cognitive radio networks: a survey, J. Inf. Secur. Appl. 38 (2018) 40–49, https://doi.org/10.1016/j.jisa.2017.11.010.

[74] S. Athmani, D.E. Boubiche, A. Bilami, Hierarchical energy efficient intrusion detection system for black hole attacks in WSNs, in: 2013 World Congr. Comput. Inf. Technol. WCCIT 2013, 2013, pp. 0–4, https://doi.org/10.1109/WCCIT.2013.6618693.

[75] S. Kumar, K. Dutta, A. Garg, FJADA: friendship based JellyFish attack detection algorithm for mobile ad hoc networks, Wirel. Pers. Commun. 101 (4) (2018) 1901–1927, https://doi.org/10.1007/s11277-018-5797-z.

[76] A.W. Min, K.H. Kim, K.G. Shin, Robust cooperative sensing via state estimation in cognitive radio networks, in: 2011 IEEE Int. Symp. Dyn. Spectr. Access Networks, DySPAN 2011, 2011, pp. 185–196, https://doi.org/10.1109/DYSPAN.2011.5936205.

[77] W. Wang, H. Li, Y. Sun, Z. Han, Securing collaborative spectrum sensing against untrustworthy secondary users in cognitive radio networks, EURASIP J. Adv. Signal Process. 2010 (2010), https://doi.org/10.1155/2010/695750.

[78] S. Bhagavathy Nanthini, M. Hemalatha, D. Manivannan, L. Devasena, Attacks in cognitive radio networks (CRN)—a survey, Indian J. Sci. Technol. 7 (4) (2014) 530–536, https://doi.org/10.17485/ijst/2014/v7i4.18.

[79] I. Ohaeri, O. Ekabua, B. Isong, M. Esiefarienrhe, M. Motojane, Mitigating intrusion and vulnerabilities in cognitive radio networks, Adv. Comput. Sci. 4 (3) (2015) 1–10.

[80] R. Singh, J. Singh, R. Singh, WRHT: a hybrid technique for detection of wormhole attack in wireless sensor networks, Mob. Inf. Syst. 2016 (2016), https://doi.org/10.1155/2016/8354930.

[81] C. Cervantes, D. Poplade, M. Nogueira, A. Santos, Detection of sinkhole attacks for supporting secure routing on 6LoWPAN for Internet of Things, in: Proc. 2015 IFIP/IEEE Int. Symp. Integr. Netw. Manag. IM 2015, 2015, pp. 606–611, https://doi.org/10.1109/INM.2015.7140344.

[82] M. Yazdani, F. Jolai, Lion optimization algorithm (LOA): a nature-inspired metaheuristic algorithm, J. Comput. Des. Eng. 3 (1) (2016) 24–36, https://doi.org/10.1016/j.jcde.2015.06.003.

[83] S. Doss, et al., APD-JFAD: accurate prevention and detection of jelly fish attack in MANET, IEEE Access 6 (c) (2018) 56954–56965, https://doi.org/10.1109/ACCESS.2018.2868544.

[84] V. PalSingh, A.S. Anand Ukey, S. Jain, Signal strength based hello flood attack detection and prevention in wireless sensor networks, Int. J. Comput. Appl. 62 (15) (2013) 1–6, https://doi.org/10.5120/10153-4987.
[85] S. Bhattacharjee, S. Sengupta, M. Chatterjee, Vulnerabilities in cognitive radio networks: a survey, Comput. Commun. 36 (13) (2013) 1387–1398, https://doi.org/10.1016/j.comcom.2013.06.003.
[86] J. Guenin, IEEE SCC41 standards for dynamic spectrum access networks, Networks (2008) 1–24.
[87] E.Z. Tragos, S. Zeadally, A.G. Fragkiadakis, V.A. Siris, Spectrum assignment in cognitive radio networks: a comprehensive survey, IEEE Commun. Surv. Tutorials 15 (3) (2013) 1108–1135, https://doi.org/10.1109/SURV.2012.121112.00047.
[88] F. Granelli, et al., Standardization and research in cognitive and dynamic spectrum access networks: IEEE SCC41 efforts and other activities, IEEE Commun. Mag. 48 (1) (2010) 71–79, https://doi.org/10.1109/MCOM.2010.5394033.
[89] S. Bhandari, S. Joshi, Cognitive radio technology in 5G wireless communications, in: 2018 2nd IEEE Int. Conf. Power Electron. Intell. Control Energy Syst. ICPEICES 2018, 2018, pp. 1115–1120, https://doi.org/10.1109/ICPEICES.2018.8897345.

CHAPTER 7

Evaluation and analysis of clustering algorithms for heterogeneous wireless sensor networks

Rashmi Mishra[a,b], Rajesh K. Yadav[b], and Kavita Sharma[c]
[a]Krishna Engineering College, Ghaziabad, UP, India
[b]Delhi Technological University, Delhi, India
[c]Galgotias College of Engineering & Technology, Greater Noida, India

1. Introduction

The wireless sensor network (WSN) is unique among the promising know-hows in the decades. It consist thousands of small sensor nodes named ordinary sensor nodes and progressive sensor nodes, of which the progressive sensor nodes participate in cluster head selection, and in base stations with a certain level of energy. These sensor nodes are distributed based on applications and on power [1]. These sensor nodes gather information from the environment, aggregate it by the cluster head, and send it to the base station. In the recent decade many applications such as agriculture, medical, rural areas, etc., faced many challenges due to the limitation of the WSNs, such as battery consumption by the sensor nodes to intellect the data and, in addition, to aggregate it; for this, the sensor nodes have to stay awake for a long time in the network. Senor nodes have to monitor real-time data/streaming data to minimize human involvement and the atmosphere effect [2]. Many authors have proposed an algorithm for balancing the network lifetime by considering the parameters of cluster head selection, number of node alive, quantity of dead nodes in the system, packet delivery ratio, etc. The clustering algorithm is well effective in raising the system time to provide the system's scalability. In the existing literature, authors have proposed many clustering algorithms for homogeneous and heterogeneous networks. To increase the network lifetime, the heterogeneous network plays a crucial role because it does not demand much energy for communication but is aimed at the assortment of the cluster head; the nodes waste their energy in choosing the right cluster head, which affects the recital of the procedure. The paper's author has analyzed the algorithms on the foundation of the remaining energy, number of nodes alive, and the first node death [3].

WSN is a combination of homogeneous and heterogeneous networks [3]. The routing algorithm plays a significant part in limiting resources such as energy consumption.

Comprehensive Guide to Heterogeneous Networks
https://doi.org/10.1016/B978-0-323-90527-5.00012-5

The disadvantages of sensor nodes are that their energy is limited; sensor node requires processing power to sense the network or information. So, to improve the system lifetime of the WSN, the network requires powerful routing algorithms that increase the time and growth of the scalability and reliability of the system [4].

1.1 Types of heterogeneous resources

Fig. 1 shows types of heterogeneous resources in the WSN. The maximum of the routing methods/procedures for the HWSN has been classified into three heterogeneities: energy, link, and computation. For the heterogeneous energy environment, the nodes are considered as per their energy levels: two-level, three-level, and multilevel. In most representative situations, the sensor nodes have different initial energy levels. In a multi-level scenario, the sensor nodes' energy level lies in one of the multilevel energy models and the two-level energy model; the sensor nodes fall into the one of the two energy levels. To optimize the network, the routing algorithm assigns energy-intensive tasks to the high-energy sensor nodes. For the link heterogeneity network, the sensor nodes have different communication capabilities such as bidirectional and unidirectional. Many algorithms use link heterogeneity to improve performance and to improve network lifetime and network delay. For computational heterogeneity, the operation requiring more computation is delivered to the sensor nodes having a powerful hardware platform. Traffic heterogeneity is also considered in computational heterogeneity [5].

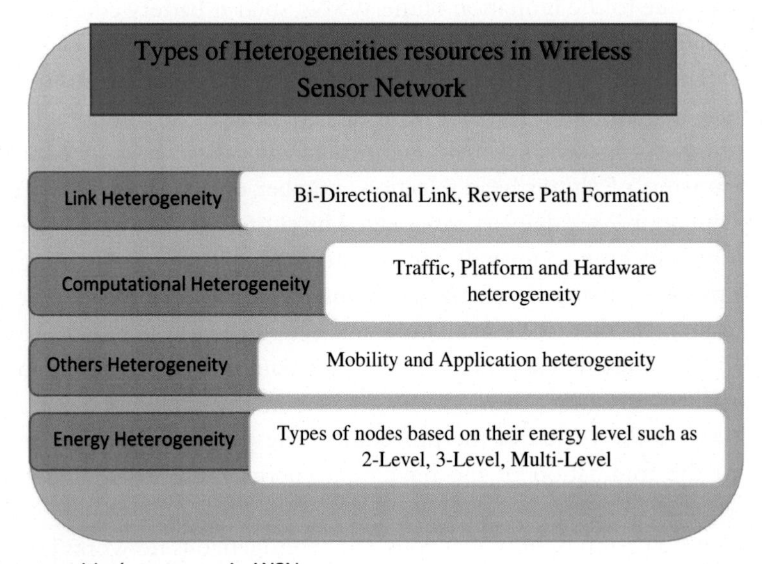

Fig. 1 Heterogeneities' resources in WSN.

1.2 Impact of heterogeneity on WSN

This section describes the impact of a heterogeneous environment on wireless sensor networks. Heterogeneous wireless sensor networks increase the network lifetime compared to homogeneous networks. Heterogeneity is totally dependent on link heterogeneity, computational heterogeneity, and the energy level of the nodes. The energy level of the nodes is divided into multiple levels. Fig. 2 shows the impact of heterogeneity on a WSN.

1.3 Performance measures

Fig. 3 displays the performance measures of a heterogeneous environment. The network is measured by calculating the number of alive nodes, throughput, number of cluster heads selected per round, number of packets transferred to the base station, and the stability period of the network [6,7]. The stability period shows after which round the first node dies. The system period shows the solidity date of the system, which means the duration of the sensor nodes in the system. The number of alive nodes displays the alive nodes per round based on the remaining energy. The throughput shows the number of packets delivered from the normal node to the cluster heads and from the cluster head to the sink nodes [8]. The number of cluster heads is calculated on the foundation of the message received by the base station. Sensor nodes with higher remaining energy near the sink station can be selected as cluster heads [9].

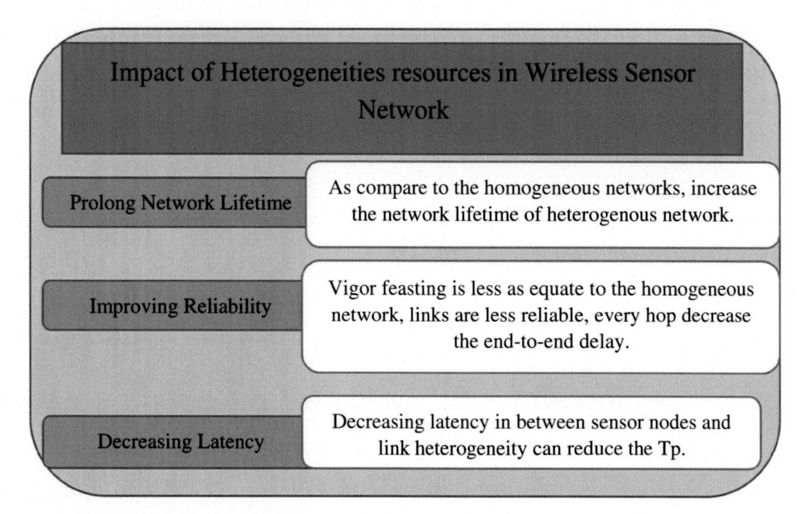

Fig. 2 Impact of heterogeneity on WSN.

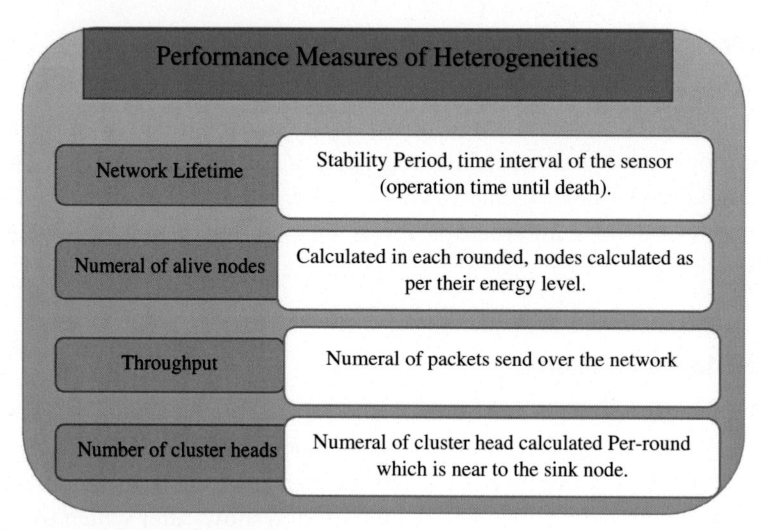

Fig. 3 Performance measures of heterogeneities.

1.4 Types of node heterogeneity of HWSN

Node heterogeneity is essential for increasing the throughput of the sent data from the source to the destination and decrease the latency. There exist multiple parameters according to which cluster-based heterogeneous algorithms are classified. The advantage of the heteronomous environment is that the hop count is more diminutive between the source and the destination; therefore, the end-to-end delivery rate is higher in the heterogeneous environment compared to the homogeneous environment [6,7]. Fig. 4 shows the node heterogeneity in WSN.

The chapter is systematized as follows: Section 2 represents the hierarchical routing protocols' evaluation process; Section 3 describes clustering and its requirement; Section 4 describes the clustering scheme; Section 5 displays the simulation results based on the death round of the first node, number of alive nodes, residual energy, number of packets transmitted to the base station; and Section 6 concludes the chapter and also discusses future work.

2. Hierarchical routing protocols' evaluation process

The essential impartiality of the hierarchical routing procedure is to provide a collision-free system using MAC protocols, minimize sensor energy or residual energy, prevent link failure, and extend the network lifetime by diminishing the dismissed data transmission. The node capabilities assign the task to the sensor nodes to stabilize the load among

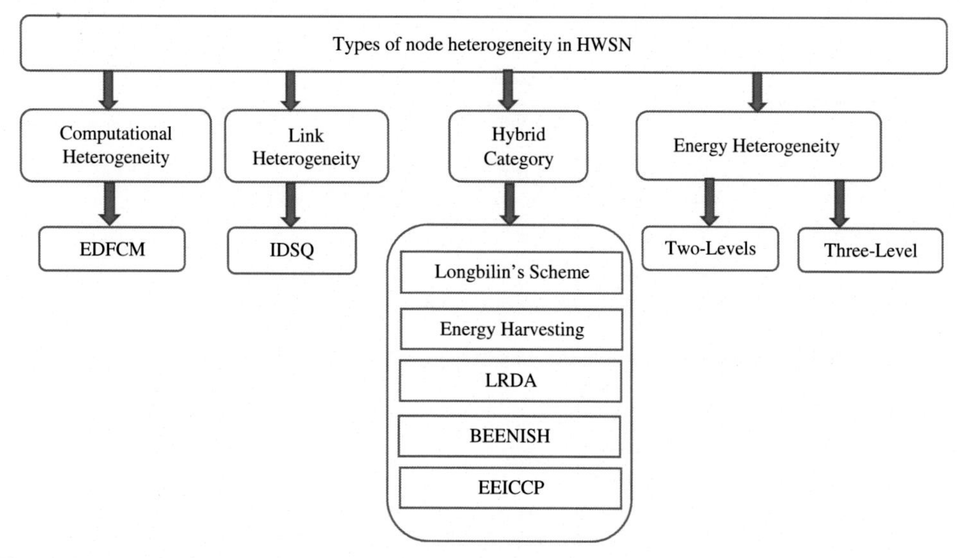

Fig. 4 Taxonomy of various heterogeneous routing schemes.

them [10]. Here, we present a detailed taxonomy of the hierarchical routing approach in Fig. 5. It is categorized into routing method, control manner, path establishment, itinerant component, agility pattern, system manner, protocol operation, radio model, communication paradigm, protocol objective, and applications. The hierarchical routing protocol is evaluated using delay, scalability, energy efficiency, and network size [11].

(i) *Delay:* Delay in a network is defined as the time spent by the packet in a network, or delay is defined as the period occupied by the packet to spread after foundation to the sink. If delay is minimum, then that link has less congestion. The delay will offer a near estimate of the concrete expectancy/latency [51,52].

(ii) *Scalability:* In terms of the WSN, scalability is defined as the routing protocol of accumulating new-fangled sensor nodes in the system. This is because in a WSN, thousands of nodes are situated communicating with each other. A single point of letdown will lead to disturbing the full network functionality [51,52].

(iii) *Energy efficiency:* The routing protocol's main objective is to save the sensor nodes' energy by sending packets to different routes. The utmost of the heterogeneous wireless network is designed to save the energy of the sensor nodes so that the network lifetime will increase [51,52].

(iv) *Network size:* Routing protocols are defined to work for both small-size networks and extensive-size networks. Networks are formed so that there is a stable network and network lifetime increases [51,52].

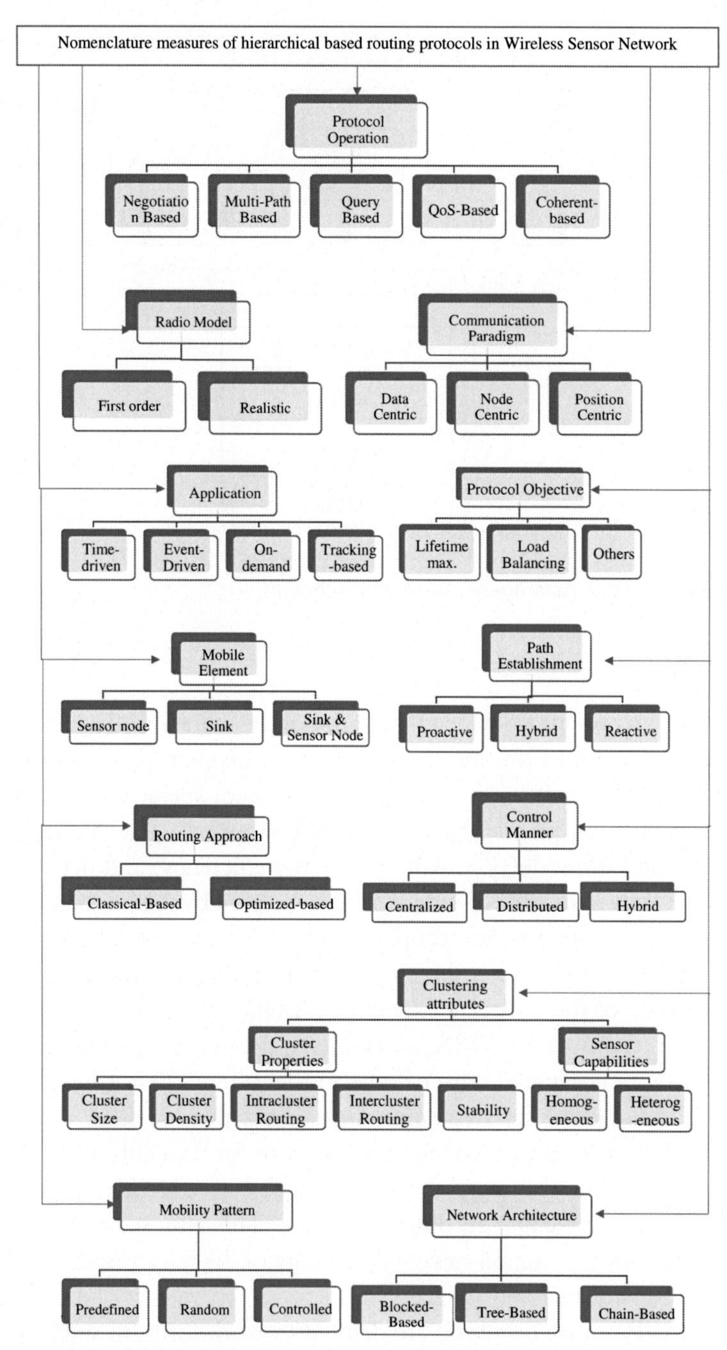

Fig. 5 Taxonomy of the hierarchically based routing protocols in a wireless sensor network.

2.1 Routing technique

The routing technique is used to direct the information from one location to the other in a WSN to increase the system lifetime, scalability, and quality of service (QoS). Founded on the applications, the path establishment is divided into many categories [12].

2.1.1 Proactive routing

This is also recognized as a table-driven routing protocol. Apiece nodes uphold the reliable and informed data preserved for the sensor nodes in their attention zone [13]. As in the ad hoc networks, topology changes from time to time so that every sensor node exchanges the updated routing table with its neighbor nodes. Some of the examples are:

(i) *Dynamic destination-sequenced distance:* The routing table contains all the links between the sensor nodes and costs related to every path. This routing protocol switches over the sensors at regular time intervals to exchange the updated routing table. These protocols can be applied periodically in cases such as: sensing the temperature, humidity level, etc.

2.1.2 Reactive routing

In reactive routing protocols, updated directions remain nonpreserved by every sensor node; they will be updated when the route is required. For creating the routes, the sender sends the route detection procedure to discover the route to the end point1 [14]. The algorithm is divided into three phases: route discovery, route maintenance, and data transmission. This type of routing protocol will interrupt the path discovery process; the actual message will be transmitted after route discovery. Examples of these types of protocols are used in intrusion detection systems and in explosion detection. AODV, DSR, and DYMO are examples of reactive routing protocols.

2.1.3 Hybrid routing

Hybrid routing utilizes proactive and reactive routing properties and will give better results. The entire system is divided into multiple zones, and one protocol run in one zone will not be used in another zone. Another protocol is used between two zones for the communication process [15]. Examples of hybrid routing are the zone routing protocols, TORA and HSLS. For the route discovery process within a zone, hybrid routing uses proactive routing algorithms, and it will use the advantages of reactive routing protocols for communication with another zone's sensor nodes.

2.2 Routing approach

Routing approaches are used to save the energy of the sensor node and increase the network lifetime, and it is categorized into two forms:

(i) *Classical-based routing*: In this approach, cluster heads will be designated randomly proceeding on the basis of the time function. Due to this, the uneven traffic is controlled by different cluster heads. There are many challenges faced by the classical-based routing approach, such as scalability, robustness, load balancing, exposure, and connectivity [16].

(ii) *Optimized-based routing:* In this approach, cluster heads remain designated on the basis of multicriteria to obtain QoS obligations. The following optimized techniques, particle swarm optimization, fuzzy logic, genetic algorithm, and artificial bee colony are used to obtain scalability, robustness, fault tolerance, load balancing, coverage, and connectivity [17].

2.3 Control manner

Routing protocols are divided into three categories:

(i) *Centralized approaches*: In the centralized approach, a cluster is formed, and the head node/sink node is selected for apiece collection. The cluster head requires the entire statistics of the nodes on the cluster, such as energy level and geographical position).

(ii) *Distributed approaches*: Without any global information, sensor nodes will collaboratively decide the route for communication between cluster heads. Each node will run an algorithm to decide to become the cluster head.

(iii) *Hybrid approaches*: This particular approach will associate the properties of the central and disseminated approaches.

2.4 Mobility pattern

Routing protocols are based on the mobility pattern of submission of the sensor nodes and the sink node. Protocols can be segregated based on the origin of the sensor node or the origin of the sink node, or the origin of the mobility pattern of the sensor node and the sink node. Different mobility patterns depend on the application and the network size1 [8].

(i) *Predefined mobility pattern*: This pattern is used for the mobile sink. Using predefined paths, mobile sinks will move from one node to another and perform their task at a predefined position.

(ii) *Random mobility pattern:* The random mobility pattern is a rummage sale for both the nodes and the base station. Within the sensor fields, mobile nodes move randomly. The random mobility outline is used in the random waypoint mobility model (RWP) and the reference point group mobility model (RPGM).

(iii) *Controlled mobility pattern*: A controlled mobility pattern is used to attain improved outcomes. In this pattern, mobility nodes will follow the routes defined by routing protocols. This movement is built on approximate constraints such as energy levels and connectivity. It is also used to control the energy hole or the hotspot problem [18].

2.5 Network architecture

The fundamental system construction plays a significant role in the purpose of HWSN routing protocols. This is classified into three categories [19]:

(i) *Block-based hierarchical routing*: As the name suggests, block-based hierarchical routing, the whole system is alienated into clusters and its cranium is known as cluster head (CH) [20]. CH is answerable for accumulating and gathering the statistics from its mobile nodes and forwarding them to the sink node. However, the problem with this routing is how to choose the cluster head and how to limit the range of the sensor nodes that directly attach to the sink node [21].

(ii) *Tree-based hierarchal routing*: In this routing scheme, an origin bush is designed between the sensor nodes and the bush's origin, and all the other nodes are the leaves of the root node [22]. All the nodes direct the information to their parent node; and then the parent node will collect the information and forward it to the parent node or base station [23].

(iii) *Chain-based hierarchical routing*: In a chain-based hierarchical routing protocol, a chain is formed between the nodes and one cluster head, also called the spearhead of the chain. The node, which is very remote from the sink node, resolves the package to its upper layer; this procedure continues until the information reaches the leader node and the sink node [24]. Nevertheless, the disadvantage of this routing is that if one node fails, the entire chain will break. The second disadvantage is that a packet has to travel for a very long time so that packet delay will increase, which decreases the reliability; and this routing is less vigorous as the disappointment of a single node will lead to the restraint break; therefore, there is the probability that packets will lead to damage [25].

3. Clustering

The main objective of forming a cluster is to increase the system lifetime. In the cluster, the complete system of a WSN is separated into a group of small chunks or small virtual groups by using some predefined rules and protocols. In this scenario, the sensor nodes are divided into two clusters, cluster members and cluster heads [25]. The cluster head is chosen by using geographical properties such as energy level, replacement from the sink node, and cluster head [26]. The cluster head gathers information from the different sensor nodes and passes all the information in one packet to the sink node to reduce overheads [24–26]. The advantage of cluster formation is that it will reduce energy utilization by improving bandwidth consumption and reducing extravagant energy utilization by dipping the network overhead. Most of the protocols have been reported for exploiting the system lifetime by allocating loads midst the sensor nodes and matching the energy utilization by choosing a cluster head so that the farthest nodes can convey the information to multiple nodes [26].

3.1 Why WSN clustering is required

It is observed that forming a cluster head will improve the performance of a WSN. Fig. 6 shows the residual energy consideration in a homogenous and in a heterogeneous environment. It shows that the sensor nodes expend more energy if the proper cluster head selection algorithm is not considered, but if the WSN considers the proper algorithm for cluster head selection, more energy is saved [25].

It will enhance network capacity by altitudinal reprocess of resources; it will rise the system period by choosing the CH. For inter-cluster routing or announcement, cluster head procedure a practical pillar from one cluster head to another cluster head [26]. Routing will grow the presentation of the WSN. Founded on different categories of the network architecture such as ad hoc or cellular network, formation of the cluster is very difficult. Due to the huge amount of sensor nodes, therefore, it is infeasible to find the sensor location, and difficult to sense the data; as ad hoc networks are self-deployed in nature, nodes have to connect with other sensor nodes [26].

- *Cluster properties*: Clusters are formed based on some properties such as network lifetime which will be increased; there should be load balancing and energy saving.
- *Cluster size*: *Cluster size is divided into two categories:* equivalent and inadequate cluster size. In equivalent clusters, all the clusters are of the same size whereas in inadequate clusters, all the clusters are inadequate. Inadequate clusters are formed because of the different energies of sensor nodes for load balancing and solving the energy hole or hotspot problem.
- *Cluster density*: Cluster density is distinguished as the number of sensor nodes in the cluster. There are two types of cluster protocols present: a stationary cluster and a dynamic cluster. In the static cluster, the number of sensor nodes is fixed in size whereas in the variable-size cluster, the number of sensor nodes is different in size. The density of the cluster affects the energy level of the cluster head [23].
- *Intra-cluster routing: Intra-cluster routing is the procedure of announcing the sensor nodes and forming* sensor nodes to the cluster head. The announcement in the intra cluster is of dual types: one-hop announcement and multihop announcement. In the one-hop announcement, the sensor nodes will directly connect or direct the information to the cluster head, whereas the data is delivered to the cluster head via relay sensor nodes [24].
- *Inter-cluster routing: Inter-cluster routing is the process of announcement among the sensor nodes, cluster head, and* base nodes. The announcement in the inter-cluster is of dual categories, one-hop announcement and multihop announcement. In the one-hop announcement, the sensor nodes and cluster head will directly contact or direct the material to the sink node, whereas in the multihop announcement, the material is delivered to the base node via relay sensor nodes and cluster head [25].
- *Stability*: The stability of any cluster depends on the cluster's density. The stability varies based on the cluster density. If the size of clusters is not static, then stability is also fixed. However, in the case of variable size clusters, the stability fluctuates during the routing procedure.

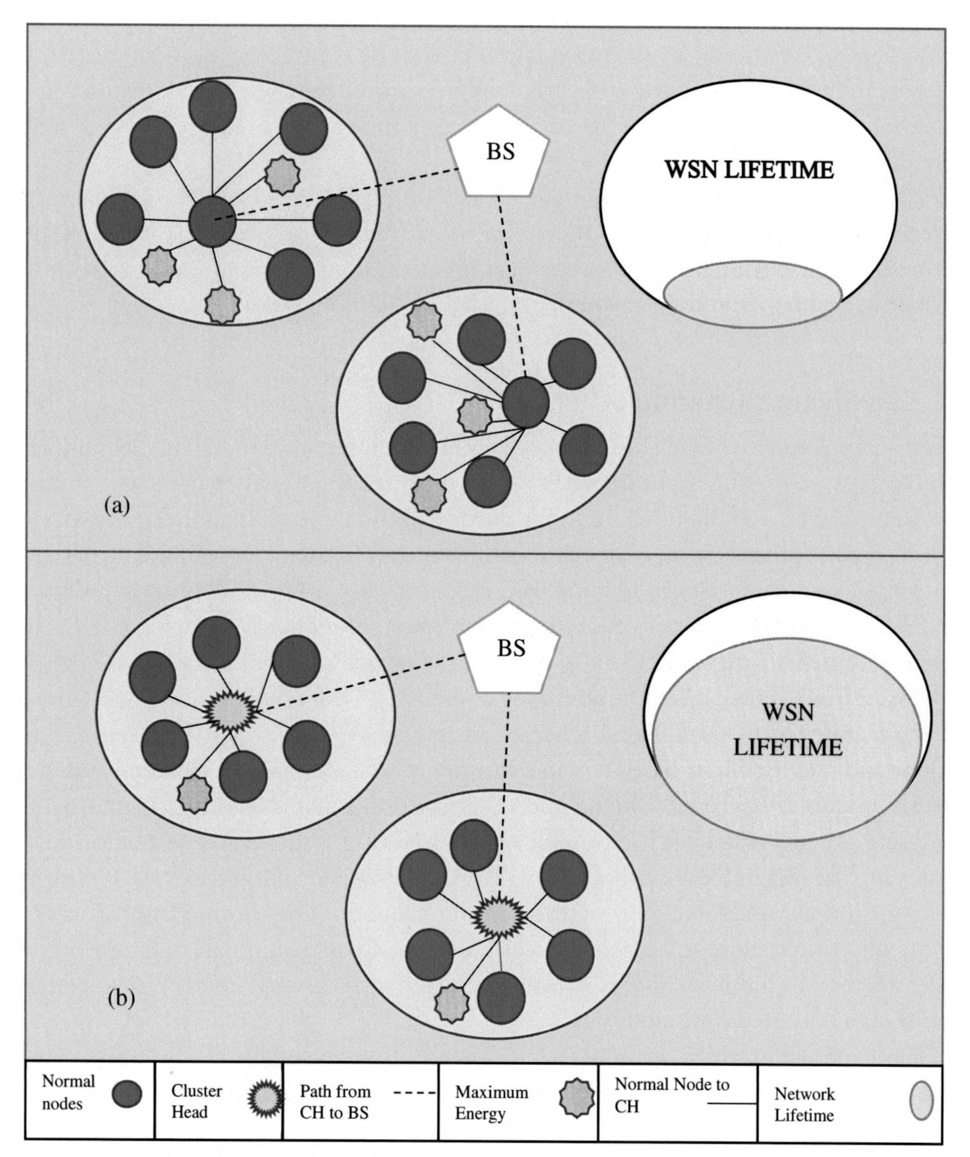

Fig. 6 (A) Cluster head selection without considering residual energy and (B) cluster head selection considering residual energy.

- *Sensor capabilities*: Founded on the sensor nodes or the resource provided by the sensor nodes, the network is divided into two parts.
- *Homogeneous network*: In this form of network, the cluster is shaped based on the source of the energy level of the nodes, computation power, and communication criteria, and also the cluster heads are designated on the foundation of some instructions and protocols or in a random fashion [45].

- *Heterogeneous network*: In this form of network, clusters are formed randomly. It does not depend on energy level, computation power, or communication criteria. All the nodes in heterogeneous networks have different competencies, and also cluster heads are selected based on the source of higher capabilities compared to the other sensor nodes in the system.
- *Path establishment*: It is rummage sale to identify the routes from the sender to the receiver. Based on the different hierarchy-based routing protocols, these routing protocols will be distinguished. Protocols are divided into three types: proactive routing, reactive routing, and hybrid routing.

3.2 Classifying clustering schemes

The main impartiality of the clustering algorithm remains toward reducing the number of control messages, reducing hotspots' problem, maintaining network coverage, utilizing sleeping schemes, avoiding collision, minimizing delay, increasing connectivity, obtaining scalability, fault tolerance, statistics accumulation, load balancing, constancy of system topology, maximizing system lifetime, and reducing energy consumption [26]. Clustering schemes are defined as the protocols used to form cluster heads in such a way that the system lifetime will grow and the node carrying advanced remaining energy will develop the cluster head, requiring minimum node count [27]. Different protocols will work on different strategies. Cluster-based schemes are divided into two parts: clustering-based scheme and methodology-based scheme. Further, the clustering-based scheme is divided into macroclustering-based scheme and microclustering-based scheme. In the macroclustering scheme, Fanin [1] explained macroclustering as general information of the cluster such as hierarchical methods (two-level, three-level, and multilevel), CH assortment technique, and objective of the routing scheme. The essential impartiality of macrocollecting scheme is to acquire scalability, load balancing, data aggregation, and fault tolerance, to diminish energy consumption, to make the most of the system lifetime, and to maintain network stability [27].

The microclustering parameters are application based such as time-driven, event-driven, query-driven, whether cluster size is controlled or uncontrolled, intercluster communication type such as one hop or multihop, intracluster communication such as one hop or multihop, methods (used for communication such as distributed, centralized, hybrid), nature of the nodes in terms of static or dynamic and homogeneous or heterogeneous, and whether the cluster head selection methods use rotation strategy or not [28].

Methodology-based schemes are further divided into the classical approach, metaheuristics founded, hybrid metaheuristics, and uncertain founded [28]. In the modern classical routing approach, the protocols mainly focus on competencies, boundaries, parameter premeditated, the technique's exact resolution, and the reproduction consequence. In the uncertain founded method, subsequent are measured as a structure such as competencies, restrictions, uncertain contribution criticisms, uncertain yield restrictions, defuzzification technique, valuation method of uncertain guidelines, uncertain regulation situation technique, uncertain reason, and imitation situation [29,46,49].

The metaheuristic-based approach uses the following parameters for evaluation: competencies, restrictions, limits deliberate in expansion, how to do the expansion procedure, expansion procedure, the persistence, expansion procedure, and simulation setting [30]. The hybrid metaheuristic approach and uncertain method use the limits for the assessment such as competencies, restrictions, uncertain contribution restrictions, uncertain production restrictions, defuzzification scheme, assessment process of fuzzy rules, fuzzy rule setting method, how to do the expansion process, the purpose of using fuzzy judgment, expansion procedure, limitations deliberate in terms of expansion, the resolution of using an expansion procedure, and imitation situation [31]. Table 1 depicts the evaluation chart of the heterogeneous WSN protocol [57].

3.3 Residual energy consideration model

To enhance the system time of a wireless sensor network, a routing algorithm is chosen to stabilize the consignment among the sensor nodes [31]. Therefore, the routing algorithm should consider the residual energy for communicating the information from the source to the base. Some factors are chosen for implementing the routing protocol for the dynamic and static WSN networks (Fig. 7).

4. Clustering scheme

The clustering scheme is classified into the following categories: The clustering scheme is divided into two categories: the homogeneous environment and the heterogeneous environment. In the following section, the protocols used for the heterogeneous environment are described, and a comparison is made based on the number of alive nodes, number of dead nodes, and number of cluster heads selected per round. Fig. 8 shows the flowchart of all the protocols mentioned in the paper. In all the heterogeneous algorithms, the sensor nodes are deployed in an environment where human intervention is not difficult. So, to improve a network's lifetime, we have to preserve the energy of the nodes. Many authors have discussed algorithms such as hybrid energy-efficient distributed clustering (HEED), distributed weight-based energy-efficient hierarchical clustering (DWEHC), hybrid clustering approach (HCA), energy-efficient heterogeneous clustered scheme (EEHCS), distributed election clustering protocol (DECP), dissipation forecast and clustering management (EDFCM), energy-efficient unequal clustering (EEUC), distributed energy-efficient clustering algorithm for HWSN (DEEC), energy-efficient clustering scheme (EECS), and multihop routing protocol with unequal clustering (MRPUC). All the algorithms are discussed in detail in the following section.

Table 1 Evaluation chart of heterogeneous WSNs protocol [43,50,53].

Algorithm	No. of clusters	No. of CHs	Intracluster	Intercluster	Overhead of cluster	Balance	Delay	Cluster stability	Location awareness	Complexity
CRDP	V	*	$\hat{s}$		L	Y	L	M	N	M
LEACH-DT	V	*	$\hat{s}$		M	N	M	M	N	M
EECR-PSO	V	*	$\hat{s}$		H	Y	H	M	Y	M
DWEHC	V	*	k-hop		H	N	H	Y/H	Y	H
TEEN	V	*	$\hat{s}$		H	N	H	Y	Y	H
EEUC	V	*	$\hat{s}$	k-hop	L	N	L		Y	H
EECS	V	*	$\hat{s}$	$\hat{S}$	M	N	L	L	Y	V/H
HCIC	V	*	$\hat{s}$	$\hat{S}$	H	N	M	M	Y	M
WCA	V	*	$\hat{s}$	$\hat{S}$	H	N	H	N	N	M
DEEC	V	*	$\hat{s}$	$\hat{S}$	M	Y	M	M	N	M
SDEEC	V	*	$\hat{s}$	$\hat{S}$	H	Y	H	M	N	M
EDFCM	V	*	$\hat{s}$	$\hat{s}$	H	Y	H	Y	Y	M
DEBC	V	*	$\hat{s}$	$\hat{s}$	H	Y	H	N	N	M
BCDCP	F	*	$\hat{s}$		H	Y	H	L	N	V/H
C4SD	V	*			M	N	H	N	N	M
SEP	V	*	$\hat{s}$	$\hat{s}$	L	N	L	M	N	M
DEECIC	V	*		$\hat{s}$	H	N	L	M	N	M
EHE-LEACH	V	*		$\hat{s}$	N	L	N	H	Y	H
WBCHN	V	*	$\hat{s}$	$\hat{s}$	H	N	H	M	N	H

Asterisk denotes variable.

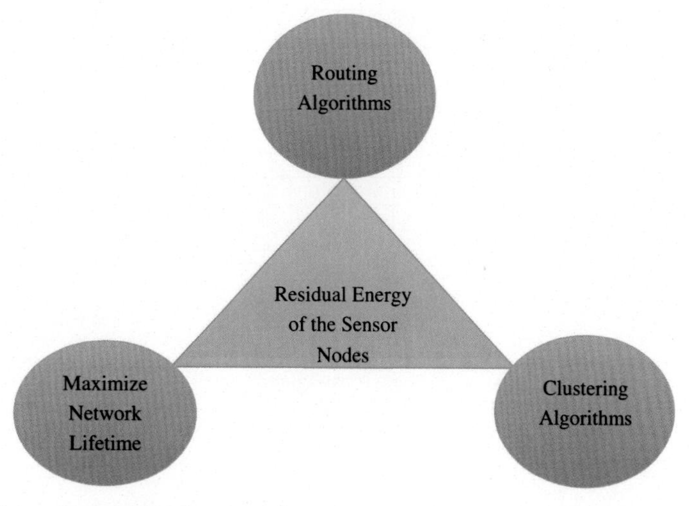

Fig. 7 Residual energy consideration model.

4.1 Hybrid energy-efficient distributed clustering (HEED)

Younis et al. [32] have published a HEED protocol. It is mainly founded on the hybrid, distributed, and energy effectual method. The essential impartiality of this etiquette is to maximize the system time by distributing the energy utilization. After some constant iterations, the clustering process is terminated. It will minimize the control message and select cluster heads by using well-established distributed methods. This protocol features that the sensor nodes regulate their communication control. The HEED protocol is different from the classical routing protocols because, in classical routing protocols, the sensor nodes can either be a node or a cluster head; they will not work as both. However, in HEED, the sensor node will act like a server/cluster head and the source node from time to time or as per application requirements.

For load balancing, if the sensor node is a cluster head and its energy gets depleted, then another node with a higher energy level will be designated as the CH. The critical overarching aim of the procedure is to grow the system lifetime utilizing the remaining energy for the selection of the cluster head and to consider some primary parameters such as energy consumed in the transfer of bits, processing, and communication; and secondary parameters are intra-cluster communication costs such as the purpose of neighbor nearness or cluster mass. Significant parameters are used for the cluster heads among the sensor nodes, and secondary parameters are used for discontinuity draw among the cluster heads; draw means that the sensor node is a variety of supplementary rather than single gathering. If the power level is cast off for the intra-cluster announcement, then the rate such as node degree will be used, if the necessity is to dispense the weight among cluster heads; and 1/node degree, if the necessity is to form a condensed cluster. The minimum power required to interconnect with the cluster head is calculated as

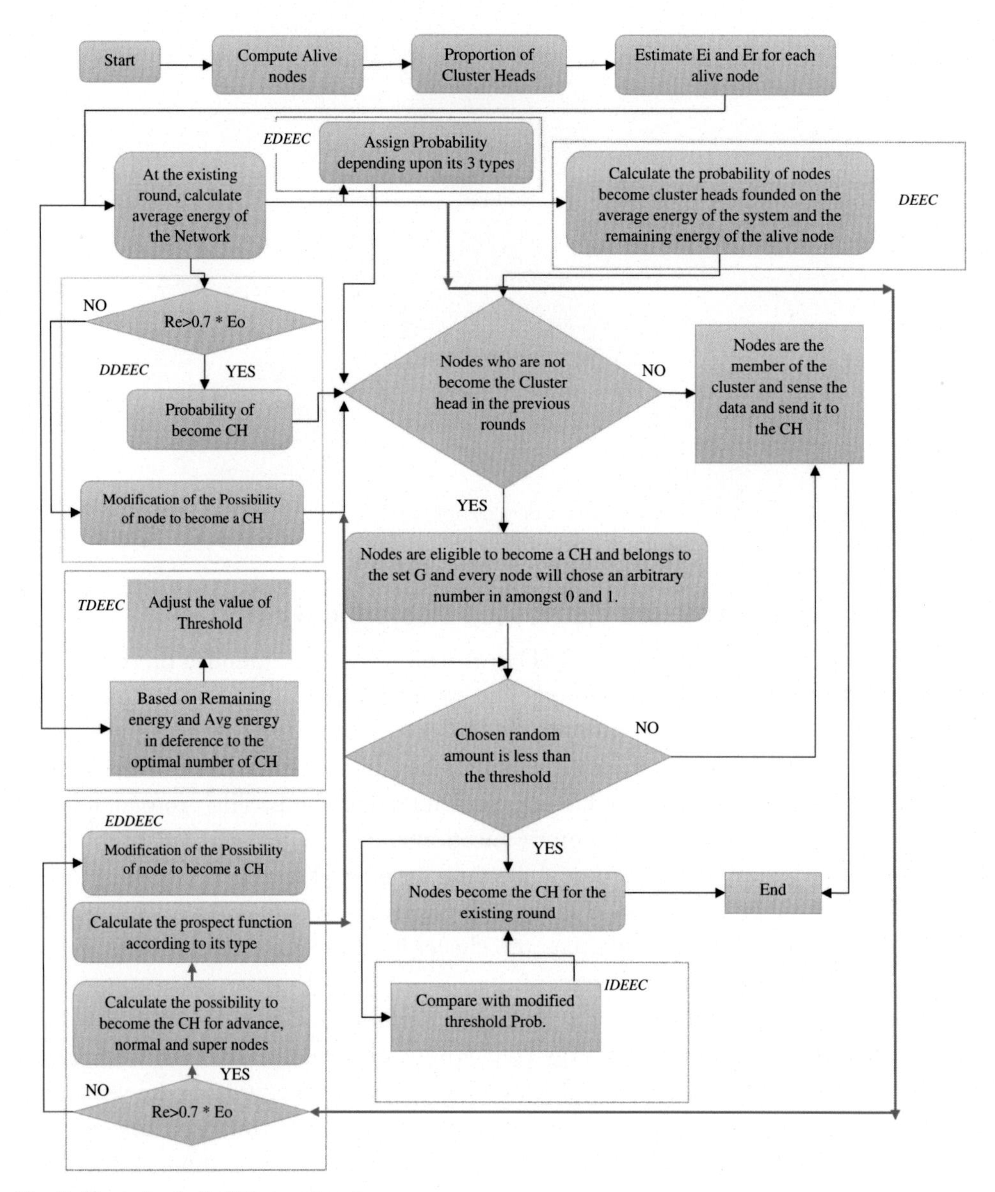

Fig. 8 Flow chart of all the protocols.

$$\text{AMP} = \frac{\sum_{i=1}^{M} min\ (p)_i}{M} \quad (1)$$

where $(p)_i$ is the minimum power level essential for apiece sensor nodes, M.

The cluster head is selected with some probability, such as

$$CH_{prob} = C_{prob} * \frac{E_{residual}}{E_{max}} \tag{2}$$

To edge the original announcements by the cluster head, C_{prob} is only used, $E_{residual}$ is the present estimated remaining energy of the nodes, and E_{max} shows the highest energy level of the nodes in a cluster. In the next phase, sensor nodes will go into the iteration until they get their new cluster head that uses the minimum communication power. Further, sensor nodes will conclude their current status. Either they will join the cluster as a member, or they will become the cluster head.

HEED is wholly distributed, in which either node is a cluster head relating to its CHprob or a cluster member within a range of clusters. There will be a significantly less chance that two nodes will become the cluster head. There will be at least one CH in an area of $(2 + \frac{1}{\sqrt{2}})R_c \times (2 + \frac{1}{\sqrt{2}})R_c$. The cluster heads of the two clusters will communicate with each other if $R_t \geq 6R_c$. Moreover, these nodes are never neighbors of each other. The process of HEED is terminated in the Niter=O (1) iteration. The worst case of the HEED protocol dispensation time complication is O(n) per node, where O(n) is the system's size concerning the number of nodes. O (1) shows that the message exchange complexity is the worst case.

4.2 Distributed weight-based energy-efficient hierarchical clustering (DWEHC)

Ding et al. [33] have proposed this method after studying the HEED protocol. The authors mentioned that the HEED etiquette fails to maintain the minimum energy utilization in the intra-cluster announcement, and the clusters created by HEED are also not stable. Like HEED, the DWEHC also does not consider properties such as the density and size of the cluster. In DWEHC, the protocol is implemented on each node till seven rounds of iteration. After completing the protocol, each node is either a usual node or a CH. The formation of clusters is in the hierarchal form, and the range of the cluster is defined in such a way that each sensor node will reach the cluster head. Within the collection of nodes, the announcement takes place using time division multiple access (TDMA), and polling is rummage sale for the announcement of information from the subordinate level to the advanced one. For intercluster communication, IEEE 802.11 is used among the cluster heads, so that the information can be communicated to the base node [34]. For the implementation purpose, the authors assumed that the nodes are discrete in a 2-dimensional interplanetary, quasi-stationary state. Sensor nodes in a system are conscious of site material such as signal strength, direction, and distance, and the authors state that energy consumption between the nodes is not even. The authors also mentioned that the following assumptions are not considered, such as the size and mass of the system, the spreading of the sensor nodes, synchronization, and sensor nodes' probability that they become the cluster head. Using an enclosure graph (i.e., a planer graph), DWEHC generates the minimum topology for every cluster.

DWEHC uses the following steps for completion of the algorithm:

(i) *Relay*: By means of distance, the algorithm mainly focuses on the path loss due to the reliability of the sensor nodes.

If

$$\left\|sd\right\|\alpha + c > \left\|sr\right\| \alpha + c + \left\|rd\right\| \alpha + c \tag{3}$$

Then, a smaller amount of energy is transmitted via node r utilization.

(ii) *Relay region*: Every sender node will transmit the data through the r node. Therefore, less energy will be consumed in the transmission. The transmission or transmit area Rα,c(s,r) with respect to the node r is defined as

$$R\alpha, c(s, r) = \left\{x\left\|\text{such that}\right\|sx\right\| \alpha > \left\|sr\right\| \alpha + \left\|rx\right\| \alpha + c\} \tag{4}$$

(iii) *Enclosure region*: Eα,c(s,r) is an enclosure region of the sensor node by r, where Eα, c(s) is the accompaniment of Rα,c(s):

$$E\alpha, c(s) = \bigcap_{r \in T(s)} E\alpha, c(s, r) \tag{5}$$

(iv) *Neighbors*: Nα,c(s) shows the neighbors of the node s. When the nodes transmit data to their neighbor node, they rely on r:

$$N\alpha, c(s) = \left\{u\left\|\text{such that } u \in T(s), u \in E\alpha, c(s)\right\}\right. \tag{6}$$

(v) *Cluster range*: Cluster range is defined as the cluster radius. It is immovable for the whole system and is defined for the farthest node exclusive to the collection of the sensor nodes.

(vi) *Weight used in cluster head election*: The weights will be calculated by the local knowledge for the cluster head selection. Weights in DWEHC can be calculated as

$$W_{weight}(s) = \sum_{u \in N_{(\alpha,c)}} (s) \frac{(R-d)}{6R} * \frac{E_{residual}{}^{(s)}}{E_{initial}{}^{(s)}} \tag{7}$$

(vii) *Levels of the cluster*: The level of the cluster head in DWEHC is multilevel. The level of the cluster depends on the child nodes.

(viii) *Myrange and Mydis*: Myrange is the remoteness from the sensor node of the cluster-head and is calculated as

$$\sqrt{(x_{CH} - x_s)^2 + (y_{CH} - y_s)^2} \tag{8}$$

where (x,y) is the cluster head, and the node synchronizes.

Mydis shows the minor energy way to the cluster head, and it is calculated as

$$(x_{CH} - x_s)^2 + (y_{CH} - y_s)^2 + (x_s - x_{s1})^2 + (y_s - y_{s1})^2 \qquad (9)$$

In DWEHC, nodes are deployed in two-dimensional interplanetary and quasi-stationary states. This protocol works on local information. Energy consumed by each node is not uniform. The author does not consider the following assumption, such as the size of the system, density of the network, the likelihood of the node developing the cluster head, synchronization, and delivery of energy utilization and nodes. The complexity of the procedure to find the neighbor node is O(min(dGt(u)dGe(u), dGt(u)logdGt(u))), where dGr(u) is within the cluster range, the degree of node u and dGt(u) is within the enclosure region and the gradation of node u. The probability of getting two nodes to become the cluster head equivalent to the cluster is minuscule. The clusters are formed in DWEHC in at most seven iterations. Broadcast message complexity and becoming cluster head complexity of each node is O(1). A node with the conducting range, R_c, is dispersed consistently and arbitrarily in $R=[0,l]2$; here, R shows that the plane is 2-dimensional, and each area is divided into squares $\left(\frac{R_c}{\sqrt{2}} \times \frac{R_c}{\sqrt{2}}\right)$. If in a cluster, there is only one node, then the following condition will be there: $R_c^2N = kL^2(in\ l)$ *for some* $k>0$. Two cluster heads will be communicating with each other if $R_t \geq 4R_c$ *where* R_t *is* min *transmitting range*. It will also create multihop clusters.

4.3 Hybrid clustering approach (HCA)

In 2011, Neamatollahi et al. [35] had reported the HCA. The essential impartiality of the projected procedure is to reduce the nodes' energy utilization and maximize the system period. HCA is founded on the distributed clustering algorithm. Nodes n are dispersed throughout the scheme; the initial determination of the system is to exploit the system period. In this procedure, the clustering algorithm is not achieved in apiece round; it will minimize the disadvantages of the static and dynamic clustering algorithms. At the end of the set-up phase, the CH will conserve their energy in the memory as a variable named ECH. The residual energy is in the range of $0<\alpha<1$. If the Eresidual is lesser than αECH, then a particular packet is sent to the sink position in the time division multiple access (TDMA) frame. After receiving this message, the base station will understand that the nodes will clutch the set-up phase on the commencement of the subsequent round, and then the base station will send a synchronization pulse to all the sensor nodes in a network. J. Greunen in 2003 showed that these pulses have vanished from the network. After receiving the synchronization pulse, each node will be ready for the clustering process. Afterwards, the assortment of the CH, a new cluster, will be formed on an on-demand basis. Furthermore, the overhead introduced by the set-up phase will be reduced tremendously [36].

- HCA accomplishes fit in relations of the period as related to other etiquettes excluding HEED. The clustering procedure in the commencement of individual rounds executes lots of operating costs on the system. Related to LEACH and HEED, HCA stretches 30% extra adaptability in rapport with system time or lifetime of the network.

In HCA, nodes are positioned using an arbitrary method in the four-sided field of (0,0) and (100,100) m, and the overall number of nodes in the square ground is 100 and there is an equivalent quantity of preliminary vigor (2J). The sink station is situated separately from the supervised area. Cluster heads are designated based on tremendous energy, and CHs perform the data aggregation. The clusters are formed when the CHs consume some predefined energy. After this new set-up phase, clustering will be performed. HCA is evaluated on HEED and LEACHs protocols using MATLAB. The evaluation process shows that HCA is 30% more efficient than both protocols. It enhances the system time since the gathering process is executed on-demand.

4.4 Energy-efficient heterogeneous clustered scheme (EEHCS)

De Freitas et al. [37] had reported EEHCS. Using the following assumptions, such as the sensor nodes placed in the system are additional energy resources, sensor nodes are motionless and organized haphazardly in the system, proportions of the sensor node and the organization of the sink nodes are already recognized. According to the protocol, the outstanding energy nodes are elected as a cluster head. The performance of the clustering protocol is measured on the foundation of system lifetime, amount of cluster head apiece rounded, the number of alive nodes rounded, and throughput. EEHCS considers the heterogeneous system, and the cluster heads are designated on the foundation of the province of contiguous energy when the sensor nodes are dispersed consistently. The energy dissipation ratio model for the network is defined using the signal-to-noise ratio of L bits transmitted in the message over a distance d.

$$E_{Tx}(L,d) = \begin{cases} L^*E_{elec} + L^*\varepsilon_{fs}{}^*d^2, & if\ d \le d_0 \\ L^*E_{elec} + L^*\varepsilon_{mp}{}^*d^4, & if\ d \ge d_0 \end{cases} \tag{10}$$

where E_{elec} is represented as energy degeneracy apiece while running the transmitter, ε_{fs} and ε_{mp} are dependent on the prototypical transmutation and d shows the remoteness among the source and destination.

EEHCS is used for straight hop announcement in the area with the base station, and it is used for the multihop announcement when the sink station is not in the detecting area. The base station appoints the cluster head on the basis of the remaining energy, and the centralized cluster head selection process will choose the number of associate nodes.

As per the simulation process, EEHCS extends the system time period up to 27.63% as compared to LEACH-C and it is two times much better in terms of first node death compared to LEACH-C.

4.5 Distributed election clustering protocol (DECP)

In 2007, Wang et al. proposed DECP [38]. DECP is a dispersed selection clustering procedure. It is meant for two-level heterogeneous WSNs. By using the remaining liveliness and the announcement cost, the probability of election is calculated. In the case of a balanced system or cluster, the node with the higher communication cost is designated by way of the CH, and in the case of the imbalanced cluster where energy dispersal is not stable, the node through the higher remaining liveliness is designated as the CH. DECP will provide the load balancing and provide the long stable region compared to classical protocols such as LEACH and SEP. In the protocol, Wang and Zhang show that the node, through advanced energy, has an additional chance to convert to the cluster head. The node having a lesser energy level will never develop the cluster head until the advanced energy node is within its investigator range. The average power dissimilarity is getting through to measure the power raze of node I, i.e.,

$$APD_i = 1 - \frac{\sum_{j \in NBR_i} E_j^{current}}{|NBR_i| * E_i^{current}} \tag{11}$$

where $E_i^{current}$ is the existing remaining power of node I, $|NBR_i|$ shows the total number of nodes, and NBR_i shows the number of neighbor nodes of i, which is in the range of node i. APD_i shows the power difference between the two neighbor nodes. If calculated ADP_i is more significant than zero; therefore, it has more energy than the neighbor nodes and a higher chance/probability of becoming the cluster head; but, if the calculated ADP_i is smaller than zero, then the node has less energy and the node resolves and will not be chosen as the cluster head. In the heterogeneous network, ADP is considered, and the communication cost is considered for cluster head selection. The announcement rate is premeditated using the following equation:

$$mCost_i = (1 - APD_i) * \frac{\sum_{j \in NBR_i} d_{i,j}^2}{|NBR_i|} = \frac{\sum_{j \in NBR_i} E_j^{current} * \sum_{j \in NBR_i} d_{i,j}^2}{|NBR_i|^2 * E_i^{current}} \tag{12}$$

$d_{i,j}$ is remoteness between the informant node and the destination node. A system is divided into clusters, and the node i occupied as a cluster head is represented as chi, and $\forall$chi is represented as CH. In the clustering procedure, the nodes in a cluster transmit their power information to all the neighbor nodes and receive all the energy information of the neighbor node. Moreover, if the nodes have all the data about their neighbor, such as communication cost and present power, then the node calculates the mCost and broadcasts the mCost to its neighbor. All the nodes select the minimum mCost and direct the poll communication to the candidate nodes. Due to the maximum received poll message, nodes become the CH and the no-CH nodes choose the nearest cluster head as member nodes.

In DECP, nodes are dispersed in a four-sided area of 100*100 using MATLAB and the sink station is situated in the middle of the square zone. As per the author, the original power of the usual node is 0.1 J and the power of the advance node is $0.5(a+1)$ J. DECP is associated with the LEACH and SEP protocols for the heterogeneous environment where the value of $m=0.2$ and value of $a=4$. DECP enhances the stable period of the network lifetime of the LEACH protocol by 102.16% and of the SEP protocol by 49.17%. The average power of the cluster head and the normal node is advanced as compared to LEACH and SEP.

4.6 Dissipation forecast and clustering management (EDFCM)

For extended time and reliable broadcast of data in heterogeneous networks, Zhou et al. [9] have projected a gathering procedure named EDFCM. The CH election process is wholly founded on the one-step power consumption forecast; it considers the energy residual and energy consumption rate, making it different from other protocols. For obtaining the longer lifetime of the system, the algorithm balances power utilization in each round, and the management node shows a significant role in the assortment of the cluster head so that an optimum number of cluster heads will be present. The node with advanced residual energy in the next round has more likelihood of being designated as a cluster head. Consequently, the likelihood of the cluster head assortment will be defined as

$$P_i(r+1) = \begin{cases} \dfrac{p}{1+\propto .m}\left(\dfrac{E_i r - E_{PR_{TO}}(r)}{\overline{E}(r+1)}\right), \text{if node } i \text{ is a type.0 node} \\ \dfrac{p}{1+\propto .m} \times (1+\propto)\left(\dfrac{E_i r - E_{PR_{TO}}(r)}{\overline{E}(r+1)}\right), \text{if node } i \text{ is a type.1 node} \end{cases} \tag{13}$$

The proposed algorithm will balance the energy round by round, which improves the generation of the system. Due to the computational heterogeneity of the node, the selected cluster head from the previous round has more remaining energy as related to other nodes in the system; the cluster head might expire or consume extra energy/power in the subsequent round of the operation. The EDFCM network operation is separated and hooked on dual stages: cluster construction and data-gathering stages. In EDFCM, the cluster creation stage is similar to that in the LEACH protocol, but two different parameters are considered in EDFCM. Firstly, the selection procedure likelihood is calculated using the above formula and the weighted function. Secondly, the organization nodes are presented in the first stage. Management nodes are introduced to select the optimum number of CH (cluster heads) in each round. The management node will communicate with the sensing node during this phase for the cluster head assortment procedure [39]. Once the node is selected as the cluster head, the management node will communicate the small data with the neighbor management node. The management nodes keep this small data and forward it to their neighbor management node. All the management nodes will broadcast the information of the cluster head node as rapidly as possible. Once the limit of the cluster head exceeds the limit, that is ($\geq$2 kopt), then the management node terminates the first phase and enters the second phase. Similarly, if the limit is ($\leq$0.5 kopt), the management nodes will randomly select the cluster heads rendering the information deposited in the memory of the CH. In the second phase, the data collecting phase, the node will send the data regarding its current remaining energy and the existing energy dissipation. The current energy dissipation information is added to the header of the data. Afterwards, getting the information from the cluster heads wholly, the base station calculates the value of energy dissipation, EPR_type0(r) and EPR_type1(r) by using the following formula:

$$\mathrm{EPR\_T0(r)} = \frac{1}{N_{type0}} \cdot \sum_{i=1}^{N_{type0}} E_{CH_{T0(i)}}(r) \tag{14}$$

$$\mathrm{EPR\_T1(r)} = \frac{1}{N_{type1}} \cdot \sum_{j=1}^{N_{type1}} E_{CH_{T1(j)}}(r) \tag{15}$$

The complexity of choosing the CH found through the merge sort and the worst/nastiest time complexity is O(n log n). Apart from the cluster head selection processes, the time complexity for the rest is O(n). The time complexity of the overall protocol is O(n2). EDFCM is compared with the LEACH and LEACH-based algorithms. EDFCM improves the most extended constancy period cluster management process and increases the base station's communication distribution.

4.7 Energy-efficient unequal clustering (EEUC)

Li et al. [40] have proposed an algorithm named "energy-efficient unequal clustering" for wireless sensor networks. The author of the paper tries to resolve the hotspot problem, which occurs in multihop routing when the cluster head is near the sink station. Cluster heads nearer to the sink station are loaded with the information and rapidly lose their energy compared to the cluster heads beyond the sink station. To resolve the hotspot problem, the author planned an algorithm in which clusters nearer to the sink stations are lesser in dimensions so that they will consume lesser energy through the intra-cluster announcement. After cluster formation, the base station sends the hello communication wholly toward the nodes, so that all the nodes will estimate the distance from the base station, which formally assists the procedure for creating clusters of unequal size [32]. In the statistics gathering process, the cluster head rotates the sensor nodes and gathers information regarding the system's energy utilization transversely. As the distance decreases from the cluster heads to the sink station, the extent of the cluster is also reduced. The algorithm's shows that the unequal cluster size recovers the system's time and balances energy consumption over LEACH and HEED.

EEUC is compared with LEACH and HEED. The communication complexity of the EEUC cluster creation is O(N). As compared with HEED, the communication burden of EEUC is limited. It avoids the cluster head message iteration and in HEED, the upper bound of the message complexity is Niter × N, where Niter is the number of repetitions. There will be very little chance that two nodes will become the cluster head in the protocol. The hot spot problem is resolved in EEUC by minimizing the interval between the first node death and the last node death in multihop routing in a clustering method. The unequal clustering mechanism improves the network lifetime in contrast as concluded from LEACH and HEED.

4.8 Distributed energy-efficient clustering algorithm for HWSN (DEEC)

Qing et al. [41] have proposed an algorithm named "energy-efficient protocol" for heterogeneous wireless sensor networks. By use of the probability function, the cluster head is designated. The likelihood function is grounded on the node's residual energy and the system's normal power. The chief goal of the CHs is to receive data from all the sensor nodes in their area and direct them to the sink station. The probability function is calculated by apiece node in the system; a node with the sophisticated likelihood function will be designated as the cluster head. DEEC uses the concept of the LEACH algorithm. DEEC works well for multilevel heterogeneous networks. The cluster head gathers all the statistics from the sensor nodes, calculates them, and sends them to the sink station. To improve the network lifetime, DEEC uses the dual level of heterogeneous nodes: normal nodes and advanced nodes. The energy level of normal nodes is low compared to the advanced nodes. E0 denotes the original energy, and m shows the number of advanced nodes. Advanced nodes have more energy than normal nodes. So, the total number of normal nodes is $(1-m)N$, and the normal nodes' energy is E_0. The entire number of advanced nodes in a system is mN, and the power associated with the advanced nodes is $E_0\ (1+a)$. Therefore, the total energy is calculated as: $E_{total}=N(1-m)E_0+NmE_0(1+a)=NE_0\ (1+am)$. The paper's author shows that the nodes are dispersed consistently, and the sink station is situated in between the internal and external network. The average energy of the node is calculated as $\overline{E}(r)=\frac{1}{N}\sum_{i=1}^{N}E_i(r)$, where $E_i(r)$ is the incredible energy of the node. Each node computes its prospect for selecting the cluster head and is calculated as

$$P_i = P_{opt}\left[1 - \frac{\overline{E}(r) - E_i(r)}{\overline{E}(r)}\right] = P_{opt}\frac{E_i(r)}{\overline{E}(r)} \tag{16}$$

The number of cluster heads selected for each round is calculated depending on the remaining energy and the reference energy ($\overline{E}(r)$).

$$\sum_{i=1}^{N} P_i = \sum_{i=1}^{N} P_{opt}\frac{E_i(r)}{\overline{E}(r)} = P_{opt}\sum_{i=1}^{N}\frac{E_i(r)}{\overline{E}(r)} = P_{opt}N \tag{17}$$

The verge probability $T(si)$ decides whether the Si node will be designated for the cluster head or not, and it is calculated as

$$T(si) = \begin{cases} \dfrac{P_i}{1 - P_i\left(r \bmod \dfrac{1}{P_i}\right)} & \text{if } S_i \epsilon\, G \\ 0 \text{ otherwise} \end{cases} \tag{18}$$

G shows the set of nodes eligible in every round for the election of the cluster head by use of the probability function.

The DEEC protocol is equated with the LEACH, SEP, and LEACH-E protocols, and for the multilevel heterogeneous networks, the prolonged protocols of LEACH and SEP are used by the author of the DEEC protocol. The stable time of the DEEC prolongs the lifetime period of the SEP and LEACH-E protocols; and SEP is the better protocol as compared to LEACH and the unbalanced zone of the SEP protocol is significantly greater than the DEEC procedure and the stable period of LEACH is the same as that of DEEC and is a 10% longer period than that of SEP. DEEC increased the stability period by 15% as compared to SEP. DEEC does not require any global knowledge of the system; it uses regular energy as the reference energy. Compared to SEP and LEACH, DEEC executes well in a multilevel heterogeneous network [44].

4.9 Energy-efficient clustering scheme (EECS)

For the periodical information assembly, Ye et al. [42] have proposed a procedure for energy proficiency and load-balanced grouping in wireless sensor networks. The cluster head is elected on the basis of outstanding power. Throughout the cluster head selection procedure, the applicant nodes are chosen and they have to contest among themselves to become the head of the cluster. The cluster head election feature is comparable to the LEACH protocol. At the cluster formation phase, the sink station transmits the hello communication to all its nodes so that all the nodes compute the remoteness from the sink position on the basis of the established indication asset.

In the assortment stage of the cluster head, the elected nodes for the cluster head are known as CANDIDATE nodes with a likelihood, *T*. Afterward, the node becomes the CANDIDITE node, broadcasts a COMPETE_HEAD_MSG to all the other nodes that are currently inside the radio variety, i.e., Rcompete. After receiving the COMPETE_HEAD_MSG, all the competing nodes compare the remaining energy with the received remaining power of the node; its received residual energy is more significant, then the node itself quits after the rivalry when it is deprived of getting the complete COMPLETE_HEAD_MSG. Else, the node will be chosen as the cluster head. For selecting the cluster head, this process uses the resident transistor announcement founded on the remaining energy.

In the cluster establishment stage, the cluster-head node broadcasts the HEAD_AD_MSG to essential nodes based on received HEAD_AD_MSG. The plain node decides to connect to the cluster on the foundation of distance. Cluster heads are dispersed equally crossways in the entire system. As compared to LEACH, EECS enhances the network lifetime by 35%.

EECS mainly focuses on the cluster set-up algorithm and not the data transmission phase. It is fully distributed in nature and the cluster heads are dispersed erratically in the system. The approach is used to scatter the load amid the cluster heads. As per the author, without the iteration process, the competition process among the cluster heads is localized; therefore, there is less message overhead and there will be well-distributed cluster heads in EECS as calculated by the weighted function. The EECS augments the imitation outcome which states that the system lifetime is 135% compared to the LEACH protocol, and energy utilization in LEACH is 53% and in EECS is 93%. The complexity of the control message through the process is O(N), where *N* is the number of nodes in the system. In every Rcomplete, at most one cluster head will be obtained.

4.10 Multihop routing protocol with unequal clustering (MRPUC)

Toward improving the network lifetime, Gong et al. [8] planned a procedure termed a disseminated, multihop routing protocol for uneven clustering. MRPUC is divided into three phases: cluster set-up, intercluster multihop routing formation, and information communication. After the network formation, the base station broadcasts the BS_ADV communication to all the nodes in the system that are conceived equivalent and apiece node will calculate the estimated remoteness d to the sink place. R_{max} and R_{min} (cluster radii) are predefined for cluster formation, and d_{max} is the maximum remoteness from the base place and the base station. The cluster range Ri of node i is calculated as

$$\mathrm{Ri} = \frac{d(i, BS).(R_{max} - R_{min})}{d_{max}} + R_{min} \tag{19}$$

R_{max} shows the maximum radius of the cluster; therefore, the nodes that exist in the R_{max} range are neighbors of each other. Each node within the R_{max} range transmits the Hello(ID, E) communication to all its neighbors. All the nodes gather all the correlative data in the table. If the node has more significant remaining energy, it will develop into the cluster head and that node will transmit the HEAD_MSG(ID, Ri, E) communication to all the nodes. After the announcement, all nodes calculate the distance $d(i,j)$ to their neighbor nodes. If $d(i,j) < \text{Ri}$, then, the node j adds the node i to the applicant cluster head set as Si. Nodes are separated into cluster head, cluster associate, and unidentified node. If the node status is unknown, that node j will intersect the collection with the lowest rate by distributing JOIN_MSG (ID, CH_ID), where CH_ID is the cluster head's ID. Using the following equation, the rate of connection of the cluster head k to join the cluster is calculated.

$$\text{Cost}k = w.\frac{d(j,k)}{R^k} + (1-w).\frac{E_{max} - E(k)}{E_{max}} \tag{20}$$

where $E_{max} = max\{E(k)\}, k \epsilon S^j$, where w is the biased factor and S^j is the candidate cluster head.

The cluster heads send the control communication to all the other cluster heads and save the replay information in their table for the intercluster announcement. Afterward, getting together all replays, all the cluster heads calculate the approximate distance and choose the parent cluster head. Then, the intercluster bush is shaped for the multihop announcement to save power. After the intercluster tree formation, all the nodes turn off the radio and awake at the time allocated to them and refer statistics to the cluster head. All the cluster heads collect the information from their nodes, convert it into one packet, and send it to the head cluster head. The head node then communicates the message to the sink station [47].

As per the author of the paper, MRPUC has improved the results of HEED and MRPEC and has also mentioned that, in HEED, CHs are very distant from the sink. Due to this, CHs have to transmit the information packets a long way, leading to enormous power consumption by the nodes. MRPUC suffers from the hotspot problem, i.e., the cluster head near the sink station expires first as an associate to the other cluster head, which is distant from the cluster head because of the burden of aggregate data, which has come from different nodes. The network lifetime is reduced after the first node dies; so as per the author, MRPUC improves the node deaths as per the rounds. It achieves 251.7% improvement over HEED and 34.4% improvement over MRPEC. The control communication complexity of MRPUC is O(N). Here, the nodes are categorized into two parts: a cluster head and a member node. There will be one individual collection head in a cluster radius, R_{min}. The first node dies at the 211th round, and 50% of the nodes expire at the 238th round. In MRPEC, the first node dies at the 157th round, and 50% of the nodes die at the 233rd round. In HEED, the first node dies at the 60th round and as an improvement, over 50% of the nodes die at the 225th round. MRPUC has improved over MRPEC in the context of the opening node death by 34.4% and improved 50% node death is 2.2%. MRPUC has improved as the first node dies compared to the HEED with 251.7% enhancement and as an improvement, over 50% of the node deaths are at 5.8%.

5. Simulation and results

The given section assesses the presentation of the numerous procedures in heterogeneous WSNs. We use MATLAB as the simulation tool. As shown in the table, we have considered $N=200$ sensor nodes arbitrarily dispersed in a network of 250×250. Despite damage due to simplification, the sink station is placed between the center of the cluster, i.e., (125×125). The preliminary supremacy of the sensor nodes is different, and the limited quantity of forecasting energy is 0.50.

Along with this, we have ignored the signal rear-ender and interfered in the wireless channel. With these system constraints, various different protocols of the HWSN are considered for evaluating the performance, such as in the context of the number of nodes alive, the remaining energy of the nodes, and the number of the departed nodes in the first round. The simulation parameters are shown in Table 2.

5.1 Evaluation of the death round of the first node

In WSNs, the system's stability and performance depend on the node's life, and it deteriorates when the first node dies. The expiry of the opening node shows that the system goes into an irregular period, and gradually, the network's performance goes into the decline mode. The network with the greater number of nodes will have a slightly increased network lifetime as compared to those networks having a lesser number of nodes; this is because the greater number of nodes will increase the burden on the cluster head but increase the network lifetime. So, for balancing the network's life, many authors have proposed many solutions for choosing the cluster head by considering the

Table 2 Simulation parameters.

Parameters	Values
Network field	250×250
Number of nodes	200
Packet size	4000 Bits
Eelec	50 nJ/bit
Efs	10 nJ/bit/m^2
Eamp	0.0013 pJ/bit/m^4
EDA	5 nJ/bit/signal
D0 (Thershold interval)	70 m
E0 (Original energy of the normal nodes)	0.5 J

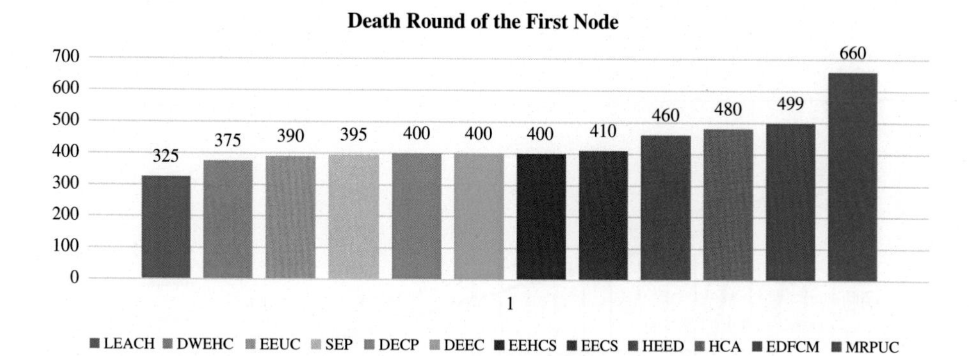

Fig. 9 Death round of the first node.

Table 3 Data values first node die.

Protocol	Round
LEACH	325
DWEHC	375
EEUC	390
SEP	395
DECP	400
DEEC	400
EEHCS	400
EECS	410
HEED	460
HCA	480
EDFCM	499
MRPUC	660

tremendous energy, weight, cost, location of the nodes, and computation power. As per Fig. 9, MRPUC, EDFCM, HCA, and HEED. Secondly, EECS, EEHCS, DEEC, DECP, SEP, and EEUC are close, and thirdly, DWEHC and LEACH performance is poor compared to others. Fig. 9 and Table 3 show the smoothed death of the primary sensor node.

5.2 Evaluation of number of nodes alive

Fig. 10 depicts that the number of alive nodes indicates the network's lifetime. In the assessment of LEACH and the other procedures, HEED and EEUC work well regarding the number of alive nodes. Other protocols such as EECS, HCA, and MRPUC work well in the beginning rounds, but the number of alive nodes ratio is decreased in further rounds. HCA is 30% better than the other protocols except HEED in terms of network lifetime.

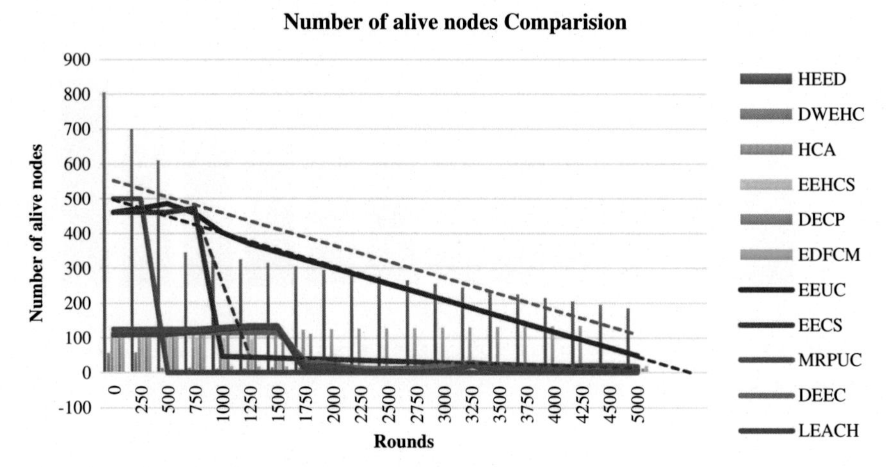

Fig. 10 Number of alive nodes

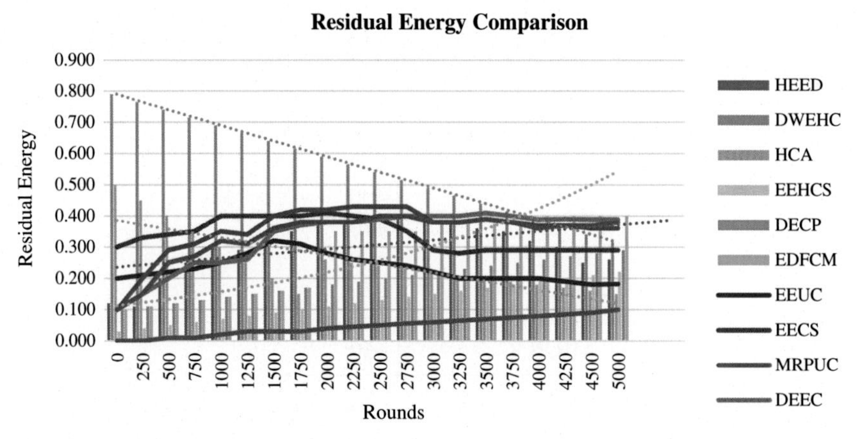

Fig. 11 Residual energy.

5.3 Evaluation based on residual energy

Fig. 11 and Table 4 illustrate that DWEHC achieves better residual energy in terms of linear decrease in energy, and DEEC, HEED, and EDFCM achieve higher residual energy but not linearly. DWEHC is good in the beginning phase but gradually decreases, whereas HEED, DEEC, and EDFCM work well after 5000 rounds. As we can see, HCA is the only algorithm that is not good for residual energy saving and reaches very low values, whereas the other algorithms work well at a certain level.

6. Performance analysis

Table 5 demonstrates the assessment/performance analysis of the protocols proposed by the different authors. The analysis is done among the procedures regarding cluster

Table 4 Data values residual energy comparison.

Rounds	HEED	DWEHC	HCA	EEHCS	DECP	EDFCM	EEUC	EECS	MRPUC	DEEC
0	0.12	0.79	0.5	0.03	0.1	0.1	0.2	0.3	0	0.1
500	0.2	0.79	0.45	0.05	0.18	0.1	0.22	0.34	0.01	0.2
1000	0.3	0.79	0.39	0.15	0.2	0.2	0.25	0.4	0.02	0.25
1500	0.299	0.78	0.34	0.18	0.18	0.33	0.32	0.4	0.03	0.35
2000	0.37	0.42	0.26	0.2	0.2	0.5	0.28	0.41	0.04	0.38
2500	0.39	0.43	0.2	0.19	0.23	0.5	0.25	0.39	0.05	0.3899
3000	0.4	0.44	0.15	0.14	0.23	0.5	0.22	0.29	0.06	0.4
3500	0.36	0.48	0	0.12	0.25	0.5	0.2	0.29	0.07	0.41
4000	0.32	0.45	0	0.11	0.25	0.5	0.2	0.29	0.08	0.38888
4500	0.25	0.43	0	0.1	0.25	0.5	0.18	0.29	0.09	0.38888
5000	0.26	0.49	0	0.05	0.25	0.5	0.182	0.29	0.1	0.38888

Table 5 Performance comparison and analysis [48,54].

Clustering routing protocol		LEACH	HEED	EECS	DECP	HCA	EEUC	DWEHC	DEEC	EDFCM	MRPUC
Cluster characteristics	Uniformity of cluster sizes	Moderate	Moderate	Uneven	Even	Random	Uneven	Even	Even	Uniform	Uneven
	Variability of cluster count	Variable	Variable	Variable	Variable	Variable	Variable	Variable	Variable	Variable	Variable
	Intercluster routing	Single-hop	Single-hop Multiple-hop	Single-hop	Single-hop Multiple-hop	Single-hop Multiple-hop	Multiple-hop	Single-hop	Single-hop	Single-hop	Single-hop Multiple-hop
	Intracluster routing	Single-hop	Single-hop	Single-hop	Single-hop	Single-hop	Single-hop	Multiple-hop	Single-hop	Single-hop	Single-hop
	Number of cluster head	Undetermined	Undetermined	Undetermined	Undetermined	Undetermined	Undetermined	Undetermined	Undetermined	Undetermined	Undetermined
	Balance	Yes	No	No	No	No	No	No	Yes	Yes	No
	Delay	Low	Moderate	Low	High	Moderate	Low	High	Moderate	Moderate	Moderate
Clustering process	Topology	Distributed	Hybrid	Distributed	Distributed	Hybrid	Hybrid	Distributed	Distributed	Distributed	Distributed
	Execution nature	Probabilistic and threshold	Iterative	Probabilistic	Probabilistic	Iterative	Probabilistic	Iterative	Probabilistic	Probabilistic	Deterministic
	Control manners	Distributed	Distributed	Distributed	Distributed	Distributed	Distributed	Distributed	Distributed	Distributed	Distributed
	Parameters for CH election	Adaptive	Adaptive	Adaptive	Adaptive	Adaptive	Adaptive	Adaptive	Adaptive	Adaptive	Adaptive
	Convergence time	Constant	Constant	Constant	Constant	Constant	Constant	Constant	Constant	Constant	Constant
	Objectives	Load balancing	Load balancing	Load balancing Periodical data communications	Load balancing	Average power distinction and communication cost	Load balancing	Load balancing	Load balancing	Balance energy consumption, lifetime of the network, reliable transmission service	Longevity and energy balance
Selection of CH	Proactivity	Proactive	Proactive	Proactive	Proactive	Proactive	Proactive	Proactive	Proactive	Proactive	Proactive
	Classification	Probabilistic	Probabilistic	Neighbor based	Neighbor based	On-demand	Architecture distance	Probabilistic	Probabilistic	Probabilistic	Neighbor based
	Parameters	Energy, communication cost	Energy, communication cost	Energy, distance	Residua energy, communication cost	Energy	Energy, distance	Weight	Residual energy, initial energy	Weight function, energy	Energy
Consider remaining energy for the CH selection		Yes	Yes	Yes	Yes	Yes	Yes	Yes	Yes	Yes	Yes
Overhead in the CH selection		Less	High	Less	Less	Less	Less	High	Medium	Less	Less
Energy efficiency		Good	Moderate	Good	Average	Good	Moderate	Moderate	Average	Low	High
Load balancing		Low	Moderate	Moderate	Very good	Very good	Good	Very good	Moderate	Moderate	Very good
Scalability		Low	High	Low	Very low	Medium	High	Moderate	Good	Very good	High
Cluster stability		Low	High	High	Very good	Very Good	High	High	Moderate	Very good	High
Delivery delay		Less	Moderate	Small	Moderate	High	moderate	Moderate	Moderate	Moderate	Less
Algorithm complexity		Low	High	Very high	Low	More	Moderate	More	Moderate	More	Moderate O(N)
Power proficiency		Moderate	Moderate	Moderate	Low	More	High	More	Moderate	More	Moderate

characteristics, clustering process, selection of cluster head, considering the remaining energy for the CH selection, overhead in CH selection, energy efficiency, load balancing, scalability, cluster stability, delivery delay, algorithm complexity, and power proficiency.

7. Conclusions and future work

To improve the system period and the stability of wireless sensor networks, well-organized energy-efficient routing protocols. Due to specific challenges such as limited energy, computational power, and packet delivery ratio of the sensor nodes, routing is challenging for the wireless sensor network. The main difference between the homogeneous and heterogeneous network is to ensure the energy utilization of the sensor nodes. Clustering methods decrease the number of messages received by the sink nodes in a very large-scale network. Simulation results show the effect of cluster head selection in a heterogeneous environment and show the best protocol compared to the others. An overall comparison regarding the cluster characteristics, delay, stability, algorithm complexity, performance, etc., is described in the summary table.

Simulation results show that HEED and EEUC are better protocols in the context of numbers of alive nodes than other protocols, and DWEHC is improved in terms of residual energy for the heterogeneous environment.

As per the market and market research (Fig. 12), approximately 40,000 United States Dollar (USD) million sensors would be produced very powerfully and employed in the upcoming years in various application areas in cost-effective ways such as in building automation, underwater acoustic sensors, wearable devices, sensing-based cyber-physical

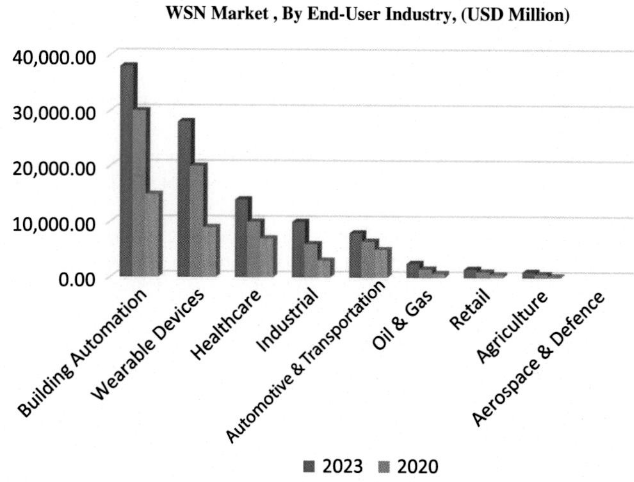

Fig. 12 The WSN market, by end-user industry (USD Million).

systems, health care, and in security and privacy management [56, 57]. So, in the future, we can propose another algorithm to secure the network's lifetime. Also, we can compare and contrast the rest of the protocols with the mentioned protocols. We can add more parameters for the comparison.

References

[1] F. Fanian, M.K. Rafsanjani, Cluster-based routing protocols in wireless sensor networks: a survey based on methodology, J. Netw. Comput. Appl. 142 (2019) 111–142, https://doi.org/10.1016/j.jnca.2019.04.021.

[2] X. Zhou, Multi-hop routing protocol with unequal clustering for wireless sensor networks, in: IEEE ISECS International Colloquium on Computing Communication Control and Management, August 2008, ISBN: 978-0-7695-3290-5, https://doi.org/10.1109/CCCM.2008.99.

[3] A.K. Tripathy, S. Chinara, Staggered clustering protocol: SCP an efficient clustering approach for wireless sensor network, in: IEEE World Congress on Information and Communication Technologies, January 2013, https://doi.org/10.1109/WICT.2012.6409209.

[4] S. Shruthi, Proactive routing protocols for a MANET—a review, in: International Conference on I-SMAC (IoT in Social, Mobile, Analytics and Cloud) (I-SMAC), October 2017, https://doi.org/10.1109/I-SMAC.2017.8058294.

[5] C. Li, Y. Mao, G. Chen, W. Jie, An energy-efficient unequal clustering mechanism for wireless sensor networks, in: IEEE International Conference on Mobile Adhoc and Sensor Systems Conference, November 2005, ISBN: 0-7803-9465-8, https://doi.org/10.1109/MAHSS.2005.1542849.

[6] J. Amudhavel, D. Rajaguru, S. Sampath Kumar, S.H. Lakhani, T. Vengattaraman, K.P. Kumar, A chaotic krill herd optimization approach in VANET for congestion free effective multi-hop communication, in: Proceedings of the 2015 International Conference on Advanced Research in Computer Science Engineering & Technology (ICARCSET 2015) - ICARCSET '15, March 2015, pp. 1–5. Article no. 27 https://doi.org/10.1145/2743065.2743092.

[7] H. Jafari, M. Nazari, S. Shamshirband, Optimization of energy consumption in wireless sensor networks using density-based clustering algorithm, Int. J. Comput. Appl. (2018), https://doi.org/10.1080/1206212X.2018.1497117.

[8] B. Gong, L. Li, S. Wang, X. Zhou, Multi-hop routing protocol with unequal clustering for wireless sensor networks, in: IEEE ISECS International Colloquium on Computing, Communication, Control, and Management, August 2008, ISBN: 978-0-7695-3290-5, https://doi.org/10.1109/CCCM.2008.99.

[9] H. Zhou, Y. Wu, Y. Hu, G. Xie, A novel stable selection and reliable transmission protocol for clustered heterogeneous wireless sensor networks, Comput. Commun. 33 (15) (2010) 1843–1849, https://doi.org/10.1016/j.comcom.2010.06.001.

[10] C. Gherbi, Z. Aliouat, M. Benmohammed, An adaptive clustering approach to dynamic load balancing and energy efficiency in wireless sensor networks, Energy 114 (2016) 647–662, https://doi.org/10.1016/j.energy.2016.08.012.

[11] P. Neamatollahi, H. Taheri, M. Naghibzadeh, M. Yaghmaee, A hybrid clustering approach for prolonging lifetime in wireless sensor networks, in: 2011 International Symposium on Computer Networks and Distributed Systems (CNDS), Tehran, 2011, pp. 170–174, https://doi.org/10.1109/CNDS.2011.5764566.

[12] H. Khamfroush, Lifetime increase for wireless sensor networks using cluster-based routing, in: International Association of Computer Science and Information Technology - Spring Conference, April 2009, pp. 14–18, https://doi.org/10.1109/IACSIT-SC.2009.99.

[13] F. Shang, D. Ren, A delay-based clustering algorithm for wireless sensor networks, in: International Workshop on Computer Science for Environmental Engineering and EcoInformatics, 158, Springer Science and Business Media LLC, 2011, pp. 463–468, https://doi.org/10.1007/978-3-642-22694-6_66 (Chapter 66).

[14] S. Arjunan, S. Pothula, A survey on unequal clustering protocols in wireless sensor networks, J. King Saud Univ. – Comput. Inform. Sci. 31 (3) (July 2019) 304–317, https://doi.org/10.1016/j.jksuci.2017.03.006.
[15] T. Gao, J.-Y. Song, J.-H. Ding, D.-Q. Wang, Fuzzy weight cluster-based routing algorithm for wireless sensor networks, J. Control Sci. Eng. (2015), https://doi.org/10.1155/2015/583092.
[16] Y.-F. Wen, T.A.F. Anderson, D.M.W. Powers, On energy-efficient aggregation routing and scheduling in IEEE 802.15.4-based wireless sensor networks, Wirel. Commun. Mob. Comput. (2012) 232–253, https://doi.org/10.1002/wcm.1249.
[17] P.S. Mehra, M.N. Doja, B. Alam, Enhanced stable period for two level and multilevel heterogeneous model for distant base station in wireless sensor network, in: Proceedings of the Second International Conference on Computer and Communication Technologies, Springer Science and Business Media LLC, January 2016, pp. 751–759, https://doi.org/10.1007/978-81-322-2517-1_72 (Chapter 72).
[18] H. Zhou, Y. Wu, G. Xie, EDFM: a stable election protocol based on energy dissipation forecast method for clustered heterogeneous wireless sensor networks, in: 2009 5th International Conference on Wireless Communications, Networking and Mobile Computing, Beijing, 2009, pp. 1–4, https://doi.org/10.1109/WICOM.2009.5304152.
[19] M.R. Dhage, S. Vemuru, Routing design issues in heterogeneous wireless sensor network, Int. J. Electr. Comput. Eng. 8 (2018), https://doi.org/10.11591/ijece.v8i2.pp1028-1039.
[20] A.K. Sohal, A.K. Sharma, N. Sood, Enhancing coverage using weight based clustering in wireless sensor networks, Wirel. Pers. Commun. 98 (2018) 3505–3526, https://doi.org/10.1007/s11277-017-5026-1.
[21] D. Agarwal, A. Gupta, Energy efficient clustering based 3-rank heterogeneous network model for wireless sensor network, in: 2015 International Conference on Advances in Computer Engineering and Applications, Ghaziabad, 2015, pp. 183–188, https://doi.org/10.1109/ICACEA.2015.7164692.
[22] V. Srividhya, T. Shankar, Energy proficient clustering technique for lifetime enhancement of cognitive radio– based heterogeneous wireless sensor network, Int. J. Distrib. Sens. Netw. (2018), https://doi.org/10.1177/1550147718767598.
[23] S. Chhabra, D. Singh, Hybrid energy efficient clustering based on residual energy, node degree and distance to base station (HEEC-RND) in heterogeneous WSNs, in: International Conference on Computing, Communication & Automation, Noida, 2015, pp. 342–347, https://doi.org/10.1109/CCAA.2015.7148441.
[24] J. Wang, Y. Yin, J. Kim, S. Lee, C. Lai, A mobile-sink based energy-efficient clustering algorithm for wireless sensor networks, in: 2012 IEEE 12th International Conference on Computer and Information Technology, Chengdu, 2012, pp. 678–683, https://doi.org/10.1109/CIT.2012.142.
[25] G. Chen, C. Li, M. Ye, et al., An unequal cluster-based routing protocol in wireless sensor networks, Wirel. Netw 15 (2009) 193–207, https://doi.org/10.1007/s11276-007-0035-8.
[26] X. Li, A routing protocol for balancing energy consumption in heterogeneous wireless sensor networks, in: International Conference on Mobile Ad-Hoc and Sensor Networks, vol 4864, 2007, pp. 79–88, https://doi.org/10.1007/978-3-540-77024-4_9.
[27] R. Pachlor, D. Shrimankar, LAR-CH: a cluster-head rotation approach for sensor networks, IEEE Sensors J. 18 (23) (2018) 9821–9828, https://doi.org/10.1109/JSEN.2018.2872065.
[28] T. Qi, Q. Bing, An energy-efficient protocol architecture with multiple clustering for wireless sensor networks, in: 2010 International Conference on Intelligent Computing and Integrated Systems, Guilin, 2010, pp. 898–901, https://doi.org/10.1109/ICISS.2010.5657046.
[29] H. Taheri, P. Neamatollahi, M.H. Yaghmaee, M. Naghibzadeh, A local cluster head election algorithm in wireless sensor networks, in: 2011 CSI International Symposium on Computer Science and Software Engineering (CSSE), Tehran, 2011, pp. 38–43, https://doi.org/10.1109/CSICSSE.2011.5963987.
[30] K. Matrouk, Prolonging the system lifetime and equalising the energy for heterogeneous sensor networks using RETT protocol, Int. J. Sens. Netw. (2009) 65–77, https://doi.org/10.1504/IJSNET.2009.029015.
[31] V.K. Chaurasiya, S.R. Kumar, S. Verma, G.C. Nandi, Traffic based clustering in wireless sensor network, in: 2008 Fourth International Conference on Wireless Communication and Sensor Networks, December 2008, https://doi.org/10.1109/WCSN.2008.4772687.

[32] O. Younis, S. Fahmy, HEED: a hybrid, energy-efficient, distributed clustering approach for ad hoc sensor networks, IEEE Trans. Mob. Comput. 3 (4) (2004) 366–379.
[33] P. Ding, J. Holliday, A. Celik, Distributed energy-efficient hierarchical clustering for wireless sensor networks, in: Proceedings of the 1st IEEE International Conference on Distributed Computing in Sensor Systems (DCOSS '05), July 2005, pp. 322–339.
[34] V.M. Galshetwar, A. Jeyakumar, Energy efficient and reliable clustering algorithms HEED and ADCP of wireless sensor networks: a comparative study, in: 2014 International Conference on Communication and Signal Processing, Melmaruvathur, 2014, pp. 1979–1983, https://doi.org/10.1109/ICCSP.2014.6950190.
[35] P. Neamatollahi, H. Taheri, M. Naghibzadeh, M.H. Yaghmaee, A hybrid clustering approach for prolonging lifetime in wireless sensor networks, in: Proceedings of the International Symposium on Computer Networks and Distributed Systems (CNDS '11), February 2011, pp. 170–174.
[36] G. Kumar, H. Mehra, A.R. Seth, P. Radhakrishnan, N. Hemavathi, S. Sudha, An hybrid clustering algorithm for optimal clusters in wireless sensor networks, in: 2014 IEEE Students' Conference on Electrical, Electronics and Computer Science, 2014, https://doi.org/10.1109/sceecs.2014.6804442.
[37] E.P. De Freitas, T. Heimfarth, C.E. Pereira, A.M. Ferreira, F.R. Wagner, T. Larsson, Evaluation of coordination strategies for heterogeneous sensor networks aiming at surveillance applications, in: Proceedings of the IEEE Sensors Conference (SENSORS '09), October 2009, pp. 591–596.
[38] X. Wang, G. Zhang, Decp: a distributed election clustering protocol for heterogeneous wireless sensor networks, in: Proceedings of the 7th International Conference on Computational Science (ICCS' 07), 2007, pp. 105–108.
[39] J.N. Al-Karaki, A.E. Kamal, Routing techniques in wireless sensor networks: a survey, IEEE Wirel. Commun. 11 (6) (2004) 6–28, https://doi.org/10.1109/MWC.2004.1368893.
[40] C. Li, M. Ye, G. Chen, J. Wu, A energy-efficient unequal clustering mechanism for wireless sensor networks, in: Proceedings of the 2nd IEEE International Conference on Mobile Ad-hoc and Sensor Systems (MASS '05), November 2005, pp. 604–611.
[41] L. Qing, Q. Zhu, M. Wang, Design of a distributed energy-efficient clustering algorithm for heterogeneous wireless sensor networks, Comput. Commun. 29 (12) (2006) 2230–2237.
[42] M. Ye, C. Li, G. Chen, J. Wu, EECS: a energy efficient clustering scheme in wireless sensor networks 10a.2, in: Proceedings of the 24th IEEE International Performance, Computing, and Communications Conference (IPCCC '05), April 2005, pp. 535–540.
[43] S. Ben Alla, A. Ezzati, A. Mouhsen, A. Beni Hssane, M.L. Hasnaoui, Balanced and centralized distributed energy efficient clustering for heterogeneous wireless sensor networks, in: 2011 3rd International Conference on Next Generation Networks and Services (NGNS), Hammamet, 2011, pp. 39–44, https://doi.org/10.1109/NGNS.2011.6142539.
[44] M. Pramanick, C. Chowdhury, P. Basak, M.A. Al-Mamun, S. Neogy, An energy-efficient routing protocol for wireless sensor networks, in: 2015 Applications and Innovations in Mobile Computing (AIMoC), 2014, pp. 815–822. Volume 12, Issue 4 https://doi.org/10.1016/S1665-6423(14)70097-5.
[45] L. Jin, Research on hierarchical routing protocol based on optimal clustering head for wireless sensor networks, Int. J. Model. Identif. Control. (April 2012) 331–337, https://doi.org/10.1504/IJMIC.2012.046413.
[46] X. Zou, C. Yang, Effects of inertia weight on DPSO-based single-hop routing protocol for wireless sensor networks, in: 2008 7th World Congress on Intelligent Control and Automation, Chongqing, 2008, pp. 6707–6710, https://doi.org/10.1109/WCICA.2008.4593944.
[47] S. Mo, H. Chen, Y. Li, Clustering-based routing for top-k querying in wireless sensor networks, EURASIP J. Wirel. Commun. Netw. (2011) 2–13, https://doi.org/10.1186/1687-1499-2011-73.
[48] D. Sharma, A. Ojha, A.P. Bhondekar, Heterogeneity consideration in wireless sensor networks routing algorithms: a review, J. Supercomput. (2018), https://doi.org/10.1007/s11227-018-2635-8.
[49] G. Devika, D. Ramesh, A.G. Karegowda, A study on energy-efficient wireless sensor network protocols, in: Nature-Inspired Computing Applications in Advanced Communication Networks, IGI Global, 2020, https://doi.org/10.4018/978-1-7998-1626-3.ch007 (Chapter 7).
[50] M.A. Al Sibahee, S. Lu, M.Z. Masoud, Z.A. Hussien, M.A. Hussain, Z.A. Abduljabbar, LEACH-T: LEACH clustering protocol based on three layers, in: 2016 International Conference on Network and

Information Systems for Computers (ICNISC), Wuhan, 2016, pp. 36–40, https://doi.org/10.1109/ICNISC.2016.018.
[51] S. Singh, S. Kumar, A. Nayyar, F. Al-Turjman, L. Mostarda, Proficient QoS-based target coverage problem in wireless sensor networks, IEEE Access 8 (2020) 74315–74325.
[52] S. Tanwar, N. Kumar, J.J.P.C. Rodrigues, A systematic review on heterogeneous routing protocols for wireless sensor network, J. Netw. Comput. Appl. 53 (2015) 39–56, https://doi.org/10.1016/j.jnca.2015.03.004.
[53] G.V. Selvi, R. Manoharan, Unequal clustering algorithm for WSN to prolong the network lifetime (UCAPN), in: 2013 4th International Conference on Intelligent Systems, Modelling and Simulation, Bangkok, 2013, pp. 456–461, https://doi.org/10.1109/ISMS.2013.134.
[54] R. Sheikhpour, S. Jabbehdari, A. Khadem-Zadeh, Comparison of energy efficient clustering protocols in heterogeneous wireless sensor networks, Int. J. Adv. Sci. Technol. 36 (2011) 27–40.
[55] Munir, Cluster based routing protocols: a comparative study, in: 2015 Fifth International Conference on Advanced Computing & Communication Technologies, Haryana, 2015, pp. 590–594, https://doi.org/10.1109/ACCT.2015.10.
[56] https://www.marketsandmarkets.com/Market-Reports/wireless-sensor-networks-market-445.html seen on January 2022.
[57] G. Shrivastava, M. Khari, Recent research in network security analytics, Int. J. Sens. Wirel. Commun. Control 8 (1) (2015) 2–4.

CHAPTER 8

Analysis of energy-efficient cluster-based routing protocols for heterogeneous WSNs

Pradeep Bedi[a], S.B. Goyal[b], Jugnesh Kumar[c], and Shailesh Kumar[d]
[a]Department of Computer Science and Engineering, Galgotias University, Greater Noida, Uttar Pradesh, India
[b]Faculty of Information Technology City University, Petaling Jaya, Malaysia
[c]St. Andrews Institute of Technology and Management, Gurgaon, India
[d]BlueCrest College, Freetown, Sierra Leone, West Africa

1. Introduction

Over the past few decades, new advancements in electronic components with limited energy capabilities and universal sensors applied for wireless sensor networks (WSNs) have shown important technological advancements. In several scenarios, these units are constructed into compact electronic components called sensor nodes by using WSNs to provide sensing and data analysis [1]. To perceive different environmental conditions, these nodes are spread in a random way with a high density over geographic locations. These environmental conditions are temperature, pressure, humidity, seismic activity, sound, and moisture. The sensed data are transferred from one hop to another and finally to sink nodes for data analysis and processing [2]. Recently, WSNs have shown their efficiency in different fields such as home automation, smart applications, military protection, health surveillance, underwater transportation, and environmental surveillance. Researchers have contributed their efforts for the incorporation of advanced computer intelligence technologies for different WSN application domains [3]. Some application or types of WSN are illustrated in Fig. 1. An example is a sensor network used for the transmission of video, audio, and images, especially for evaluation and surveillance applications; these are termed wireless multimedia sensor networks (WMSNs). Another example is wireless industrial sensor networks (WISNs) within warehouses or industrial sectors for machine performance surveillance and process automation. The network could be identified as a wireless body area network (WBANs) whenever utilized for healthcare services. In contrast, whenever sensor nodes are installed beneath the water to enable underwater navigation monitoring, emissions control, and disaster management, they are referred to as wireless underwater sensor networks (WUSNs). Sensor networks that are mobile can be called wireless mobile sensor networks (WMSNs). Such implementation domains have issues regarding their design, topology, and specifications

Comprehensive Guide to Heterogeneous Networks
https://doi.org/10.1016/B978-0-323-90527-5.00003-4

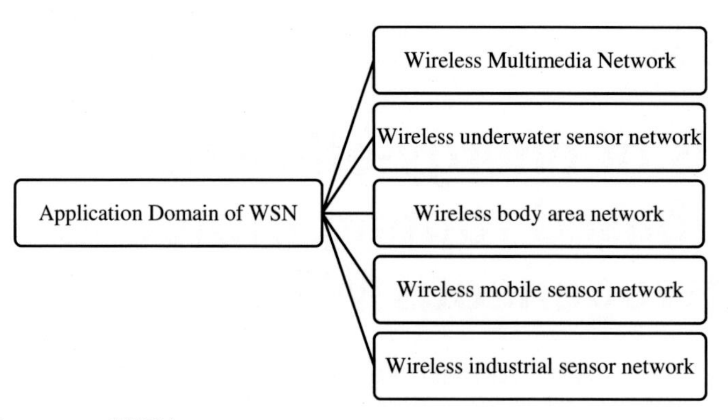

Fig. 1 Application areas of WSNs.

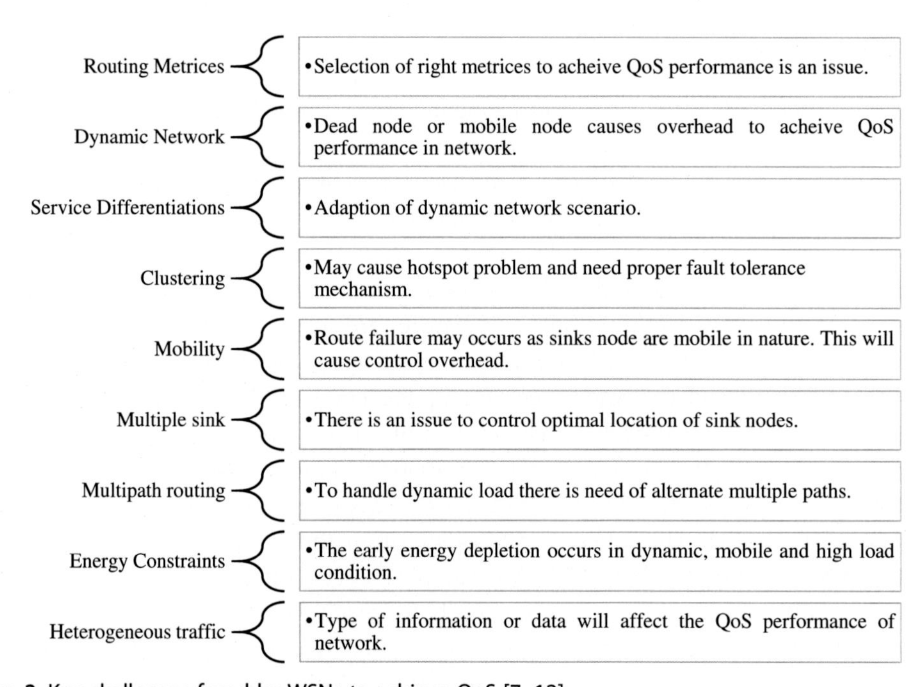

Fig. 2 Key challenges faced by WSNs to achieve QoS [7–12].

to achieve quality of service (QoS) performance such as energy efficiency, bandwidth usage, throughput, scalability, network longevity, etc. [3–6]. Offering QoS security in an environment that has limited resources, is, therefore, one of the vital difficulties tackled by either changing established WSN routing protocols or by imposing new QoS provisioning strategies (Fig. 2).

1.1 Problem definition

As WSNs are designed for the adoption of different surveillance and monitoring applications, there are many hardware and communications device characteristics that induce certain problems in providing QoS. The following are the most critical issues that impact QoS in WSN (Fig. 2).

A WSN forms a short-lived network that has no permanent infrastructure in which all nodes are free to travel and all nodes are configured [13–16]. In a WSN, even though the network topology can also shift rapidly, each node behaves similarly as a router as well as a host. Below, some of the algorithms are mentioned. The following are the classifications of routing protocols based on functions:

- Proactive: In a proactive routing protocol, a routing table from source to destination is maintained at each node that needs to be updated periodically. Optimized Link State Routing (OLSR) Protocol is a popular proactive routing protocol utilized in WSNs. Using selected nodes known as multipoint relays (MRPs), this protocol allows for the effective flooding of control messages throughout the network. The MPR nodes solve the problem of control message flooding in the network. Each node selects MRPs and serves to transmit control messages, resulting in a distributed protocol operation. In addition, OLSR is proactive and suitable for communications networks where paths are continuously changing. This approach is better for traffic conditions in which a large group of nodes communicates with some other large group of nodes, and the source and destination couples change over time. This technique is suitable for systems that are vast and congested.
- Reactive: There is no preservation of the routing table in the reactive routing protocol. In such algorithms, routes are dynamically selected whenever needed. Ad-hoc On-demand Distance Vector (ADOV) is a type of reactive routing protocol. The method is available in attempts to discover paths, which means that only when a source node requires data packets to be transmitted is a route determined. Route discovery and route maintenance are the two fundamental functions of AODV. To discover and manage routes, AODV uses route reply (RREP), route error (RERR), and route request (RREQ) messages. When a source node needs a route to a destination node for which it does not have one, it broadcasts an RREQ packet across the network in route discovery. The destination IP address, source sequence number, source IP address, request ID, hop count, and destination sequence number are all included in an RREQ packet. When a node gets a request packet with the identical source address and request ID fields as previous route request packets, the packet is rejected. When a link breakage in an active route is discovered during route maintenance, the node alerts the source node by sending an RERR message. If it still has data to send, the source node will reboot the route discovery process.

- Hierarchical protocol: Such a protocol combines the proactive and reactive protocols. A cluster-based routing protocol is a type of hierarchical protocol. This routing protocol categorizes the various nodes as a cluster. The node with high residual energy is selected as the head node for the entire cluster. The clustering algorithms are more energy efficient compared to proactive and reactive protocols. But they also face some problems that are mentioned below:
  - **i.** Consumption of more energy by CHs as they are involved in tasks such as intracluster communication, data aggregation to a base station or other CHs, etc.
  - **ii.** The energy efficiency of the entire network is reduced as some CHs consume more energy due to regular data forwarding.
  - **iii.** There is regular updating regarding CH selection, which also consumes energy.

1.2 Scope and motivation

Clustering has proven to be one of the most powerful methods for increasing network scalability and designing a WSN routing protocol that is energy efficient. Many sensors can even save energy by modifying the sampling rate of each node. For topology modulation, coverage preservation, or localization, these methods manipulate sampling rates using mathematical optimization models or heuristic models. Moreover, model-based WSN management has some drawbacks, as mentioned herein. The routing algorithm is designed to follow various QoS criteria to provide better efficiency and to increase the lifetime of WSNS challenges and problems considered by the network. This chapter, therefore, discusses the recently proposed routing protocols for cluster-based HWSN. With recent developments in machine learning [17–19], several algorithms are developed that provide application-specific assurance for QoS that is mainly focused on optimization algorithms. There are many existing optimization algorithms such as particle swarm optimization (PSO), genetic algorithm (GA), grey wolf optimization (GWO), etc., to provide QoS in heterogeneous WSN application areas. This is because the basic objective of a WSN is to give versatility and strength to network failure, dynamic network topology, and the conditions of the variable channel in the WSN. Thus, computational intelligence (machine learning and bio-inspired optimization)-based routing protocols are analyzed along with their performance and limitations in this chapter.

1.3 Key contributions

This section highlights the key contributions of the chapter:

- The major portion of this chapter presents a state-of-the-art analysis of existing cluster-based routing protocols for WSNs along with applications and challenges. This chapter also provides an overview of the heterogeneity of WSNs and clustering algorithms used with their advantages as well as disadvantages.

- This chapter also presents a literature review on the application of bio-inspired optimization algorithms as well as machine learning for enhancement of network lifetime.
- This chapter also proposes a machine learning-based framework with a bio-inspired algorithm to predict the lifetime of the network and to find the best optimal path in multihop as well as multipath mobile and heterogeneous scenarios. This chapter also provides the theoretical benefits of the proposed algorithm.

1.4 Chapter organization

In this chapter, Section 2 provides an overview of heterogeneous WSNs with different heterogeneity conditions. Section 3 gives an overview of cluster-based routing protocols and reviews some of the advancements in recent years for the energy efficiency of cluster-based routing protocols. Section 4 highlights the contributions of researchers for homogeneous as well as heterogeneous cluster-based routing protocols with the application of machine learning. Section 5 provides an analytical overview of the implementation of optimization or the evolutionary approach in finding the optimal path for data forwarding in cluster-based WSNs. After observation of the issues faced during the implementation of the heterogeneous cluster-based routing protocol by researchers, Section 6 presents a system architecture using optimization and machine learning techniques to handle QoS performance. Section 7 highlights the theoretical benefits of the proposed methodology compared with existing algorithms. Section 8 presents the challenges. And finally, Section 9 and 10 presents conclusion and future scope.

2. Overview of heterogeneous WSN (HWSN)

In most of the routing algorithms for WSNs, researchers assume the homogeneous scenario with respect to resources and capabilities. However, in practical or real-time scenarios, there is no homogeneous scenario, as at any moment a new node can get added to the existing WSN that contains a higher energy level compared to other existing nodes. Heterogeneity helps the protocol find the optimal path for data communication with improvements in network lifetime and performance. WSN heterogeneity is generally broken down into three categories, as illustrated in Fig. 3.

Computational heterogeneity: (i) The heterogeneous node has a more efficient microprocessor and more memory than its usual node, which means machine heterogeneity. (ii) The heterogeneous nodes allow complex data processing and long-term storage with their powerful computational resources. In certain sensing areas, the communication environment is unfavorable due to obstacles, so that it is sometimes appropriate to deploy the nodes with different communication ranges. (iii) Effectiveness for computation and storage. A sensor node is very constrained in its capacity for computation and storage. In certain protocols, nodes must serve as aggregation and relief nodes

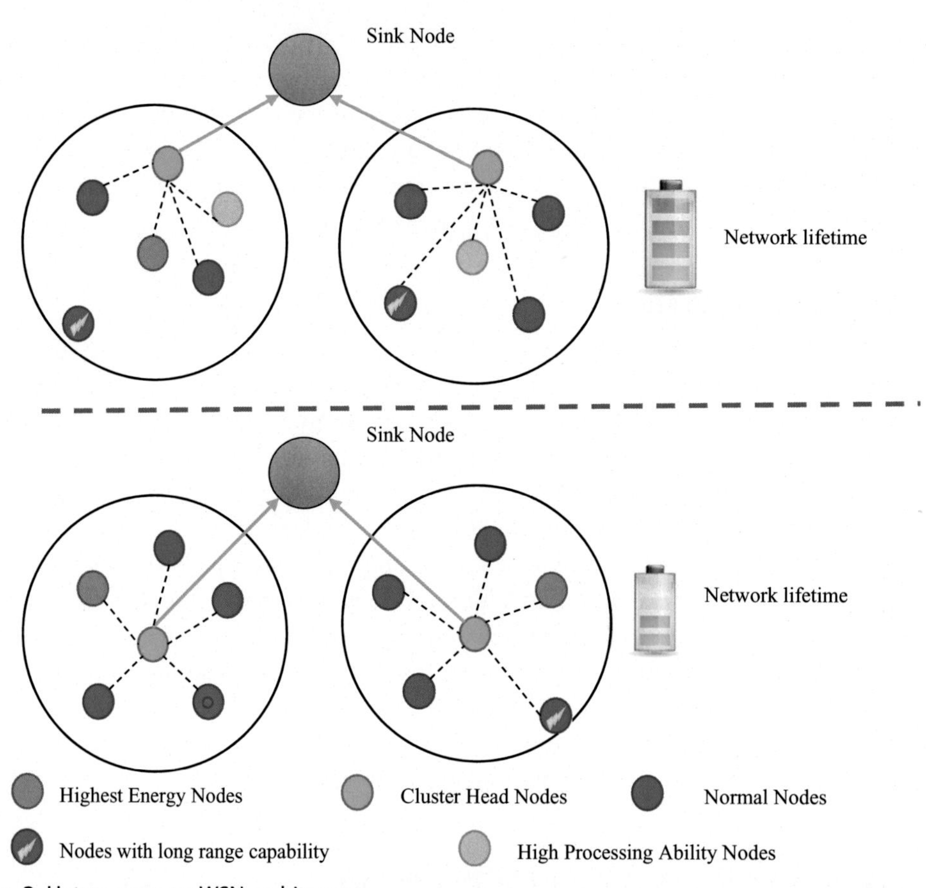

Fig. 3 Heterogeneous WSN architecture.

periodically. To satisfy this requirement, these nodes must have greater computational and storage capacity than the other nodes.

Energy heterogeneity: This term is associated with different energy levels of sensor nodes in WSNs. The WSN becomes heterogeneous when all nodes in the entire network have different energy levels. Most of the research works for HSWN are based on energy heterogeneity. When the nodes are first used on the network, they have different energy levels compared to each other. Due to a restricted power resource and several sensor nodes, battery change is a very difficult task for the nodes, and in some specific scenarios, it is often impossible. Then we use several nodes with more network energy to serve as data aggregation, processing, and transmission centers to balance energy dissipation in the entire network.

Link heterogeneity: Link heterogeneity is a high bandwidth and long-distance transceiver for the heterogeneous node as usual. More reliable data transmission can be made possible with connection heterogeneity. In such heterogeneity, some nodes have long-range communication capabilities.

Some researchers have suggested WSNs with heterogeneity, with the implementation of optimization algorithms intending to increase the network longevity under dynamic network scenarios. In principle, heterogeneous WSNs can be categorized into those based on different cluster sizes and those based on different energy levels.

3. Overview of cluster-based routing protocols

The clustering-based method was selected among the routing methods provided in the study because it is the most appropriate for WSNs. Different clusters, either equal or unequal, are formed in cluster-based routing protocols of WSNs. Each cluster is centrally coordinated by a head node termed the cluster head (CH). The intercluster communication is coordinated through CH nodes. The nodes of a cluster in this organization transmit their information immediately to their CH nodes. The CH aggregates data received from members of its cluster to transmit them to the sink straight or probably or by another cluster head [12]. The advantages of clustering a WSN are:

- It assures cluster head level information collection that can minimize energy utilization by removing unnecessary information.
- Routing can be conveniently handled because only specific nodes such as the cluster head require retaining the other regional routing structure of a cluster head, therefore requiring small routing details. Besides, this would immensely enhance the scalability of the network.
- It also preserves the bandwidth of connectivity because sensor nodes only interact with their corresponding cluster heads, hence preventing redundant information from being exchanged among them.

3.1 Classification of cluster-based routing protocols

The cluster-based routing protocol is energy efficient compared to other routing protocols in which nodes with high energy levels are selected as the head node (communicating node) and others are selected as sensing nodes that send their sensed data to the head node. These properties enhance the lifetime, stability, and robustness of the entire network. So, to achieve specific goals, cluster-based routing protocols are categorized into three categories as block, grid, and chain cluster-based protocols, as illustrated in Fig. 4. Some of the block-based cluster routing protocols are LEACH, HEED, UCS, TEEN, etc. Similarly, TTDD is a type of grid-based routing protocol and PEGASIS is a type of chain-based routing protocol. They are described in further sections. Along with that, a separate comparative analysis of these protocols is given in Table 1 with their advantages and disadvantages.

3.1.1 Low-energy adaptive clustering hierarchy (LEACH)

The LEACH protocol is a type of block-based cluster routing protocol. In LEACH, every node has the same likelihood of being a cluster head. The LEACH protocol is based

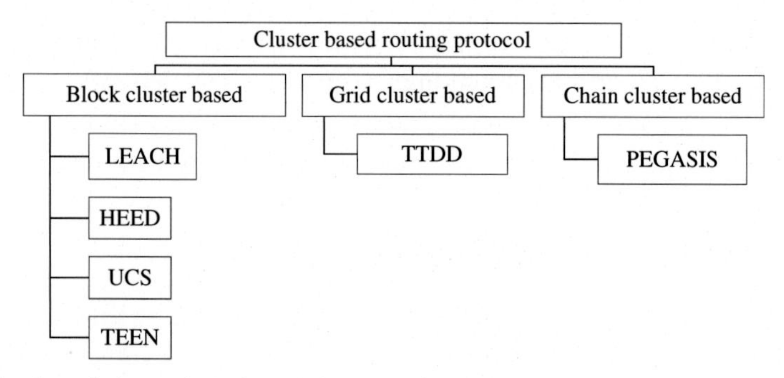

Fig. 4 Classification of cluster-based routing protocols in WSNs.

on TDMA whose aim is to improve the lifespan of clusters. For this, the algorithm is divided into two stages: the setup phase and the steady Phase. In the setup phase, the cluster is formed and in each cluster, the CH is selected. Whereas in the steady phase, data from normal nodes are aggregated at the CH and transmitted to the base station. In this, TDMA is used to protect cluster heads from unwanted collisions. On the basis of probability, the cluster head is selected. The working steps of the LEACH protocol is as follows (Fig. 5A):

```
Start
For rounds (R)= 1: max(R)
Apply Random distribution probability function Nodes --1/max(R)--> CH
Each node has probability of 1/max(R) to become CH.
If Node is not selected as CH
Join Nearest CH
CH Allot time schedule to each node inside cluster
Nodes --Communication_TDMA--> CH
End if
End for
end
```

3.1.2 Hybrid energy-efficient distributed clustering (HEED)

It is a routing mechanism that is completely decentralized. It attains load balancing and consistent dispensation of the cluster head. This accomplishes increased energy conservation and scalability by multihop interaction. This also results in inadequate energy utilization due to several cluster heads. The objective of the HEED algorithm is to enhance the network lifetime and the energy efficiency that is dependent on the CH selection process. HEED considers intercluster communication power and residual energy to decide the CH. The working process is almost similar to LEACH. The difference is that in LEACH, there may be a tie among nodes as the CH for the same cluster. So, to break

Table 1 Summary of cluster-based routing protocols.

Protocols		Advantages	Disadvantages	Energy efficiency	Stability	Complexity
Block	LEACH	Each node has chance to become CH Removes unnecessary collision due to application of TDMA	Not applicable for large-scale network	Low	Average	Low
	HEED	Achieves better load balancing and scalability	Overhead due to number of rounds	Average	High	Average
	UCS	Variable number of nodes in a cluster	Not suitable for large-scale network	Low	High	Average
	TEEN	Controlled data transmission Good for time-constrained applications	Data loss due to inconvenience in communication of CH	High	High	High
Grid	TTDD	Removes problem related with mobility	Latency is large	Low	High	Low
Chain	PEGASIS	Data redundancy is removed. Uniform distribution	Doesn't have dynamic network scenario and is not scalable Bottleneck issue	Low	Low	High

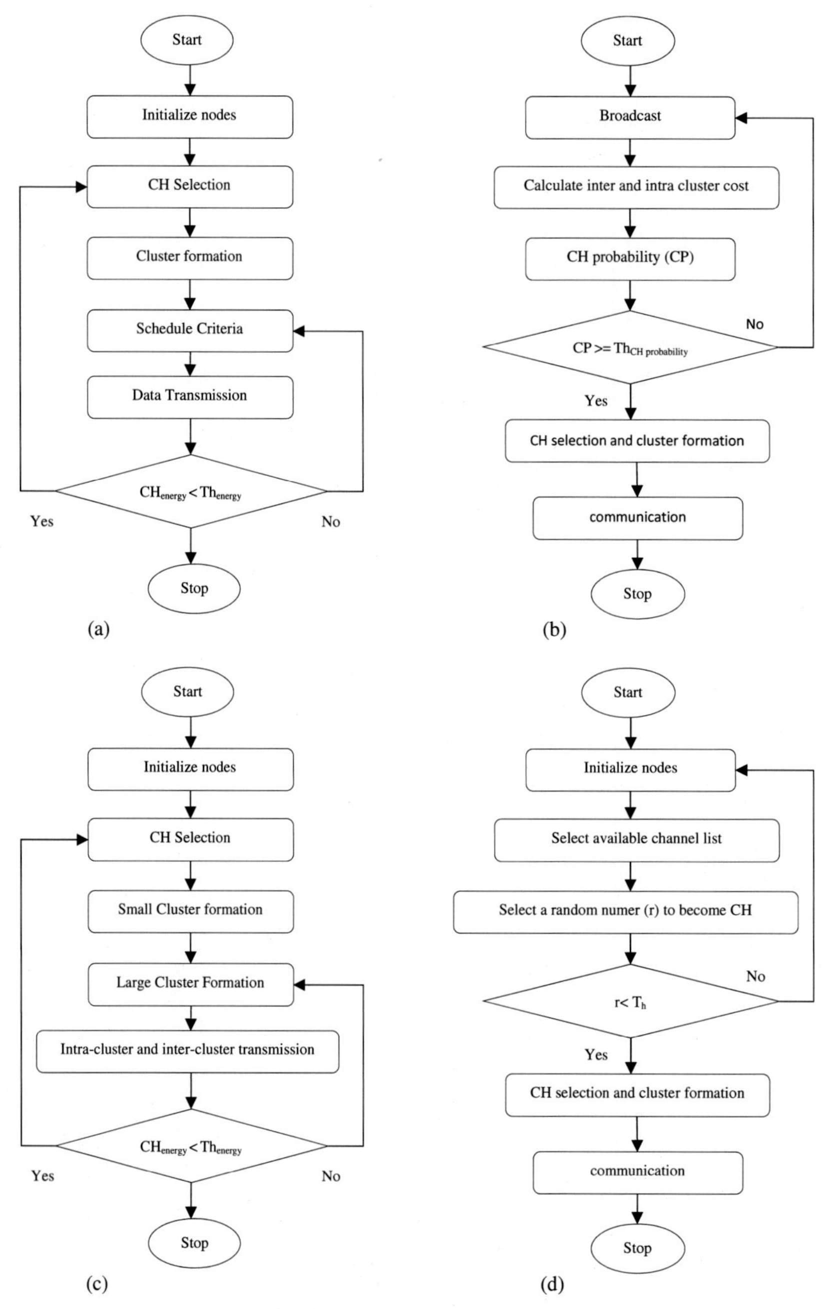

Fig. 5 Working process of cluster-based protocols. (A) LEACH protocol flow chart, (B) HEED protocol flow chart, (C) UCS protocol flow chart, (D) TEEN protocol flow chart,

(Continued)

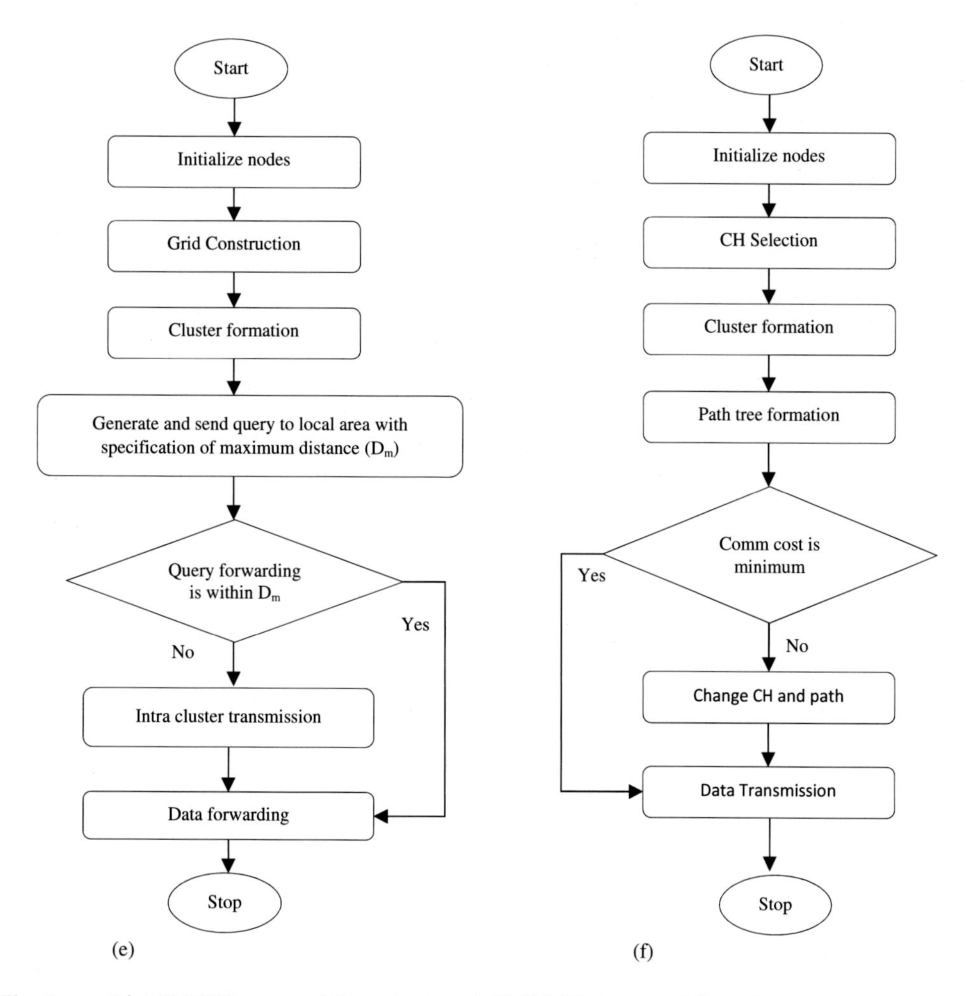

Fig. 5, cont'd (E) TTDD protocol flow chart, and (F) PEGASIS protocol flow chart.

the tie, the HEED algorithm considers the communication power to decide one single CH inside a cluster. A set of discrete power levels of SNs is decided and the radius of the cluster is determined. The working steps of HEED are as follows (Fig. 5B):

```
Start
For rounds (R)= 1: max(R)
Apply Random distribution probability function (Nodes --1/max(R)--> CH)·
Each node has probability of 1/max(R) to become CH.
If Node is not selected as CH
Join Nearest CH
```

```
elseif
CH_cluster > 1
Select CH == CH_max_comm_power
end elseif
CH Allot time schedule to each node inside cluster
          Communication_TDMA
Nodes ──────────────────────→ CH
End if
End for
end
```

3.1.3 Unequal clustering (UCS)

Nodes in the cluster could be a variable. It is also a bilayered structure with two-hop connectivity between clusters. The remaining node energy is not taken into account and is not adequate for large-scale networks. The objectives for clustering any network are dependent on different requirements. Hot spot elimination and energy saving are two main reasons to choose over unequal clustering algorithms. Some other objectives that need to be fulfilled include scalability, fault tolerance, stability, and load balancing. The UCS method uses probability-based cluster formation of different unequal sizes. Small clusters are selected near the base station, whereas large clusters are formed away from the base station. Near the base station, the CH focuses on intercluster communication, whereas far away, the CH focuses on intracluster communication. This results in elimination of hot spot problems. The flowchart of the UCS is illustrated in Fig. 5C.

3.1.4 Threshold sensitive energy-efficient sensor network protocol (TEEN)

In this, data transmission could be regulated by a variation of two thresholds. It is best suited for time-critical implementation. The node would not interact when thresholds are not accomplished. If cluster heads do not interact with one another, information might be lost. TEEN is a type of reactive protocol advanced with the LEACH protocol. In this algorithm, the CH collects data inside the cluster, then aggregates them and transmits to the base station. In TEEN, all nodes get a turn to become a CH, selected randomly, for even distribution of energy. The working steps of the TEEN algorithm are as follows (Fig. 5D):

```
Start
For rounds (n)= 1: max(n)
Node_i senses available channel list
Assign random number (r) to Node_i {0 ≤ r ≤ 1}
Evaluate threshold (T_h) such that:
```

$$T_h = \begin{cases} \dfrac{P}{1 - P(n \bmod (1/P))} & \text{for set of nodes} \\ 0 & \text{otherwise} \end{cases}$$

```
Where P= Percentage of becoming CH
```

```
If r< T_h
Node_i decides to be CH
end
CH broadcast to all Nodes
Node joins to nearest CH
CH Allot time schedule to each node inside cluster
Nodes --Communication_TDMA--> CH
End for
end
```

3.1.5 Two-tier data dissemination (TTDD)

This fixes various mobile sinks and traveling issues of sinks in enormous scale in WSNs. It is ideal for the identification of WSNs in abnormal data traffic events. It has large delays and low efficiency in terms of energy. In this, sensor nodes are used to be stationary and to be aware of the location. TTDD is based on grid topology to provide network scalability and efficiency. In this algorithm, whenever a sink node needs data, it generates a query message within its local area with specification of the maximum distance. This reduces the flooding from those nodes that are far away from the sink node or less than the predetermined maximum distance. Whenever the message reaches the nodes that are in range of the sink (called intermediate nodes), then it is forwarded to the upstream dissemination node. Then, these upstream nodes transmit to the source node for data transmission. The flowchart of the working principle of TTDD is illustrated in Fig. 5E.

3.1.6 Power-efficient gathering in sensor information systems (PEGASIS)

In this, the energy load is evenly distributed. Due to dynamic cluster design, it decreases the overhead. A node becomes a bottleneck during this time lag and the network is not very scalable. The PEGASIS algorithm is operated in two main steps (Fig. 5F):

Step 1: Tree construction

The path from the sensor node is selected on the basis of minimum hop count.

Step 2: Fine tuning

Nodes change their path to the base station as the overall cost (energy) is minimum while communicating data from source to destination.

Decisions at both steps are taken on the basis of the status of neighbor nodes.

4. Literature review

4.1 Homogeneous cluster-based routing protocols

Some of the contributions of researchers for designing cluster-based routing protocol are presented below:

The first cluster-based routing protocol was LEACH, which is hierarchical. A significant number of suggestions for energy-efficient clustering emerged after LEACH. Mahmood et al. [20] proposed a modified LEACH termed MODLEACH. The main principle of this algorithm was that if the residual energy of CH is greater than a predefined threshold, the cluster would not be altered in the upcoming stage so that the power consumption in the construction of the cluster can be saved. The experimental result presented by Mahmood et al. [20] shows the overall lifetime of the entire network was approximately 2500 rounds.

Barfunga et al. [21] proposed a comparable method of selecting and preserving cluster heads, but the coordination between the cluster head and the base system is not taken into account by all [22]. In general, the major energy use in traditional heterogeneous networks arises from the transmission processes between the cluster head and the base system. The model presented by Barfunga et al. [21] was termed the enhanced LEACH-C protocol. It was implemented using variable rounds for multihop communication. The result analysis showed the efficacy of the model in terms of energy efficiency.

Fan et al. [22] proposed a multihop LEACH in which each cluster chooses the closest cluster head in a single hop range. This has enhanced energy conservation as multihop routing requires less power than transmitting information straightforwardly to the base station, particularly in broad sensor networks. However, the key downside of this algorithm and other approaches [23,24] that optimally merge clustering and multihop communication was that cluster heads possess higher routing traffic around the sink and those nodes would die quickly, thereby reducing the lifespan of the networking. In this method, the WSN combines the clustering and multihop communication methods with nodes near and far away from the sink. The network performance is increased compared to the clustering algorithm and multihop methods.

Several approaches intend to increase the duration before the first node runs out of power, in addition to the attempt to minimize power utilization [25,26]. The issue with these max-min lifespan routing protocols is that, based on the entire network size, they do not conserve resources. Conserving energy for the entire network is much more essential in an enormous sensor network than saving energy for particular nodes. After analyzing these issues, Wang et al. [27] proposed a clustering algorithm based on fusion methodology and termed the algorithm NEAHC to enhance network lifespan. The idea is to find the optimized route from the origin to the endpoint by supporting the maximum available battery capacity, minimal multihop energy usage, and adequate sensor node equality. Simulation findings reveal that this approach doesn't just manage the total network energy usage and extend the node failure time, but also ensures more efficient data transmission.

Ullah et al. [28] proposed a clustering technique to transmit data that would effectively reduce redundant and erroneous data by applying an extreme learning machine

(ELM). The data are filtered using the Kalman filter before transmission to the CH. The result analysis was evaluated in terms of accuracy compared to other clustering algorithms.

A modified clustering methodology was proposed by Radhika et al. [29] that reduces the overhead that occurs during cluster formation and data forwarding. This work effectively scheduled the task of clustering by considering the resources of alive sensor nodes. Reclustering nodes is done by measuring the upgrade period using an inference method to achieve minimum energy consumption.

Manzoor et al. [30] designed a two-level hierarchical architecture that showed an improvement over the LEACH protocol; they called it the two-level hierarchy of low energy adaptive hierarchy (TL-LEACH). This algorithm showed its efficiency in terms of end-to-end data transmission and showed improvement over LEACH. The drawbacks of this algorithm are its central architecture and decreased robustness in large-scale WSNs. The energy efficiency of TL-LEACH was improved by the extended TL-LEACH (ETL-LEACH) for large-scale WSNs.

Madhumathy et al. [31] suggested a routing protocol based on the agent cluster that would subdivide the cluster according to the satisfaction list into separate subgroups. A node agent node may talk to the cluster's head is associated with the independent subgroups. To reduce energy consumption during data transmission from the node to the cluster head, the proposed agent cluster-based routing algorithm was developed. The node of the agent is chosen based on a satisfaction list that decreases the chances of the agent node being more confident and durable.

The current LEACH clustering protocol has been updated by Behera et al. [32] by adding a threshold limit for the selection of cluster heads while switching the power level between the nodes at the same time. The proposed improved LEACH protocol increases the number of live nodes to 1750 rounds, which can increase the lifespan of a WSN. It was shown that, compared to other energy-efficient protocols, the proposed algorithm improves stability time and network life in various scenarios of location, energy, and node density.

Razzaq et al. [33] proposed a K-means cluster-based routing protocol and found the ideal fixed packet size based on the radio parameters and channel conditions of the transceiver. This approach decreases the energy consumption of the individual node and maximizes the whole network life.

Ahmad et al. [34] proposed the location aware routing protocol for a cluster having a multihop nature. The clusters formed in this algorithm use a self-organized algorithm in which every node is selected to be the cluster node. The direct interaction between the chosen CH at the sink node can be prevented to save energy and minimize the impact of overhearing transmissions. A communication multihop system, which takes position-based routing, is then introduced. Using a wireless simulation device, the proposed solution was developed and tested.

Abidoye et al. [35] discussed that in each iteration or round, a new CH is chosen according to the LEACH protocol, which periodically involves the creation of new clusters. Due to the routing overhead, this may lead to inefficient use of resources that cannot be admitted by Internet of Things (IoT) devices. In the previous round, the CH elected not to have a sufficient energy level to communicate among nodes and there is a reasonable probability that in the next round, the CH selected must have a good energy level. Therefore, a CH replacement algorithm was used to reduce the energy overhead during cluster creation and data advertisement for CH node selection process.

Xiong et al. [36] proposed an improvement over LEACH protocol by changing the primitive threshold value and termed it the coverage-preserving CH selection algorithm (CPCHSA). The network coverage sensing capability was optimized. The major drawback of this system was that there is no fixed CH generation in each round, that is, every round generates different clusters.

Peng et al. [37], based on LEACH, proposed an improved LEACH-B protocol. First, the Balanced Iterative Reducing and Clustering using Hierarchies (BIRCH) algorithm's great convergence and global optimization capabilities will reasonably divide the entire area of the network into a multitude of subregions. Then, in the subregion, select the cluster leader, taking into account the residual energy factor.

A comparative analysis of some existing homogeneous protocols is presented in Table 2.

Table 2 Comparative analysis of cluster based routing protocol in homogeneous WSN.

Ref	Protocol	WSN	Key features	Network Lifetime
[20]	MODLEACH	Homogeneous	LEACH protocol is the base algorithm. Dual transmission power was observed after CH reselection process	~2500 rounds
[26]	Nonlinear programming	Homogeneous	Optimal function was evaluated on convex function to choose relay nodes	~2500 rounds
[28]	Extreme learning machine	Homogeneous	ELM was used to remove redundant data	–
[29]	Fuzzy logic and machine learning	Homogeneous	ML classifies the data on each node and reduces the redundant data transfer rate. Fuzzy logic was used to recluster the network	~1500 rounds
[32]	Improved LEACH	Homogeneous	The LEACH protocol was improved for IoT scenario	~1750 rounds
[34]	Position aware clustering	Homogeneous	Self-organized clusters are formed in this protocol	>4000 rounds

4.2 Heterogeneous cluster-based routing protocols

Some of the typical contributions of researchers for cluster-based routing protocols for HSWN are presented in this section.

Manchanda et al. [38] proposed a novel framework for heterogeneous WSNs that are energy efficient in nature. This proposed methodology consists of four basic steps: selection of CH on the basis of energy remaining, data forwarding from nodes present inside the cluster to the CH, aggregation of data (compression) to transmit to the sink node, and decompression of data at the sink node. A simulation was performed on MATLAB and analyzed for different network scenarios; it enhances the network longevity and stability. Similarly, different challenges faced while implementing HWSN were highlighted by Gupta et al. [39].

Guezouli et al. [40] implemented a controlled mobility technique utilizing unmanned aerial vehicles (UAVs) for heterogeneous WSNs. Thus, according to observations, the suggested deterministic and genetic approaches should deal effectively with the difficulty of mobile heterogeneous WSNs, particularly in comparison with the earlier implemented random approaches. The outcomes acquired showed that an increased transmission rate, a faster coverage time, and a faster lag can be achieved by our proposed technologies.

Dawood et al. [41] used it for many purposes such as temperature control, humidity tracking, etc. Sensor nodes are composed of a nonreplaceable battery, restricted memory, and processing abilities. The life expectancy of the sensor node and network, therefore, relies on the use of power. Due to the unsuitability of the region of deployment for manual mediation, energy investment is essential for improving the lifespan of the network. Several researchers have suggested various routing protocols for utilizing energy efficiently. Clustering is among the several essential strategies for condensing the power utilization of the network. The essential energy conservation method for clustering algorithms is the selection of the cluster head. Because excessive energy is consumed throughout data transmission, therefore coordination within the cluster is extremely important. The connection gap is important between the cluster head and the member node. A large amount of energy would be acquired by a node with a large contact gap inside the cluster. To conserve energy in the network and enhance the lifespan of the network, the suggested technique decreases the contact gap among the clusters.

Cao et al. [42] studied the information fusion method of portable heterogeneous WSNs and considered the WSN nodes as neurons to train the extreme learning machine. The ELM collects the sensory information obtained by the mobile HWSN, trains the network with this collected information, and discover the clustering route for data transmission over the network. The ELM network is optimized by applying a bat algorithm. The optimization algorithm will optimize the learning parameters of the ELM to fast convergence toward the optimal result. This proposed methodology was compared with a conventional SEP algorithm in terms of network lifetime and energy efficiency. The simulation outcomes demonstrate that the suggested BAT-ELM could efficiently decrease network traffic, conserve energy, enhance work productivity, and dramatically extend the lifespan of the network in comparison to the conventional SEP technique, the BP neural network technique, and the ELM technique.

Leu et al. [43] proposed a cooperative routing protocol for HWSN that is energy efficient. The heterogeneous sensor network is created by many WSNs installed in a similar geographic location, and the sensors transmit messages for their own WSNs as well as for other WSNs. Networking paths are statically constructed in accordance with the methods used for transferring activity data, the remaining power of embedded sensors, and nearby neighbors. In contrast, data from sensors that are substantially identical route are combined to save energy.

Priyadarshi et al. [44] discussed the limiting factor for WSNs, which has always been the small battery of the sensors. The network could be subdivided into clusters to increase network life for a prolonged time. A three-level heterogeneous clustering protocol is suggested in this study to improve the network route. The three heterogeneity levels break the sensors into three distinct classes utilizing their resources. Through using the tipping point and energy variables, the suggested technique selects the most capable nodes as the cluster head. Skilled selection of the CH causes an increase in the efficiency and functionality of the entire network route. Utilizing the MATLAB simulator, the suggested method is simulated and the outcomes are compared with traditional methods to demonstrate the benefits of the suggested method over other technologies.

Gomathy et al. [45] proposed a heterogeneous cluster-based secure WSN. In this algorithm, security is also considered along with energy efficiency. The network has identified the wormhole and black hole attacks in WSNs. The simulation result shows an improvement of 10% in energy efficiency whereas malicious node detection was improved by 96%.

Zhao et al. [46] proposed a routing protocol based on an improved grey wolf optimizer for HWSNs. In this work, the protocol defined different fitness functions used to evaluate the nodes' fitness values for optimal CH selection. The result analysis was performed in terms of network longevity and it was observed that the network was alive for approximately 2500 rounds.

Yazid et al. [47] proposed routing protocols relying on a clustering methodology. The primary concentration was done on the powerful clustered routing protocols devoted to heterogeneous WSNs, such as the secure election protocol (SEP) and its new routing algorithms. The productivity of these methods has also been evaluated based on different constraints, including energy conservation, network lifespan performance, and reliability. Especially in comparison to other SEP routing methods, the simulation outcomes demonstrate that the threshold of the SEP clustered routing protocol gives better output in terms of power, network network, and power consumption.

Verma et al. [48] proposed a routing protocol termed the improved dual hop routing technique (IDHR) and multiple data sink energy-efficient cluster-based routing protocol (MEEC). In IDHR and MEEC, the CH choice is accomplished by integrating variables

Table 3 Comparative analysis of cluster-based routing protocol in heterogeneous WSNs.

Ref	Protocol	WSN	Key features	Network Lifetime
[42]	ELM optimized with bat optimization	Heterogeneous	Redundant data removed using ELM	–
[43]	Co-operative cluster-based routing	Heterogeneous	Dynamic selection of route for communication	~2000 rounds
[44]	Three-level clustering protocol	Heterogeneous	Three diverse groups are created on the basis of energy	~6000 rounds
[46]	Modified GWO	Heterogeneous	Optimal CH selection using multiple objective functions	~2500 rounds
[49]	Energy-coverage ratio clustering protocol	Heterogeneous	Optimal number of clusters selected on the basis of minimum energy consumption and region maximization	~8500 rounds

of node density combined with additional variables, including power and distance among them. Whereas in MEEC, the multiple data sink scenario was created to interpret nodes and information associated with them for data propagation. The node density factor appears to be loyal to the power conservation of nodes by lowering the total contact distance between the nodes and the associated CH.

Zeng et al. [49] suggested an energy ratio clustering protocol (*E*-CRCP) associated with the reduction of the power utilization system and the use of the local covering ratio. Initially, the configuration of the energy model was done. Thereafter, the optimum number of clusters was computed based on the minimum energy utilization theory and the choice of the cluster head was done based on the maximization of local exposure theory. To evenly distribute the network load as much as feasible, the CH with the minimal remaining energy and the maximum power utilization was substituted during the succeeding iteration of CH choice to extend the lifetime of the network.

A comparative analysis of some existing heterogeneous protocols is presented in Table 3.

5. Application of optimization for energy efficiency in WSN

In WSNs, all sensor nodes are deployed statically over a large area and communicate through wireless channels in a cooperative and sharing way. They also have the capability to communicate with each other in a mobile condition. The sensors' function is to sense

Table 4 Comparative analysis of optimization algorithms for energy efficiency.

Ref	Optimization	Compared with	1	2	3	4	5
[46]	Modified GWO	SEP, DEEC, GWO	√	√	–	–	√
[54]	TPSO-CR	LEACH, LEACH-C, PSO, GA	√	√	–	–	√
[55]	GASONeC	With increasing nodes in specific area	√	–	–	–	√
[56]	GWO	GA, PSO, multiobjective fuzzy clustering	√	–	–	–	√
[57]	ALO	LEACH, iLEACH, E-OEERP	√	–	√	√	√
[58]	GA-ACO	LEACH, SEP	√	√	–	–	√
[59]	ACO	PSO, BCO	–	√	–	–	–
[60]	GAPSO-H	GA, LEACH, PSO	√	–	–	–	√
[61]	MCH-EOR	GWO, ALO, GA, PSO	√	–	–	√	√
[27]	ABC	LEACH-C, GWO, PSO	√	–	–	–	√

1, Homogeneous scenario; 2, heterogeneous scenario; 3, mobility support; 4, multipath support; 5, energy efficient.

the environment, transmit to the base station, and sometimes analyze the data. These computational capabilities make them a better choice in applications such as environmental monitoring, military surveillance, patient surveillance, traffic surveillance, etc. Apart from applications, some factors that need to be focused on include energy efficiency and minimum resource utilization, network coverage, low latency, load balancing, etc. Due to these issues, it is a quite challenging task to choose an optimal routing protocol. For this, clustering techniques are used in WSNs that can somehow resolve these issues. As it is known that the clustering technique distributes the entire network in small clusters in every cluster, a CH node is elected [50–53]. All available sensor nodes in a cluster forward their data to the CH, which transmit that data to the base station (BS). As the CH only transmit data to the BS, this avoids collision issue that arise due to communication channel sharing among sensor nodes. Then, in clustering protocols, another issue that arises is how to make a cluster and who to elect a cluster. In traditional clustering algorithms such as LEACH, SEP, HEED, etc., there is a need for regular updating of CH nodes, as their CH selection is based on energy. To avoid such conditions, researchers have been recently implementing bio-inspired algorithms such as particle swarm optimization, genetic algorithm, grey wolf optimization, ant colony optimization, etc. To analyze the efficiency of optimization or evolutionary approaches in WSNs, some existing algorithm contributions are shown in Table 4.

6. Evolutional optimization cluster-based routing protocol (EOCRP)

6.1 System model

Regardless of great achievements in WSNs, there are still some challenging issues related to energy efficiency. To eliminate the energy efficiency problems under cluster-based heterogeneous WSNs, this chapter proposed the bio-inspired optimization algorithm as a solution that can be easily integrated with technologies such as IoT, edge computing, etc. The present research considers randomly deploying *n* random sensor nodes in any area with a dynamic size for WSN. The objective of this chapter is to propose a methodology that can predict the lifetime of the network and design an optimized cluster-based routing protocol to enhance QoS performance.

The assumptions for the network scenario are as follows:

- The heterogeneous network contains sensor nodes that have different initial energies.
- The sensor nodes deployed are mobile.
- The processing scenario and sensor parameters are considered to be the same for all nodes.
- Bidirectional transmission.

6.2 Architecture

This work is performed in four stages:

(a) Cluster formation.
(b) Cluster head node selection on different objective functions such as distance, transmission range, remaining energy, etc., using multiobjective optimization algorithms.
(c) Prediction of the cluster as well as head node lifetime using a machine learning algorithm.
(d) Multipath data transmission.

To address the existing problems in heterogeneous WSNs, this chapter proposes a methodology to resolve the problem of network longevity under the limited energy of sensor nodes. The proposed system architecture is illustrated in Fig. 6. The architecture integrates the ensemble learning approach for lifetime prediction of the cluster as well as the head node. Along with that, swarm optimization is used to find the optimal cluster head node and evolutionary algorithms are used to make data forwarding decisions. All these modules are integrated to achieve QoS goals.

Algorithm: Proposed heterogeneous cluster-based routing protocol

Input: $\{S_1, S_2, \ldots S_n\}$, Sensor nodes, BS= Base station, O_f=Objective functions, Itr_{max} = Maximum Iteration

Output: Data transmission over best suitable paths selected from multipath

1: Begin
2: Randomly deploy $\{S_1, S_2, \ldots S_n\}$ in an random selected area.
3: Select 2 nodes randomly S_i and S_j
4: For S_i : S_j, do
5: Evaluate difference among nodes $Diff_n = S_i - S_j$, $Diff_{BS} = BS_n - S_n$
6: If $Diff_n <$ some pre-defined threshold
7: Predict S_n, BS_n lifetime in cluster using ensemble machine learning algorithm

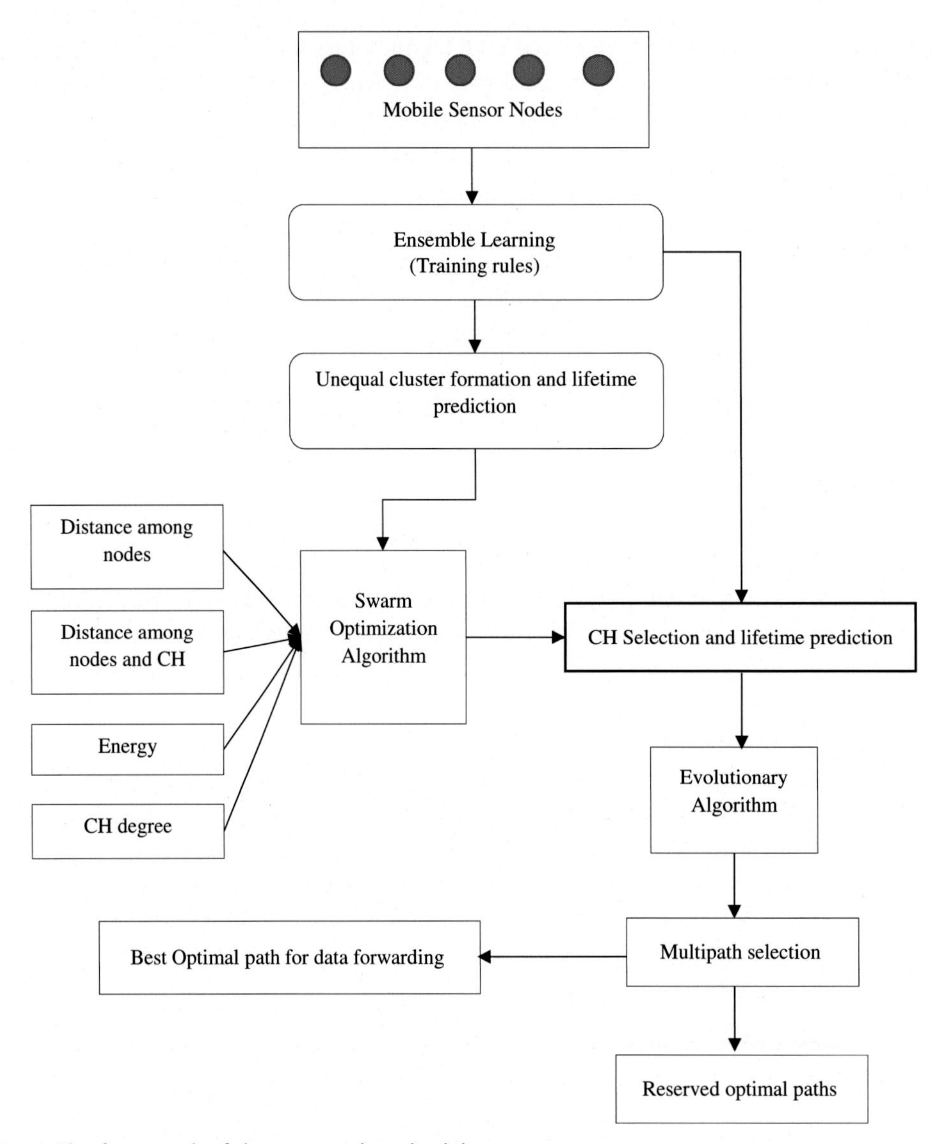

Fig. 6 The framework of the proposed methodology.

```
8:  End if
9:  End for
10: Discover neighbour nodes
11: Repeat step 4 to 10 until clusters are formed.
12: Set Objective functions O_f={f_1, f_2, ......f_n}
13: For 1 : Itr_max
14: Select CH node using swarm optimization algorithm
```

```
15: Best CH is selected and its lifetime is predicted using ensemble
    machine learning algorithm.
16: End for
17: For 1 : Itr_max
18: Generate multiple paths from source to destination using
    evolutionary algorithms
19: Return best optimal path
20: End for
21: Other paths are reserved for future data forwarding in condition of
    path failure
```

7. Theoretical benefits of proposed EOCRP

This methodology has some key points that can result in better QoS:

- The application of machine learning for network lifetime prediction will result in energy preservation. As energy is consumed while regularly updating the CH nodes, a prediction algorithm will find the optimal CH node as well as reduce unnecessary periodic routing updating.
- This framework proposes a multiobjective swarm optimization approach that finds optimal CH nodes. This optimally manages energy among the SN, CH, and BS.
- The evolutionary approach will find the best optimal path for data forwarding but at the same time, some alternate optimal paths can also be found and saved for future use if any unexpected network abnormalities occur such as node movement, dead node, malicious node, etc.

Table 5 illustrates the features included in previous research works to analyze the performance. The features and parameters included show the theoretical benefits of EOCRP with other existing algorithms such as optimization algorithm, application of machine learning, multiobjective functions, support mobility or multipath selection, intercluster communication, network longevity, heterogeneous scenario, and unequal clustering.

Table 5 Comparative features of proposed methodology with existing works.

Features	1	2	3	4	5	6	7	8	9	10
[57]	√	√	–	–	√	√	√	–	–	–
[61]	√	√	–	√	–	√	√	–	–	–
[27]	√	√	–	√	√	–	√	–	–	√
[62]	√	–	–	–	√	–	√	–	√	√
[63]	√	–	√	–	√	–	–	√	√	–
Proposed	√	√	√	√	√	√	√	√	√	√

1, Clustering; 2, optimization algorithm; 3, machine learning; 4, multiobjective; 5, multipath; 6, mobility; 7, intercluster communication; 8, network lifetime prediction; 9, heterogeneous WSN; 10, unequal cluster.

In Table 5, Feature 1 represents whether the routing protocol is a type of cluster-based protocol. Feature 2 represents the bio-inspired optimization algorithm used to find optimal results. Feature 3 represents the integration of the machine learning approach for prediction while routing to maximize the energy efficiency. Feature 4 represents whether the optimization algorithm is single objective or multiobjective in nature. Feature 5 represents whether multipath routing is supported. Feature 6 represents whether the proposed algorithm supports node mobility. Feature 7 represents whether the communication supports only intracluster or whether it also supports intercluster. Feature 8 represents the lifetime of the entire network, that is, its energy efficiency. Similarly, Feature 9 represents whether the network scenario supports heterogeneity. Feature 10 represents whether the algorithm forms an equal cluster size or whether it supports the formation of unequal clusters.

8. Challenges

8.1 Challenges faced in HWSN

In this chapter, different existing works were presented on heterogeneous cluster routing protocols using optimization and machine learning. These works generally focus on issues of heterogeneous scenarios of the WSN. But, in past research works, there was no profound description of cluster lifetime prediction in the dynamic network scenario. This section summarized techniques used for heterogeneity along with their drawbacks. This chapter also explored the performance parameters that can be focused on for network lifetime improvement in future research work. This section presented the challenges faced related to heterogeneous clustering. To better understand the challenges faced in existing works, Table 6 presents state-of-art reviews related to the heterogeneous cluster-based routing protocol. Similarly, Table 7 represents all possible parameters that need to be focused on for further research works.

Table 6 Basic concepts and limitations of existing algorithms.

Ref	Algorithm	Basic concept	Limitations
[58]	GA-ACO	For optimal data forwarding path, GA is combined with ACO	The network is stable but increases the algorithm complexity
[59]	ACO–K-mean	Modified k-mean algorithm was used to create cluster and ACO for best path selection for optimal transmission	Doesn't support mobility

Table 6 Basic concepts and limitations of existing algorithms—cont'd

Ref	Algorithm	Basic concept	Limitations
[60]	GAPSO-H	Energy heterogeneity is used to categorize sensor nodes. CH is selected using the GA algorithm	Doesn't support mobility and multiple sink node scenario
[61]	MCH-EOR	To resolve the hotspot problem, optimization is used. Sailfish optimization is used for CH selection	Doesn't support mobility for dense WSN
[62]	I-SEP	Improvement on SEP routing protocol for IoT application	Doesn't support mobility
[64]	SSEEP	State switchable routing protocol that is designed for a heterogeneous cluster	Suitable for large-scale network only
[65]	SEP	Improvement over SEP with greedy search routing	Doesn't support mobility
[66]	R-EEHC	CH is elected by finding the ratio of energy of the current and initial states	The network lifetime decreases steeply with increase in number of rounds
[67]	SCBC	Secondary cluster head nodes are selected and connected in a chain with other CHs to save energy	Simulation scenario is fixed with 100 nodes
[68]	FMCB-ER	Clusters are formed using a hierarchical grid. CH is selected using algorithm fuzzy logic and TOPSIS	Doesn't support mobility
[69]	HEESR	Closest node to CH joins the cluster. Path is selected based on advertisement for intercluster communication	Static CH and route selection. Energy hole problem
[70]	Optimized DEEC and LEECH-C	Mobile sink has to be in the center of all regions. Region distance is calculated for TDMA schedule	Need of centralized sink node to coordinate with mobile sink nodes
[63]	Neurofuzzy cluster selection	Distance, remaining energy, and degree of CH are considered to be deciding factors. Fuzzy rules are used to select CH. Neural network is used to decide the transmission path	All nodes are assumed to be trustful nodes

Table 7 Performance parameters for enhancement of lifetime of heterogeneous networks.

		Performance metrices									
Ref	**Algorithm**	**1**	**2**	**3**	**4**	**5**	**6**	**7**	**8**	**9**	**10**
[58]	GA-ACO	–	–	√	–	–	–	–	–	–	–
[59]	ACO–K-mean	–	–	–	–	–	–	–	–	–	√
[60]	GAPSO-H	–	√	√	√	√	–	–	–	–	–
[61]	MCH-EOR	√	√	√	√	–	–	–	–	√	–
[62]	I-SEP	√	√	√	–	–	–	–	–	–	–
[64]	SSEEP	–	–	√	√	–	–	–	–	–	–
[65]	SEP	–	–	√	√	–	–	–	–	–	–
[66]	R-EEHC	–	–	√	√	–	–	–	–	–	–
[67]	SCBC	–	–	√	√	√	–	–	–	–	–
[68]	FMCB-ER	–	√	√	–	–	√	√	√	√	–
[69]	HEESR	–	–	√	√	√	–	–	–	–	–
[70]	Optimized DEEC and LEECH-C	–	–	–	√	√	–	–	–	–	–
[63]	Neurofuzzy cluster selection	√	–	√	√	–	–	–	–	–	–

1, Network lifetime; 2, throughput; 3, alive nodes; 4, energy; 5, packets delivered; 6, end-to-end delay; 7, latency; 8, jitter; 9, packet delivery ratio; 10, accuracy.

9. Conclusion

The clustering routing protocol is commonly used in WSN topology as it improves the energy efficiency of the entire network. However, it may lead to problems such as low QoS, load balancing, energy degradation under heterogeneous network conditions, node mobility, and security. This chapter summarizes the objectives for future research works in the field of clustering techniques for WSNs. After analyzing the current issues and challenges of research works, this chapter suggests the implementation of a machine learning approach and optimization algorithm for cluster lifetime prediction as well as optimal path selection for future research work.

10. Future scope

The WSN is one of the infrastructure-less scenarios that may be utilized to send or receive data in a wireless situation, as shown by the aforementioned study. The most difficult task, particularly in WSNs when a significant number of nodes (either mobile nodes) are linked to transmit data, requires consideration of the restricted WSN resources. So, there is a need to manage WSN resources carefully to achieve QoS performance. Some scopes for future research directions are presented in Fig. 7.

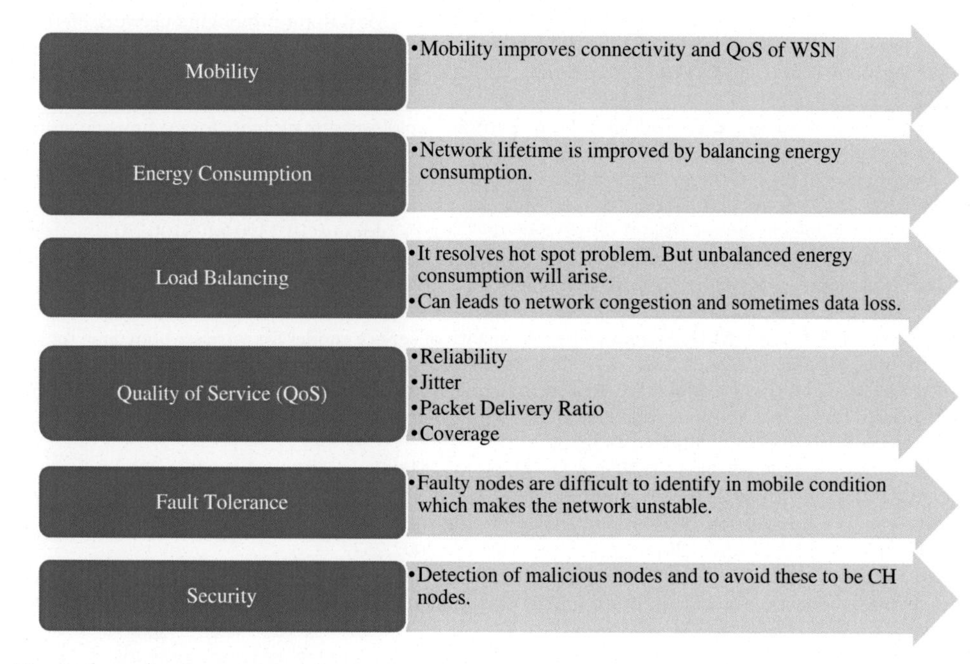

Fig. 7 Scope for future research.

References

[1] V. Bhandary, A. Malik, S. Kumar, Routing in wireless multimedia sensor networks: a survey of existing protocols and open research issues, J. Eng. (2016), https://doi.org/10.1155/2016/9608757. Hindawi Limited.

[2] L.D.P. Mendes, J.J.P.C. Rodrigues, A survey on cross-layer solutions for wireless sensor networks, J. Netw. Comput. Appl. (2011) 523–534, https://doi.org/10.1016/j.jnca.2010.11.009. Academic Press.

[3] S. Aswale, V.R. Ghorpade, Survey of QoS routing protocols in wireless multimedia sensor networks, J. Comput. Netw. Commun. 2015 (2015), https://doi.org/10.1155/2015/824619.

[4] Z. Hamid, F.B. Hussain, QoS in wireless multimedia sensor networks: a layered and cross-layered approach, Wirel. Pers. Commun. (2014) 729–757, https://doi.org/10.1007/s11277-013-1389-0. Springer.

[5] M. Asif, S. Khan, R. Ahmad, M. Sohail, D. Singh, Quality of service of routing protocols in wireless sensor networks: a review, IEEE Access 5 (2017) 1846–1871, https://doi.org/10.1109/ACCESS.2017.2654356.

[6] S. Bhatnagar, B. Deb, B. Nath, Service differentiation in sensor networks, in: Proceedings of Wireless Personal Multimedia Communications, 2001. https://www.researchgate.net/publication/2550868_Service_Differentiation_in_Sensor_Networks. (Accessed 31 January 2021).

[7] M.M. Afsar, M.H. Tayarani-N, Clustering in sensor networks: a literature survey, J. Netw. Comput. Appl. 46 (2014) 198–226, https://doi.org/10.1016/j.jnca.2014.09.005.

[8] T. Korkmaz, M. Krunz, Multi-constrained optimal path selection, in: Proceedings—IEEE INFOCOM, vol. 2, 2001, pp. 834–843, https://doi.org/10.1109/infcom.2001.916274.

[9] E.I. Oyman, C. Ersoy, Multiple sink network design problem in large scale wireless sensor networks, in: IEEE International Conference on Communications, vol. 6, Institute of Electrical and Electronics Engineers Inc., 2004, pp. 3663–3667, https://doi.org/10.1109/icc.2004.1313226.

[10] B. Nazir, H. Hasbullah, Mobile sink based routing protocol (MSRP) for prolonging network lifetime in clustered wireless sensor network, in: ICCAIE 2010–2010 International Conference on Computer Applications and Industrial Electronics, 2010, pp. 624–629, https://doi.org/10.1109/ICCAIE.2010.5735010.
[11] Z.M. Wang, S. Basagni, E. Melachrinoudis, C. Petrioli, Exploiting sink mobility for maximizing sensor networks lifetime, in: Proceedings of the Annual Hawaii International Conference on System Sciences, 2005, p. 287, https://doi.org/10.1109/hicss.2005.259.
[12] M. Radi, B. Dezfouli, K.A. Bakar, M. Lee, Multipath routing in wireless sensor networks: survey and research challenges, Sensors (Basel) (2012) 650–685, https://doi.org/10.3390/s120100650.
[13] V.C. Gungor, G.P. Hancke, Industrial wireless sensor networks: challenges, design principles, and technical approaches, IEEE Trans. Ind. Electron. 56 (10) (2009) 4258–4265, https://doi.org/10.1109/TIE.2009.2015754.
[14] Y. Liao, M.S. Leeson, M.D. Higgins, Flexible quality of service model for wireless body area sensor networks, Healthc. Technol. Lett. 3 (1) (2016) 12–15, https://doi.org/10.1049/htl.2015.0049.
[15] M. Khalid, Z. Ullah, N. Ahmad, M. Arshad, B. Jan, Y. Cao, A. Adnan, A survey of routing issues and associated protocols in underwater wireless sensor networks, J. Sens. (2017), https://doi.org/10.1155/2017/7539751. Hindawi Limited.
[16] S.A. Munir, B. Ren, W. Jiao, B. Wang, D. Xie, J. Ma, Mobile wireless sensor network: architecture and enabling technologies for ubiquitous computing, in: Proceedings -21st International Conference on Advanced Information Networking and Applications Workshops/Symposia, AINAW'07, vol. 1, 2007, pp. 113–120, https://doi.org/10.1109/AINAW.2007.257.
[17] M. Saleem, G.A. Di Caro, M. Farooq, Swarm intelligence based routing protocol for wireless sensor networks: survey and future directions, Inf. Sci. 181 (20) (2011) 4597–4624, https://doi.org/10.1016/j.ins.2010.07.005.
[18] W. Guo, W. Zhang, A survey on intelligent routing protocols in wireless sensor networks, J. Netw. Comput. Appl. (2014) 185–201, https://doi.org/10.1016/j.jnca.2013.04.001. Academic Press.
[19] J. Kumar, S. Tripathi, R.K. Tiwari, A survey on routing protocols for wireless sensor networks using swarm intelligence, Int. J. Internet Technol. Secur. Trans. (2016) 79–102, https://doi.org/10.1504/IJITST.2016.078574. Inderscience Publishers.
[20] D. Mahmood, N. Javaid, S. Mahmood, S. Qureshi, A.M. Memon, T. Zaman, MODLEACH: a variant of LEACH for WSNs, 2013 Eighth International Conference on Broadband and Wireless Computing, Communication and Applications, 2013, pp. 158–163. https://doi.org/10.1109/BWCCA.2013.34.
[21] S.P. Barfunga, P. Rai, H.K.D. Sarma, Energy efficient cluster based routing protocol for wireless sensor networks, in: 2012 International Conference on Computer and Communication Engineering, ICCCE 2012, 2012, pp. 603–607, https://doi.org/10.1109/ICCCE.2012.6271258.
[22] X. Fan, Y. Song, Improvement on LEACH protocol of wireless sensor network, in: 2007 International Conference on Sensor Technologies and Applications, SENSORCOMM 2007, Proceedings, 2007, pp. 260–264, https://doi.org/10.1109/SENSORCOMM.2007.4394931.
[23] N. Javaid, M. Aslam, K. Djouani, Z.A. Khan, T.A. Alghamdi, ATCEEC: a new energy efficient routing protocol for wireless sensor networks, in: 2014 IEEE International Conference on Communications, ICC 2014; IEEE Computer Society, 2014, pp. 263–268, https://doi.org/10.1109/ICC.2014.6883329.
[24] J. Gnanambigai, N. Rengarajan, R.J. Prarthana, An analytical approach for quadrant based leach: an energy efficient routing protocol for WSNs, in: Proceedings of the IEEE International Caracas Conference on Devices, Circuits and Systems, ICCDCS, Institute of Electrical and Electronics Engineers Inc, 2014, https://doi.org/10.1109/ICDCSyst.2014.6926117.
[25] D. Van Giang, T. Taleb, K. Hashimoto, N. Kato, Y. Nemoto, A fair and lifetime-maximum routing algorithm for wireless sensor networks, in: GLOBECOM—IEEE Global Telecommunications Conference, 2007, pp. 581–585, https://doi.org/10.1109/GLOCOM.2007.114.
[26] L. Zhang, S. Chen, Y. Jian, Y. Fang, Z. Mo, Maximizing lifetime vector in wireless sensor networks, IEEE/ACM Trans. Netw. 21 (4) (2013) 1187–1200, https://doi.org/10.1109/TNET.2012.2227063.
[27] Z. Wang, H. Ding, B. Li, L. Bao, Z. Yang, An energy efficient routing protocol based on improved artificial bee colony algorithm for wireless sensor networks, IEEE Access 8 (2020) 133577–133596, https://doi.org/10.1109/ACCESS.2020.3010313.
[28] I. Ullah, H.Y. Youn, Efficient data aggregation with node clustering and extreme learning machine for WSN, J. Supercomput. 76 (12) (2020) 10009–10035, https://doi.org/10.1007/s11227-020-03236-8.

[29] S. Radhika, P. Rangarajan, On improving the lifespan of wireless sensor networks with fuzzy based clustering and machine learning based data reduction, Appl. Soft Comput. J. 83 (2019) 105610, https://doi.org/10.1016/j.asoc.2019.105610.

[30] K. Manzoor, S.H. Jokhio, T.J.S. Khanzada, I.A. Jokhio, Enhanced TL-LEACH routing protocol for large-scale WSN applications, in: Proceedings—2019 Cybersecurity and Cyberforensics Conference, CCC 2019, Institute of Electrical and Electronics Engineers Inc., 2019, pp. 35–39, https://doi.org/10.1109/CCC.2019.00-12.

[31] I. Banerjee, P. Madhumathy, An agent cluster based routing protocol for enhancing lifetime of wireless sensor network, in: 1st International Conference on Advanced Technologies in Intelligent Control, Environment, Computing and Communication Engineering, ICATIECE 2019, Institute of Electrical and Electronics Engineers Inc, 2019, pp. 265–268, https://doi.org/10.1109/ICATIECE45860.2019.9063788.

[32] T.M. Behera, U.C. Samal, S.K. Mohapatra, Energy-efficient modified LEACH protocol for IoT application, IET Wirel. Sens. Syst. 8 (5) (2018) 223–228, https://doi.org/10.1049/iet-wss.2017.0099.

[33] M. Razzaq, D. Devi Ningombam, S. Shin, Energy efficient K-means clustering-based routing protocol for WSN using optimal packet size, in: International Conference on Information Networking; IEEE Computer Society, vol. 2018, 2018, pp. 632–635, https://doi.org/10.1109/ICOIN.2018.8343195.

[34] W. Wibisono, T. Ahmad, R. Anggoro, Position-based scheme for multi-hop routing protocol in cluster-based wireless sensor networks, in: Proceeding of 2018 4th International Conference on Wireless and Telematics, ICWT 2018, Institute of Electrical and Electronics Engineers Inc., 2018, https://doi.org/10.1109/ICWT.2018.8527735.

[35] A.P. Abidoye, I.C. Obagbuwa, Models for integrating wireless sensor networks into the internet of things, IET Wirel. Sens. Syst. 7 (3) (2017) 65–72, https://doi.org/10.1049/iet-wss.2016.0049.

[36] W. Wu, N. Xiong, C. Wu, Improved clustering algorithm based on energy consumption in wireless sensor networks, IET Netw. 6 (3) (2017) 1–7, https://doi.org/10.1049/iet-net.2016.0115.

[37] P. Li, W. Jiang, H. Xu, W. Liu, Energy optimization algorithm of wireless sensor networks based on LEACH-B, in: Lecture Notes on Data Engineering and Communications Technologies, vol. 1, Springer, 2017, pp. 391–404, https://doi.org/10.1007/978-3-319-49109-7_37.

[38] R. Manchanda, K. Sharma, A novel framework for energy-efficient compressive data gathering in heterogeneous wireless sensor network, Int. J. Commun. Syst. 34 (3) (2021), https://doi.org/10.1002/dac.4677, e4677.

[39] P. Gupta, S. Tripathi, S. Singh, Energy-Efficient Routing Protocols for Cluster-Based Heterogeneous Wireless Sensor Network (HetWSN)—Strategies and Challenges: A Review, Springer, Singapore, 2021, pp. 853–878, https://doi.org/10.1007/978-981-15-8335-3_65.

[40] L. Guezouli, K. Barka, A. Djehiche, UAVs's efficient controlled mobility management for mobile heterogeneous wireless sensor networks, J. King Saud Univ. - Comput. Inf. Sci. (2020), https://doi.org/10.1016/j.jksuci.2020.09.017.

[41] M. Sheik Dawood, S. Sakena Benazer, S.K. Vijaya Saravanan, V. Karthik, Energy efficient distance based clustering protocol for heterogeneous wireless sensor networks, Mater. Today Proc. (2020), https://doi.org/10.1016/j.matpr.2020.11.339.

[42] L. Cao, Y. Cai, Y. Yue, S. Cai, B. Hang, A novel data fusion strategy based on extreme learning machine optimized by bat algorithm for Mobile heterogeneous wireless sensor networks, IEEE Access 8 (2020) 16057–16072, https://doi.org/10.1109/ACCESS.2020.2967118.

[43] L.L. Hung, F.Y. Leu, K.L. Tsai, C.Y. Ko, Energy-efficient cooperative routing scheme for heterogeneous wireless sensor networks, IEEE Access 8 (2020) 56321–56332, https://doi.org/10.1109/ACCESS.2020.2980877.

[44] R. Priyadarshi, P. Rawat, V. Nath, B. Acharya, N. Shylashree, Three level heterogeneous clustering protocol for wireless sensor network, Microsyst. Technol. 26 (12) (2020) 3855–3864, https://doi.org/10.1007/s00542-020-04874-x.

[45] V. Gomathy, N. Padhy, D. Samanta, M. Sivaram, V. Jain, I.S. Amiri, Malicious node detection using heterogeneous cluster based secure routing protocol (HCBS) in wireless adhoc sensor networks, J. Ambient. Intell. Humaniz. Comput. 11 (11) (2020) 4995–5001, https://doi.org/10.1007/s12652-020-01797-3.

[46] X. Zhao, S. Ren, H. Quan, Q. Gao, Routing protocol for heterogeneous wireless sensor networks based on a modified grey wolf optimizer, Sensors 20 (3) (2020) 820, https://doi.org/10.3390/s20030820.

[47] Y. Yazid, I. Ez-zazi, M. Salhaoui, M. Arioua, E.O. Ahmed, A. González, Extensive analysis of clustered routing protocols for heteregeneous sensor networks, in: Proceedings of the Third International Conference on Computing and Wireless Communication Systems, ICCWCS 2019, April 24–25, 2019, Faculty of Sciences, Ibn Tofaïl University, Kénitra, Morocco, EAI, 2019, https://doi.org/10.4108/eai.24-4-2019.2284208.

[48] S. Verma, N. Sood, A.K. Sharma, A novelistic approach for energy efficient routing using single and multiple data sinks in heterogeneous wireless sensor network, Peer Peer Netw. Appl. 12 (5) (2019) 1110–1136, https://doi.org/10.1007/s12083-019-00777-5.

[49] M. Zeng, X. Huang, B. Zheng, X. Fan, A heterogeneous energy wireless sensor network clustering protocol, Wirel. Commun. Mob. Comput. 2019 (2019), https://doi.org/10.1155/2019/7367281.

[50] R. Kandpal, R. Singh, Improving lifetime of wireless sensor networks by mitigating correlated data using LEACH protocol, in: India International Conference on Information Processing, IICIP 2016—Proceedings, Institute of Electrical and Electronics Engineers Inc., 2017, https://doi.org/10.1109/IICIP.2016.7975316.

[51] S. Siavoshi, Y.S. Kavian, H. Sharif, Load-balanced energy efficient clustering protocol for wireless sensor networks, in: IET Wireless Sensor Systems, vol. 6, Institution of Engineering and Technology, 2016, pp. 67–73, https://doi.org/10.1049/iet-wss.2015.0069.

[52] A.S.D. Sasikala, P. Aravindh, Improving the energy efficiency of leach protocol using VCH in wireless sensor network, Int. J. Eng. Dev. Res. 3 (2015) 918–924.

[53] R.P. Mahapatra, R.K. Yadav, Descendant of LEACH based routing protocols in wireless sensor networks, Procedia Comput. Sci. 57 (2015) 1005–1014, https://doi.org/10.1016/j.procs.2015.07.505. Elsevier.

[54] R.S.Y. Elhabyan, M.C.E. Yagoub, Two-tier particle swarm optimization protocol for clustering and routing in wireless sensor network, J. Netw. Comput. Appl. 52 (2015) 116–128, https://doi.org/10.1016/j.jnca.2015.02.004.

[55] X. Yuan, M. Elhoseny, H.K. El-Minir, A.M. Riad, A genetic algorithm-based, dynamic clustering method towards improved WSN longevity, J. Netw. Syst. Manag. 25 (1) (2017) 21–46, https://doi.org/10.1007/s10922-016-9379-7.

[56] A. Lipare, D.R. Edla, V. Kuppili, Energy efficient load balancing approach for avoiding energy hole problem in WSN using grey wolf optimizer with novel fitness function, Appl. Soft Comput. J. 84 (2019) 105706, https://doi.org/10.1016/j.asoc.2019.105706.

[57] G. Yogarajan, T. Revathi, Improved cluster based data gathering using ant lion optimization in wireless sensor networks, Wirel. Pers. Commun. 98 (3) (2018) 2711–2731, https://doi.org/10.1007/s11277-017-4996-3.

[58] M. Wu, M. Collier, A new cluster-based evolutionary routing algorithm for extending the lifetime of heterogeneous sensor networks, in: IEEE Intl Conference on Computational Science and Engineering (CSE) and IEEE Intl Conference on Embedded and Ubiquitous Computing (EUC) and Intl Symposium on Distributed Computing and Applications for Business Engineering (DCABES), 2016, pp. 185–188.

[59] L. Lakshmanan, A. Jesudoss, V. Ulagamuthalvi, Cluster based routing scheme for heterogeneous nodes in WSN–A genetic approach, in: Lecture Notes on Data Engineering and Communications Technologies, vol. 26, Springer, 2019, pp. 1013–1022, https://doi.org/10.1007/978-3-030-03146-6_117.

[60] B.M. Sahoo, H.M. Pandey, T. Amgoth, GAPSO-H: a hybrid approach towards optimizing the cluster based routing in wireless sensor network, Swarm Evol. Comput. 60 (2021) 100772, https://doi.org/10.1016/j.swevo.2020.100772.

[61] D. Mehta, S. Saxena, MCH-EOR: multi-objective cluster head based energy-aware optimized routing algorithm in wireless sensor networks, Sustain. Comput. Inform. Syst. 28 (2020) 100406, https://doi.org/10.1016/j.suscom.2020.100406.

[62] T.M. Behera, S.K. Mohapatra, U.C. Samal, M.S. Khan, M. Daneshmand, A.H. Gandomi, I-SEP: an improved routing protocol for heterogeneous WSN for IoT-based environmental monitoring, IEEE Internet Things J. 7 (1) (2020) 710–717, https://doi.org/10.1109/JIOT.2019.2940988.

[63] K. Thangaramya, K. Kulothungan, R. Logambigai, M. Selvi, S. Ganapathy, A. Kannan, Energy aware cluster and neuro-fuzzy based routing algorithm for wireless sensor networks in IoT, Comput. Netw. 151 (2019) 211–223, https://doi.org/10.1016/j.comnet.2019.01.024.
[64] G. Zhao, Y. Li, L. Zhang, SSEEP: state-switchable energy-conserving routing protocol for heterogeneous wireless sensor networks, in: ICEIEC 2019- Proceedings of 2019 IEEE 9th International Conference on Electronics Information and Emergency Communication, Institute of Electrical and Electronics Engineers Inc., 2019, pp. 685–688, https://doi.org/10.1109/ICEIEC.2019.8784570.
[65] Y. Cao, L. Zhang, Energy optimization protocol of heterogeneous WSN based on node energy, in: 2018 3rd IEEE International Conference on Cloud Computing and Big Data Analysis, ICCCBDA 2018, Institute of Electrical and Electronics Engineers Inc., 2018, pp. 495–499, https://doi.org/10.1109/ICCCBDA.2018.8386566.
[66] P. Khandnor, T.C. Aseri, Reactive energy efficient heterogeneous cluster based (R-EEHC) routing protocol for WSN, in: Proceedings of the 2017 International Conference on Wireless Communications, Signal Processing and Networking, WiSPNET 2017, vol. 2018, Institute of Electrical and Electronics Engineers Inc., 2018, pp. 1442–1445, https://doi.org/10.1109/WiSPNET.2017.8300001.
[67] N.D. Tan, N. Dinh Viet, SCBC: sector-chain based clustering routing protocol for energy efficiency in heterogeneous wireless sensor network, in: International Conference on Advanced Technologies for Communications, vol. 2016, IEEE Computer Society, 2016, pp. 314–319, https://doi.org/10.1109/ATC.2015.7388341.
[68] D. Mehta, S. Saxena, Hierarchical WSN protocol with fuzzy multi-criteria clustering and bio-inspired energy-efficient routing (FMCB-ER), Multimed. Tools Appl. (2020) 1–34, https://doi.org/10.1007/s11042-020-09633-8.
[69] H. Qabouche, A. Sahel, A. Badri, Hybrid energy efficient static routing protocol for homogeneous and heterogeneous large scale WSN, Wirel. Netw 27 (1) (2020) 575–587, https://doi.org/10.1007/s11276-020-02473-2.
[70] D. Sethi, An approach to optimize homogeneous and heterogeneous routing protocols in WSN using sink mobility, Mapan - J. Metrol. Soc. India 35 (2) (2020) 241–250, https://doi.org/10.1007/s12647-020-00366-5.

CHAPTER 9

Imperative load-balancing techniques in heterogeneous wireless networks

Tanu Kaistha[a] and Kiran Ahuja[b]
[a]I.K.G Punjab Technical University, Jalandhar, India
[b]DAV Institute of Engineering and Technology, Jalandhar, Punjab, India

1. Introduction

A heterogeneous network (HetNet) that combines various types of cells ranging from macrocells to different small cells can solve problems in data traffic in different networks [1] in contrasts with their bandwidths, working frequencies, costs, inclusion areas, and latencies [2,3]. The principle targets of heterogeneous systems are to expand limits, load balancing, change cell edge coverage area, and effectively utilize range [4]. A heterogeneous network is used for modern mobile communication networks and refining spectral efficiency in a geographic area by using combinations of picocells, femtocells, microcells, and relay nodes [5].

Fig. 1 shows the network coverage provided by macrocells. In a heterogeneous environment, administrators give more reliable client encounters when contrasted to a homogenous environment by incorporating an assortment of advancements as well as cell layers that rely upon the inclusion area's topology. A heterogeneous network environment may be a grid, distributed, or a cloud network.

A notable issue in a heterogeneous network is the development of equitable production strategies for the distribution of tasks between different loads with different load-balancing techniques [7,8]. Load balancing is the process of allocating the load across all distributed application domains to improve performance, resource utilization, bandwidth balance, and response time while avoiding overload and underload conditions [9]. It is necessary to make good performance of all nodes with the most neglected topics with different strategies [10]. Due to the increase in traffic over the Internet to the web server, there are problems where there are many users with high access speeds. Therefore, the concept of resource integration is applied. A popular Google web server works in the same way. It distributes user requests on different networks on servers that are distributed locally in different locations. There are numerous ways to use a user request with open connected data (LOD) that require a web serve [1,2]. Load balancing plays an important role in the distribution and compatibility of computers. Split the incoming charge into smaller tasks that can be assigned to computing resources through the parallel interface. B sure to max the central

Comprehensive Guide to Heterogeneous Networks
https://doi.org/10.1016/B978-0-323-90527-5.00011-3

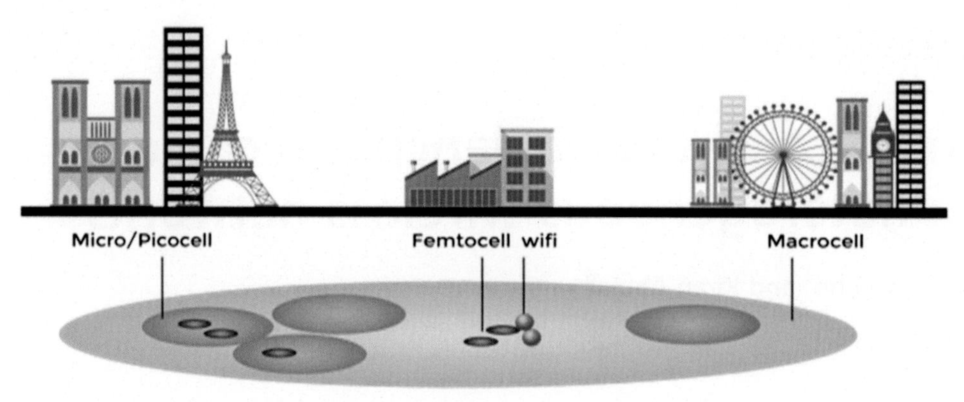

Fig. 1 Architecture of a heterogeneous network [6].

processing unit capacity, memory capacity, and network load. Load balancing in a multi-user application is processed without compromising the performance of the web server. After receiving a request from the user and determine the load on the resources available, the download manager transmits the request to the uploading server. The basic loading balancer is described in Fig. 2. It performs the following functions [6].

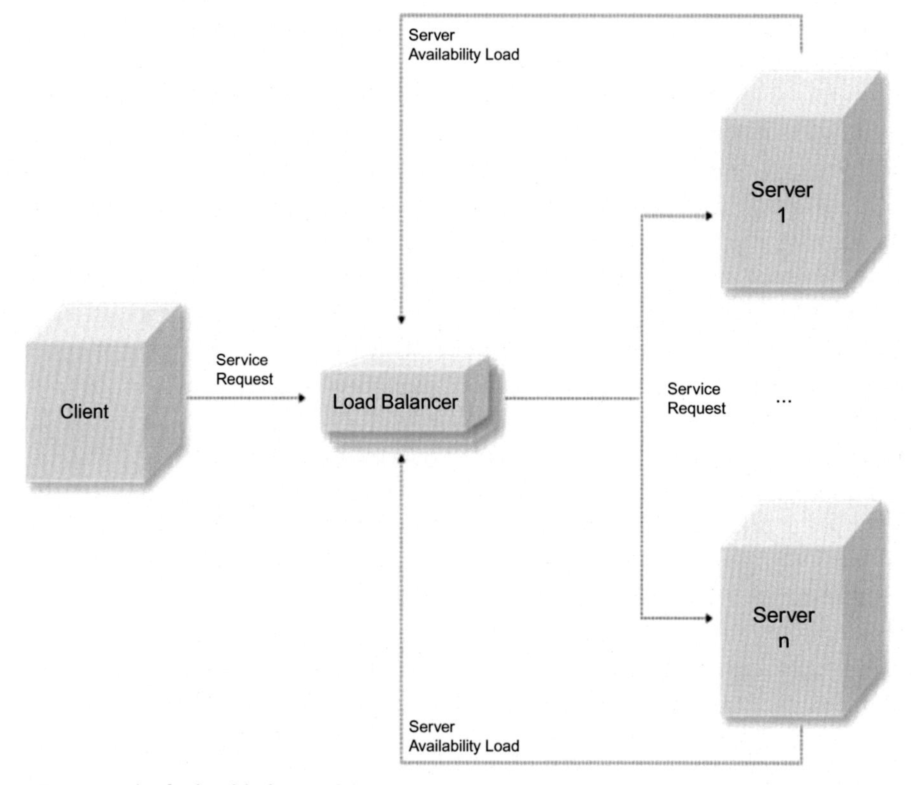

Fig. 2 Framework of a load balancer [1].

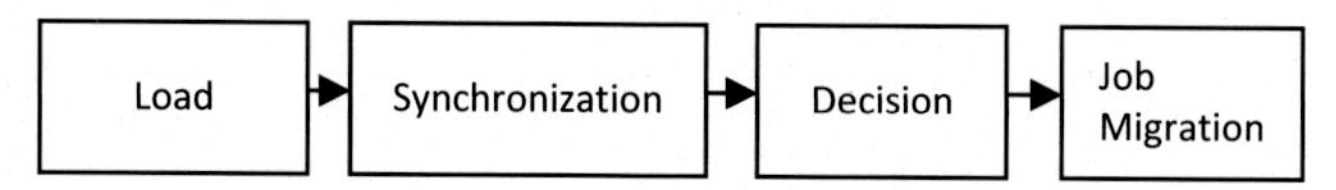

Fig. 3 Basic load-balancing steps [4].

- Distributes incoming traffic through multiple computer resources.
- Increases client satisfaction.
- Improves resource utilization.
- Provides fault tolerance and a flexible framework by adding or subtracting resources as per client demand.
- Determines resource availability and reliability for task execution.

The following steps for load balancing are described in Fig. 3 [10,11].

- *Load monitoring*: This helps to locate the load state.
- *Synchronization*: The data can be exchanged between different resources.
- *Decision making*: Figuring the new work development and settling on the work minute choice.
- *Job migration*: Progression of actual data.

Load balancing can also be characterized by the use of different policies [12].

- The information policy specifies which work information should be collected as well as when and where.
- The transfer policy specifies the exact time to be decided for the start-up of load-balancing activity.
- The location policy determines the availability and capabilities of the task to find an appropriate accomplice for a resource provider.
- The selection policy characterizes the tasks moved from an overloaded source to an idle receiver.

Due to the above advantages, load balancing has become an area of intensive research, and a number of measurement algorithms have been developed. The main objectives of this chapter are as follows:

- Introduction to a heterogeneous network in load balancing.
- Different types of load-balancing techniques.
- Comparative analysis of load-balancing algorithms.
- Artificial intelligence systems for load balancing.
- Machine learning technique for load balancing.

2. Different types of load-balancing techniques

Several load-balancing techniques work to accomplish their task on various complexity levels. Load-measurement strategies can be categorized based on the local distribution of nodes and the environment shown in Fig. 4.

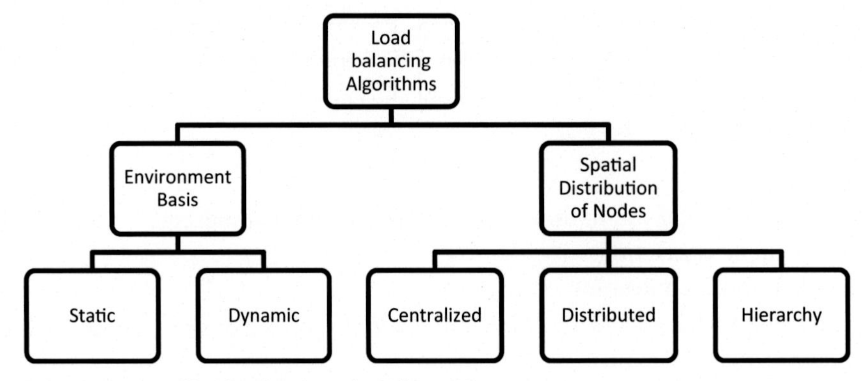

Fig. 4 Categorization of load-balancing algorithms [4].

In this section, different load-balancing methods can be divided into different classes on the basis of various criteria, such as the one shown in Fig. 4 [11,13].

2.1 Classification of static load balancing

In static load balancing, the decision of tasks to processors is taken before the execution starts, and data for that are assumed to be known before run time. In this strategy, once the load is allotted at the node, it cannot be exchanged for another load. The significant advantage of this method is that it is simple to execute with minor delays and less overhead. At the same time, this technique is less helpful when nodes are heterogeneous and tasks have a distinctive execution time. So it requires a system that should manage such conditions and overcome this issue [14,15]. Round robin, reproduced toughening, and randomized are some methods for static load balancing. It prompts the utilization of dynamic load balancing.

Among various proposed algorithms, powerful static load-balancing algorithms are as follows:

2.1.1 Round Robin algorithm

This algorithm distributes tasks uniformly to every node based on a round robin order, where equal loads are allocated to every node giving any priority, and if the last node is reached, it will come back to the first node. It is simple, easy to implement, free from starvation, and does not require any interprocess communication, as shown in Fig. 5. The use of load balances on a network will be much needed if the network is an active network and widely accessible by users. The reason is that it allows network inequality to occur. A round robin algorithm can be used for network measurement because it is a simple algorithm to schedule processes to provide efficient process performance. The main advantage of this algorithm is that it does not require process communication. This algorithm is used on web servers where HTTP applications have the same environment and are distributed evenly (Table 1).

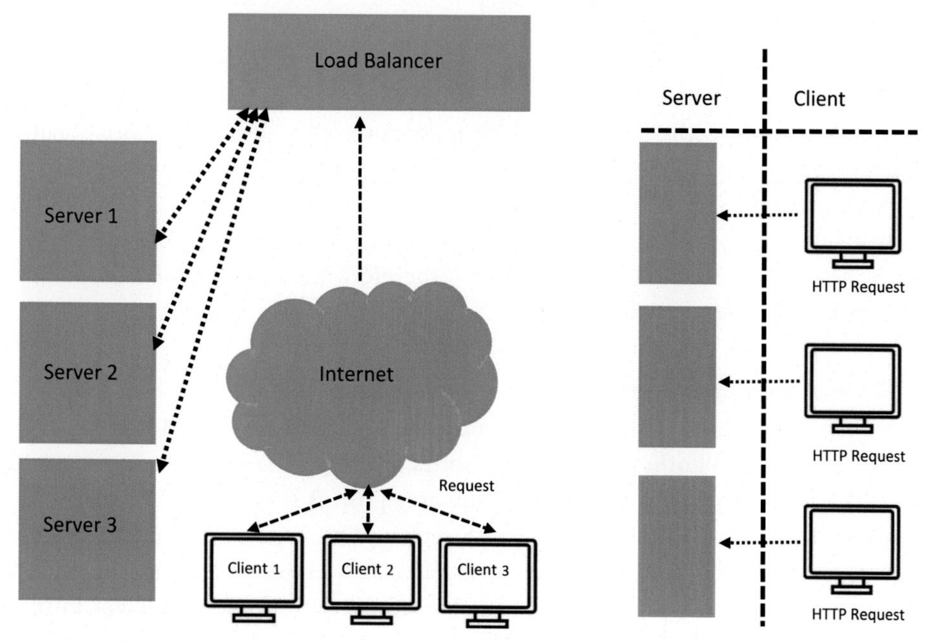

Fig. 5 Basic round robin algorithm [4].

Table 1 Different types of algorithms with information as well as pros and cons [4].

Algorithms	Information	Pros	Cons
Static	Detailed knowledge is compulsory about each node and client requirement	Used in a homogeneous environment	Has flexibility and scalability problems, and is in line with the change of the user and load requirements
Dynamic	The performance metrics for each of the nodes can also be observed in accordance with the load	Used in a heterogeneous environment	Complex and time consuming
Centralized	Each individual node is responsible for the care and maintenance of the network metrics	Useful in a small network	Not fault tolerant
Distributed	Every processor is responsible for load balancing and makes efficient load-balancing decisions, which is part of the network	Very useful in large, heterogeneous environments	Complex algorithms and communication overhead
Hierarchy	The nodes at different levels of the hierarchy interact with the nodes in order to get the information	Useful in a large- to medium-sized network in a heterogeneous environment	Complex and less fault tolerant

2.1.2 Randomized algorithm

This algorithm works with computing nodes and is randomly selected, without obtaining information about the previous or current load on the node. The random algorithm uses random pieces in the same way as helpful input to direct its operation to achieve the best performance in all possible random selections. Legally, algorithm performance will be random variables determined by random bits; therefore the duration of the operation, output, or both is a random variable.

2.1.3 Central manager algorithm

In the central manager algorithm, the central node selects the low-load slave node that keeps with it a list of all other related slave nodes. During a load change at any point, a message is sent to the slave nodes in the main area [16]. A single site is basically a type of database that is stored, but only in one location. This type of database is modified and managed from the actual location. This area is therefore basically any database system or intermediate computer program. A central location is available via an Internet connection (LAN, WAN, etc.). In this case, the central processor is able to collect all processor data and load data, where a selection based on this algorithm is likely to be made. The load manager makes load-balancing decisions based on system load information and allows the best decision when the process is built. This algorithm is expected to work better than similar applications where different hosts perform powerful tasks.

2.1.4 Threshold algorithm

This algorithm deals with allocating the load existence of computing nodes chosen locally without sending any messages and also keeps a copy for the new processor of the system load information. The load of the handling node can be described by different measurements, which are underloaded, medium, and overloaded. t_under and t_upper can be used to describe these levels.

Step 1: Underloaded(U_L):

$$U_L : \alpha < t_{under} \text{ (where } \alpha \text{ is load)}$$

Step 2: Medium(M):

$$M : t_{under} \leq \alpha \leq t_{upper}$$

Step 3: Overloaded (O_L)

$$O_L : \alpha > t_{upper}$$

First, in this case all processors are considered underloaded. When the processor upload status exceeds the upload limit, then it sends messages regarding the new upload status to all remote processors. If the location is not overloaded, the process will be redistributed locally; otherwise, an underloaded processor is selected. This algorithm has a systematic interaction with a large number of local process allocations and also decreases the overhead to improve performance. A disadvantage of this algorithm is that in this case, all processors are assigned to an area where all the remote processors are overloaded. In this case, the load on one processor can be much higher than in another overloaded processor, causing a significant disruption in the load while the duration of the processing also increases.

2.2 Classification of dynamic load balancing

Dynamic load balancing for heterogeneous networks requires changes to the system to be redistributed, monitored, and further separated into distributed and nondistributed labor [17,18]. A dynamic load-balancing algorithm is executed in a distributed system by all the nodes, and the responsibility of load balancing is shared and can be divided into cooperative and noncooperative. In a nondistributed system, load balancing is taken by some of the nodes but not with all nodes. It can also be placed in one of two forms: centralized or semidistributed. In a centralized system, the central node is responsible for the load balancing of the entire distributed system. In a semidistributed system, the nodes of the system are segmented into clusters. There has also been the development of load-balancing policies that are made to balance the load. Transfer load is mainly used to start the load-balancing task at which a load-balancing activity is to begin and deals with the allocation of nodes to the job.

Load balancing can be described by a different set of rules: location, distribution, and selection rules. The source domain for the decision function will be determined by the location principle. The distribution rule is used to determine how the load will be redistributed among the existing resources in the field. The selection rule is used to determine whether a balancing act to be performed as a preventive measure, or not.

The most important, dynamic load balancing algorithms are as follows [17,19]:

2.2.1 Central queue algorithm

The central queue algorithm works on dynamic distribution, keeping new functionality and incomplete applications in the rotating First In First Out line and the requirements for high connection between nodes. After that, the job application is accepted and the first one is expelled from the queue; if no job is requested, then the application is interrupted until a new job is created. When the processor load falls below the threshold, then the local load manager sends a new job application to the central load manager. The central load manager responds to the request as soon as a suitable function is found in the application request queue.

2.2.2 Local queue algorithm

When the load falls under the threshold limit in the local queue algorithm, all new processors with process relocation are initiated by the host. When the host gets underloaded, it requests a new job from distant hosts; after that, it checks its neighbor's list for correct jobs by the numbers obtained. The concept of a local queue algorithm is to process all new processes in an area with a migration process where the host load falls below the user-defined limit. This parameter describes the minimum number of processes that the load manager attempts to assign to each processor [2]. In the first step, new processes created on the main host are allocated on all underloaded hosts. The number of parallel activities created by the first parallel construct on the main host is usually sufficient for allocation on all remote hosts. When the host gets underloaded, the local load manager attempts to get several processes from remote hosts. It randomly sends requests with the number of local ready processes to remote load managers after receiving such a request; it compares the local number of ready processes with the received number. If the former is greater than the latter, then some of the running processes are transferred to the requester and an affirmative confirmation with the number of processes transferred is returned.

2.2.3 Least connection algorithm

The least connection algorithm selects the load distribution based on connections present and maintains the connections on each node depending on the increased or decreased number. This method works best in conditions where the loads have indistinguishable capacities.

A survey of different load balancing algorithms has been done, and their advantages and disadvantages are summarized in Table 2 [20]. The service provider can choose the optimal algorithm that best suits the purpose to give smooth, reliable, and fast service delivery to customers, which can be helpful in the future for development [21,22]. In accordance with the critical state and the steady state in the HetNet network, the maximum traffic load and the global range of the growth rate have been studied. In addition, a variety of methods have been developed to provide valuable insights into the impact of a number of important network parameters, such as multimedia content, node diversity and density, and limited caching capabilities on the performance of the system. To cope with the growing issue of wireless data transfer, innovative methods for increasing the bandwidth have been investigated, such as the heterogeneous network (HetNet). The material is elastic-network caching support.

The corresponding load-balancing mechanism must have a few properties that could make it recognizable and essential. These properties should provide higher throughput, scalability, performance, response time, efficient resource utilization, and low overhead. These are discussed as follows [23,24]:

Table 2 Summary of load-balancing algorithms [4,7,11].

Approaches	Algorithm	Advantages	Disadvantages
Static	Round robin	Straightforward and simple to actualize. Least framework overhead	Not helpful if undertakings having distinctive execution time
	Randomized	No interprocess communication	Maximum response time. Uneven load distribution
	Central manager	Helpful when dynamic undertakings are made	High interprocess communication
	Threshold	Minimize remote process allocation	Increased execution time
Dynamic	Central queue	Helpful for heterogeneous nodes. Undertakings dispensed after execution	No interprocess migration
	Local queue	Less interprocess communication	
	Least connection	Useful for nodes of similar capabilities	Not useful for tasks of different durations
	Random	Simple and easy to implement	Poor utilization of system resources
	Probabilistic	Minimized communication overhead	Algorithm is not scalable
	Centralized information and centralized decision	All resources are utilized with efficient performance results	Not scalable
	Centralized information and distributed decision	Less inclusion of message information with robust nature	Not generally up and coming. More overhead happens
	Distributed information and distributed decision	Efficient performance	More communication overhead sometimes

- *Resource utilization*: It advises whether the algorithm can use all the assets correctly to get the best performance. Static load-balancing algorithms have minimal utilization of resources as they simply try to assign tasks to processors to achieve less response time. Dynamic load balancing makes better use of resources as it takes into account the fact that the load must be evenly distributed to the processors. The proper load-balancing algorithm should have high resource utilization.

- *Process migration*: When a system chooses to export a process, process migration is provided. The algorithm is able to decide whether it should make changes in load distribution during execution of a process. The time taken in the migration of a task from one to another should be minimum to improve the performance of the system. The process migration is the overhead, which cannot be removed but should be minimized.
- *Fault tolerant*: A fault-tolerant system is a system in which the processing is not affected because of the failure of any particular processing device in the system. It enables an algorithm to work continually at the time of some failure. Dynamic load-balancing algorithms can deal with failure more than static load balancing. The load balancing should be fault tolerant.
- *Response time*: This parameter is used to measure the time taken by the load-balancing algorithm to complete a task and the minimum time in a distributed system. This parameter should be minimized. The static load-balancing algorithms have a short response time. Dynamic load-balancing algorithms may have a much higher response time.
- *Scalability*: Scalability is the capacity of the system to perform under an expanded overload. It is defined as the maximum number of nodes that an algorithm can use to provide advanced results. It is the ability of an algorithm to perform load balancing for a system with a finite number of nodes.
- *Adaptability*: This parameter tells whether the algorithm can adapt to the varying situations. Static load-balancing algorithms are not adaptive. Dynamic load-balancing algorithms are adaptive toward every situation.
- *Performance*: Performance refers to the effectiveness of the system after the complete execution of load-balancing algorithms. It describes the usefulness of the system once load balancing is done and is also used to check the efficiency of the system. This has to be improved at a reasonable cost, such as reducing task response time while keeping acceptable delays. If all the above parameters are satisfied ideally, it will maximize the entire system's performance.
- *Nature*: This parameter is related to determining the nature or behavior of load-balancing algorithms, that is, whether the load-balancing algorithm is static or dynamic, preplanned or no planning.
- *Overhead*: The overhead associated with any load-balancing algorithm supports the extra cost of integrating the different algorithms. Static load-balancing algorithms incur less overhead as once tasks are assigned to processors, no redistribution of tasks takes place, therefore there is no relocation overhead. Dynamic load-balancing algorithms incur relatively more overhead as relocation of tasks takes place. Overhead should be minimized so that a load-balancing technique can work efficiently.
- *Throughput*: It depicts the sending and receiving rates of data for the total number of tasks completed on given input at a given time, and if the throughput is high, then adaptability is high.

3. Comparative analysis of load-balancing algorithms

Table 3 summarizes the types of algorithms with information and their pros and cons. Here, we have a comprehensive list of all the above-mentioned algorithms for static and dynamic systems, as well as some other algorithms with unique features.

parameters are selected as per the need of the advanced time in Table 4 [23,24] gives us the comparative analysis of load-balancing algorithms with different parameters. There have been numerous strategies for directly enhancing the performance of heterogeneous networks described, we will discuss the significance of AI in wireless heterogeneous systems in the next section.

4. Impact of AI in wireless heterogeneous networks

Artificial intelligence (AI) is a system for solving complex problems with a system or machine that has taken steps without human intervention by witnessing complex natural processes such as learning, reasoning, and self-correction. In computer science, AI, called machine intelligence, is mechanical, unlike intelligence shown by humans. In other words, it is also used to describe machines (or computers) that mimic tasks associated with the human mind, such as learning and problem-solving [27].

Table 3 Categorization of load-balancing algorithms [25,26].

Load-balancing techniques	Criteria considered	Advantages	Disadvantages
Round robin	Time quantum	Fairness in load balancing, less complex	Quantum size overhead
First come first serve	Arrival time	Ease of implementation	Does not consider any other parameter
Central load balancer	Heterogeneous physical servers	Shorter response time	Less dynamic and robust
Heuristic based	Total turnaround time and response time	Better response time	Data locality, communication costs, and flow time are not consider
VM-assign algorithm	Least loaded VM	Better response time, fairness in load	For mixed loads (static and dynamic)
GA based	Efficiency, performance	Better performance and efficiency	More complex, long time consumption

Table 4 Comparison of load-balancing algorithms/parameters [23].

Parameters/ algorithms	Round robin	Randomized	Central manager	Threshold	Central queue	Local queue	Least connection
Nature	Static	Static	Static	Dynamic	Dynamic	Dynamic	Dynamic
Overhead	Less	Less	Less	More	More	More	More
Resource utilization	Less	Less	Less	Less	Less	More	More
Process migration	No	No	No	No	No	Yes	No
Fault tolerant	No	No	Yes	No	Yes	Yes	No
Response time	Less	Less	Less	Less	More	More	Less
Waiting time	More	More	More	More	Less	Less	Less
Adaptability	Less	Less	Less	Less	More	More	More
Scalability	More	Less	More	Less	Less	More	More
Performance	Less	Less	Less	Less	More	More	More
Stability	More	More	More	More	Less	Less	Less

Nowadays, AI has become helpful in everyday life and has greatly influenced our way of life even though employing essential computer workers and not limited to just a PC or a design industry and used in business, medical, legal, educational, manufacturing, and wireless communications [28]. Some examples of AI include spam filters, creative email input, voice elements, self-help text, good personal assistants, automated processes, online customers, security monitoring, sales, and business forecasting.

The traffic load has increased over the past few years as a result of the Internet's and mobile telecommunications' rapid development, and the heterogeneous network's optimized mobile applications and improved wireless access have helped to address a variety of problems and functionalities. There are distinct types of cells such as macrocells, base stations (BS), and NodeB (eNBs) developed in Long-Term Evolution (LTE), picocell and femtocell with different types of transmission, cover and operating systems in heterogeneous network. The biggest issue for a high-power system is developing productive techniques for the proper distribution of loads between different loads with different load-balancing techniques. Due to the heterogeneous network's capability, performance should keep improving for a variety of features, including power conversion, load balancing, mobility management, flexibility, neighbor optimization, and more user requirements in the workplace.

In the next topic, we discuss the importance of measuring load in the AI system with different techniques.

4.1 AI importance in load balancing

Load balancing is the process of allocating all workloads to all distributed system domains to optimize usage and resource utilization, analyze bandwidth and response time, and avoid overload and underload conditions. Fig. 2 shows the basic structure of the load-balancing concept. It is needed to get the best execution of each small header in different processes. Although many ways have been suggested to improve the heterogeneous network's performance, it has attracted significant interest from industry and academia by implementing artificial intelligence techniques such as neural networks and machine learning. AI techniques can automatically solve problems for vast systems, such as a heterogeneous network with intelligent and automated systems. The advanced features in self-organizing networks (SONs) can dramatically reduce human engagement in workflows, coverage, and QoS while increasing network capacity in the heterogeneous network [17]. The different load-balancing techniques in AI to get an efficient network are discussed as follows [29].

4.1.1 Neural network in a wireless heterogeneous network

The artificial neural network (ANN) is a model for preparing data inspired by biological sensors. It is similar to the brain process and contains a tremendous amount of neurons to detect specific issues. Neural network consists of numerous nodes that mimic the neurological features of the human brain connected to the joints that have been evaluated by weight to communicate between them. The final output of each node is called its node value; when it receives input information and performs a function, the result is sent to the subsequent neurons. ANNs can learn by changing weight esteems. There are two topologies of the ANN top: the feed and the feedback. A trained neural system can go "expert" in classifying data that will provide bearings in unpredictable situations. An ANN-based handoff decision algorithm is used to reduce the inactivity of the wireless infrastructure. There are other benefits of the neural network such as information capacity, readability, and real-time power; it can also be stored if the neurons fail.

4.1.2 Fuzzy logic for a wireless network

Fuzzy logic is a logic system that has more values based on degrees of truth than the usual true or false (1 or 0) Boolean logic on which a modern computer is based. It has been used to capture the concept of an empirical fact, where the actual value may differ between absolute and absolute truth. It includes 0 and 1 as cases of extreme truth (or "state of things" or "truth") but also includes various regions of truth between fuzzy logic developed in numerous areas, ranging from advanced logic to artificial intelligence. In fuzzification, mathematical input values map into fuzzy membership functions [30]. In contrast, the defuzzification parts can be used to map the output of a dense production to a "crisp" output value that may be used for decision or control purposes. An optimized logic algorithm has been proposed to improve transmission quality with minimal management delay, minimal packet loss, and poor handling management. Fuzzy logic applications have been submitted to be measured by channel, channel size, and channel layout on mobile networks [31], it can solve the problem of impartial installation, validation data, variations in the languages used, and a relatively stable system with limited time.

4.1.3 Genetic algorithm

A genetic algorithm was developed to provide optimum utilization of resources in a wireless environment. It also confirms the QoS requirement for customer service. In this case, variations in the ability to manipulate and select can be used as a future activity for further good results and presentations. Among the many evolutionary algorithms, a genetic algorithm (GA) must be discovered and designed to perform complex systems [30]. A GA adopts the process of genetic modification, which involves two key functions: the crossover, which makes easier discovery of the correct solution, and the transformation of genes. In the most open cases, the GA is skilled at solving problems where the size of the solution is too high to be fully searchable, and can be properly converted to one (or more) correct results in a short period of time. It can also solve many multipurpose problems easily. Therefore, GAs are widely used to develop heterogeneous networks, especially in cell organization and spatial induction where a large number of parameters need to be evaluated [31]. An example is the multipurpose GA to address the multipurpose communication problem of the heterogeneous network placement problem, aiming to increase communication coverage and total bandwidth while reducing cost savings.

4.1.4 Particle swarm optimization (PSO)

Particle swarm optimization (PSO) is a high-quality, human-based, global-based, and intelligence-based algorithm that finds a solution to the problem of increasing search space and social behavior. The social network is defined and empowered by the individual, and the number of people defined as random guesses in a set of problems is introduced and referred to as the candidate solution, also known as particles. Particles calculate the robustness of student solutions and recall the area in which they were most successful [32]. A good human solution is called the particles of the best space and it gets details from its neighbors. The shape of the particles is influenced by the excellent position visited by itself, that is, its experience and the shape of the leading particles in its position. When a particle cohost is aggressive, the best part of the site is called the global best particle, and the emerging algorithm is called the PSO gbest. When using small areas, the algorithm is often referred to as the best PSO. The performance of each particle is measured using a density function that varies according to the accessibility problem.

4.1.5 Artificial bee colony (ABC)

The artificial bee colony (ABC) algorithm was introduced in 2005 and is the metaheuristic algorithm used. This algorithm mimics the process of making honey and has three stages. There are employee bees, onlooker bees, and scout bees. Based on choices made in the bees section employed and probabilistic selection in the onlooker bee category, bees in the employee bee and onlooker bee categories generate local researchers' sources of available neighborhoods [33]. In the scout bee category, which symbolizes the abandonment of energy-consuming food resources in the process, the solutions no longer work when they seek to find new regions in space and have good exploration and developmental potential.

4.1.6 Markov models and Bayesian-based games

The Markov models and Bayesian methods are not AI methods, but still provide mathematical solutions for the particular networks with the ability to transform them [34] by changing dynamics of the state between the different regions in both learning, home and navigation systems on the higher-level networks can be improved based on tree depletion. Achieving the right types oflocal hidden Markov models (HMM) is often used to adjust signal strength history. This study builds on the problem of network selection in a heterogeneous network with incomplete data by combining large reaction energies. It also investigates the use of specific Bayesian methods for anomaly detection in a heterogeneous network, which can provide human interpretations.

The methods discussed above are used to handle various aspects of wireless communication such as network management, mobility management, traffic loading, load loading, etc., in the heterogeneous network. This chapter provides a comprehensive introduction to machine learning for wireless networks and a survey of the latest advances in ML applications for performance improvement to address various challenges in future wireless networks. The next section includes the necessary fundamentals for the reader to understand the concepts of machine learning.

5. Need of machine learning in a wireless heterogeneous network

Machine learning (ML) was introduced in 1959, and it can be defined as a numerical method that can make predictions or improvements based on input data. The practical applications of ML are to predict unforeseen items or data and to learn and build effective algorithms for these predictions. Examples of input data are products, information, or electronic data taken from the Internet; however, ML requires a large amount of data to ensure the maximum accuracy of the prediction [35].

ML emerged from pattern recognition studies, computational education in AI environments, and studies of the human brain. It explored the design and survey of algorithms while also making algorithms able to read and adapt to change settings in a wireless network [36] and it describes a machine-learning method called Q-learning to detect a supply chain problem on parallel networks. It makes the system fit to take in and improve from its encounters and starts with perceptions or information. It creates a scientific model of first-hand data, known as "training data," to solve forecasts or selections and enables investigating vast amounts of data with immediate and consistent results. The collaboration of ML with AI and the advances in use can make it compelling to analyze large amounts of data. There are various methods of ML such as supervised, unsupervised, semisupervised, and reinforcement learning (RL). The benefits of ML include reducing over-signing, providing better performance, avoid previous errors, improving algorithm complexity, compute power, and overcoming network information shortages [35,37–39].

5.1 Supervised learning

Supervised learning is an ML activity input from supervised training data. The training details contain a collection of training examples. In unsupervised learning, each example is a pair that contains the input object (usually the vector) and the required output value (also called the control signal). The machine's purpose is to read the correct output given the new input. In the tested texts, the following supervised reading methods are used.

5.1.1 Support vector machine

The basic support vector machine (SVM) model is a linear classifier, which aims to separate p-dimensional data points using the ap-1 hyperplane. The best hyperplane is the one that leads to the greater division or limit between the two given categories, and is called the maximum-margin hyperplane. However, the dataset is usually not evenly distributed in the first space. In this case, the actual space can be mapped to a much higher scale by including kernel functions, such as polynomial or Gaussian kernels, resulting in nonlinear separation.

5.1.2 K nearest neighbors

K nearest neighbors (KNN) is a nonparametric lazy learning algorithm that is not a parameter for classification and regression, where no assumptions about data distribution are required. Taking the classification function as an example, KNN's basic goal is to determine the point of the test in the form of a vector feature based on the majority vote of its K nearest neighbors. The class training data can control the prediction of test data, a weighted approach can be adopted by including the weight of each neighbor equal to its distance to the test site. In addition, one of the keys to using KNN is trading with parameter K, and selecting a value can be addressed.

5.2 Unsupervised learning

Unsupervised learning is an ML activity that aims to read the task of explaining the hidden structure from unlabelled information. In the experimental activities, the following unsupervised learning methods are used.

5.2.1 K-means clustering algorithm

In K-means clustering, the purpose is to divide data points into clusters, and each data point belongs to clusters that have a close meaning. The most common type of K-means algorithm is based on iterative refinement. Initially, the methods were performed randomly. Then, in each iteration, each point of data is given exactly one set of data, the meaning of which has a much smaller Euclidean distance than the data point, and the meaning of each set is elevated. The algorithm continues to work until the members of each group do not change.

5.2.2 Principal component analysis

Principal component analysis (PCA) is a dimension-reduction tool that can be used to reduce a large set of variables to a small set that still contains most of the information in the large set. It is mathematically defined as an orthogonal linear transformation that transforms the data to a new coordinate system.

5.3 Reinforcement learning

Reinforcement learning differs from the supervised learning problem in that correct input/output pairs are never presented. Specifically, the following reinforcement learning algorithms are applied in surveyed studies.

5.3.1 Q-learning

One of the most common strength-enhancing learning algorithms is Q-learning. Specifically, an RL agent interacts with nature to read Q values, depending on whether the agent is taking action. The Q value is defined as the discounted accumulative reward, starting at a tuple of a state and an action and then following a certain policy. Once the Q values have been read after a reasonable amount of time, the agent can make a quick decision under the current situation by taking action with a large Q value. More details about Q reading can be passed on to you. In addition, to manage continuous state spaces, it can be used to read fuzzy Q.

5.3.2 Multiarmed bandit learning

In a multistakeholder (MAB) model with a single agent, the agent, in turn, takes action and receives a random reward created by consistent distribution, aimed at maximizing the combined reward, trade-off between taking the current, best price (exploitation) and collecting data for the next big reward. In this case, the model is expected to gain other stable or equitable regions.

5.3.3 Actor-critic learning

The actor-critic algorithm for learning is made up of the character, the critic, and the environment in which the character interacts. In this algorithm, the character begins to choose the action according to the current plan and earns the cost immediately. After that, the critic revives the function of state value according to the time variance error, and then the character will review the policy. As for the plan, it can be reviewed based on the policy taught at Boltzmann's distribution. When each action is updated indefinitely in each region, the algorithm will change to the correct state values.

5.3.4 Joint utility and strategy estimation learning

In this algorithm, each agent carries an expected service rating, the review of which is based on immediate reward. The opportunity to select each action, named a strategy, is updated in the same way according to the help rating. The main advantage of this algorithm lies in the fact that it is fully distributed where the reward can be calculated directly locally, as for example the amount of data between the transmitter and its receiver is paired.

5.3.5 Deep reinforcement learning

In this author propose deep NN called a deep Q network (DQN) to approximate optimal Q values, which allows the agent to learn from the high-dimensional sensory data directly. Reinforcement learning based on DQN is known as deep reinforcement learning (DRL) [36,40].

The methods discussed above are used to handle various aspects of machine learning in the heterogeneous network. However, these new tools require further research to fulfill their full potential in (practical) wireless networks.

6. Conclusion and future scope

It has been surveyed that load balancing is a significant concern in a heterogeneous network. Load balancing has been intensively explored in past decades; a vast number of related studies and results have been introduced on the topic. Load balancing is required to get good performance with the most negligible overhead. This chapter has taken a survey of various load-balancing algorithms and strategies with different parameters. Computing resources are rapidly developing and growing with different issues of load imbalance. Here, we show that load balancing distributes the load equitably among different loads, expanding in general execution. The problem is to efficiently manage the multidimensional nature of the heterogeneous systems using transformational strategies. The heterogeneous network will be characterized by a high degree of capillarity, population, and high bit rate. With this research and analysts, AI-based strategies are explored. They have demonstrated the ability to develop and master the best development of different load-balancing techniques in the heterogeneous network as well as the technical challenges and research problems.

In the future, by introducing different optimization techniques, offloading, and caching techniques, the efficiency of the network can be improved for different parameters. This investigation is further helpful in planning a new load-balancing algorithm.

References

[1] T.V. Gopal, N.K. Nataraj, C. Ramamurthy, V. Sankaranarayanan, Load balancing in heterogenous distributed systems, Microelectron. Reliab. 36 (9) (1996) 1279–1286.

[2] Y. Zhang, H. Kameda, S.L. Hung, Comparison of dynamic and static load-balancing strategies in heterogeneous distributed systems, IEE Proc. Comput. Digit. Tech. 144 (2) (1997) 100–106.

[3] K. Zhu, D. Niyato, P. Wang, Network selection in heterogeneous wireless networks: evolution with incomplete information, in: 2010 IEEE Wireless Communication and Networking Conference, IEEE, 2010, pp. 1–6.

[4] Y. Li, Z. Lan, A survey of load balancing in grid computing, in: International Conference on Computational and Information Science, Springer, Berlin, Heidelberg, 2004, pp. 280–285.

[5] M. Haddad, S.E. Elayoubi, E. Altman, Z. Altman, A hybrid approach for radio resource management in heterogeneous cognitive networks, IEEE J. Sel. Areas Commun. 29 (4) (2011) 831–842.

[6] W. Shi, B. Li, N. Li, C. Xia, A network architecture for load balancing of heterogeneous wireless networks, J. Networks 6 (4) (2011) 623.
[7] W. Tangtrongpairoj, A. Jansang, A. Phonphoem, Ripple algorithm: an LTE load balance mechanism with distributed and heuristic approach, in: The 2013 10th International Joint Conference on Computer Science and Software Engineering (JCSSE), IEEE, 2013, pp. 111–115.
[8] M. Alam, G. Rabiul, C. Biswas, N. Nower, M.S.A. Khan, A reliable semi-distributed load balancing architecture of heterogeneous wireless networks, arXiv (2012). preprint arXiv:1202.1918.
[9] A. Mäkelä, S. Siikavirta, J. Manner, Comparison of load-balancing approaches for multipath connectivity, Comput. Netw. 56 (8) (2012) 2179–2195.
[10] Y. Bejerano, S.J. Han, Cell breathing techniques for load balancing in wireless LANs, IEEE Trans. Mob. Comput. 8 (6) (2009) 735–749.
[11] S. Goswami, A. De Sarkar, A comparative study of load balancing algorithms in computational grid environment, in: 2013 Fifth International Conference on Computational Intelligence, Modelling and Simulation, IEEE, 2013, pp. 99–104.
[12] R. Mukhopadhyay, D. Ghosh, N. Mukherjee, A study on the application of existing load balancing algorithms for large, dynamic, heterogeneous distributed systems, in: Proceedings of 9th WSEAS International Conference on Software Engineering, Parallel and Distributed Systems (SEPADS), 2010, pp. 238–243.
[13] S. Boudko, W. Leister, S. Gjessing, Heterogeneous wireless network selection: load balancing and multicast scenario, Int. J. Adv. Netw. Serv. 6 (3&4) (2013).
[14] S. Begum, C.S.R. Prashanth, Review of load balancing in cloud computing, Int. J. Comput. Sci. Issues 10 (1) (2013) 343.
[15] I. Demirci, Ö. Korçak, Cell breathing algorithms for load balancing in Wi-Fi/cellular heterogeneous networks, Comput. Netw. 134 (2018) 140–151.
[16] Q. Ye, B. Rong, Y. Chen, M. Al-Shalash, C. Caramanis, J.G. Andrews, User association for load balancing in heterogeneous cellular networks, IEEE Trans. Wirel. Commun. 12 (6) (2013) 2706–2716.
[17] S.F. El-Zoghdy, S. Ghoniemy, A survey of load balancing in high-performance distributed computing systems, Int. J. Adv. Comput. Res. 1 (2014).
[18] S.T. Milan, L. Rajabion, H. Ranjbar, N.J. Navimipour, Nature inspired meta-heuristic algorithms for solving the load-balancing problem in cloud environments, Comput. Oper. Res. 110 (2019) 159–187.
[19] N. Rathore, I. Chana, Load balancing and job migration techniques in grid: a survey of recent trends, Wirel. Pers. Commun. 79 (3) (2014) 2089–2125.
[20] W. Chen, H. Li, Q. Ma, Z. Shang, Design and implementation of server cluster dynamic load balancing in virtualization environment based on OpenFlow, in: Proceedings of the Ninth International Conference on Future Internet Technologies, 2014, June, pp. 1–6.
[21] K. Ahuja, B. Singh, R. Khanna, Network selection based on available link bandwidth in multi-access networks, Digit. Commun. Netw. 2 (1) (2016) 15–23.
[22] D. Mithbavkar, H. Joshi, H. Kotak, D. Gajjar, L. Perigo, Round robin load balancer using software defined networking (SDN), Capstone Team Res. Proj. 5 (2016) 1–9.
[23] M. Kushwaha, S. Gupta, Various schemes of load balancing in distributed systems-a review, Int. J. Sci. Res. Eng. Technol. 4 (7) (2015) 741–748.
[24] Y. Jiang, A survey of task allocation and load balancing in distributed systems, IEEE Trans. Parallel Distrib. Syst. 27 (2) (2015) 585–599.
[25] L. Maftahi, S. Rakrak, S. Raghay, Automated SLA negotiation: a novel approach for optimizing cloud data overload, in: International Symposium on Ubiquitous Networking, Springer, Singapore, 2015, pp. 249–257.
[26] R.A. Haidri, C.P. Katti, P.C. Saxena, A load balancing strategy for cloud computing environment, in: 2014 International Conference on Signal Propagation and Computer Technology (ICSPCT 2014), IEEE, 2014, pp. 636–641.
[27] E. Charniak, Introduction to Artificial Intelligence, Pearson Education India, 1985.
[28] N.J. Nilsson, N.J. Nilsson, Artificial Intelligence: A New Synthesis, Morgan Kaufmann, 1998.
[29] X. Wang, X. Li, V.C. Leung, Artificial intelligence-based techniques for emerging heterogeneous network: state of the arts, opportunities, and challenges, IEEE Access 3 (2015) 1379–1391.

[30] M. Alkhawlani, A. Ayesh, Access network selection based on fuzzy logic and genetic algorithms, in: Advances in Artificial Intelligence, 2008, p. 2008.
[31] W.X. Shi, S.S. Fan, N. Wang, C.J. Xia, Fuzzy neural network based access selection algorithm in heterogeneous wireless networks, J. Commun. 31 (9) (2010) 151–156.
[32] P. Muñoz, R. Barco, I. de la Bandera, Optimization of load balancing using fuzzy Q-learning for next generation wireless networks, Expert Syst. Appl. 40 (4) (2013) 984–994.
[33] D.N. Miku, P. Gulia, Improve performance of load balancing using artificial bee colony in grid computing, Int. J. Comput. Appl. 86 (14) (2014).
[34] X. Chen, J. Wu, Y. Cai, H. Zhang, T. Chen, Energy-efficiency oriented traffic offloading in wireless networks: a brief survey and a learning approach for heterogeneous cellular networks, IEEE J. Sel. Areas Commun. 33 (4) (2015) 627–640.
[35] C. Jiang, H. Zhang, Y. Ren, Z. Han, K.C. Chen, L. Hanzo, Machine learning paradigms for next-generation wireless networks, IEEE Wirel. Commun. 24 (2) (2016) 98–105.
[36] M. Kulin, T. Kazaz, E. De Poorter, I. Moerman, A survey on machine learning-based performance improvement of wireless networks: PHY, MAC and network layer, Electronics 10 (3) (2021) 318.
[37] M. Chen, U. Challita, W. Saad, C. Yin, M. Debbah, Machine learning for wireless networks with artificial intelligence: a tutorial on neural networks, arXiv 9 (2017). preprint arXiv:1710.02913.
[38] M.G. Kibria, K. Nguyen, G.P. Villardi, O. Zhao, K. Ishizu, F. Kojima, Big data analytics, machine learning, and artificial intelligence in next-generation wireless networks, IEEE Access 6 (2018) 32328–32338.
[39] R. Amiri, H. Mehrpouyan, L. Fridman, R.K. Mallik, A. Nallanathan, D. Matolak, A machine learning approach for power allocation in HetNets considering QoS, in: 2018 IEEE International Conference on Communications (ICC), IEEE, 2018, pp. 1–7.
[40] T. Kaistha, K. Ahuja, Load balancing in wireless heterogeneous network with artificial intelligence, in: Artificial intelligence, CRC Press, 2021, pp. 187–197.

CHAPTER 10

Intelligent intersystem handover delay reduction algorithm for heterogeneous wireless networks

Topside E. Mathonsi[a], Okuthe P. Kogeda[b], and Thomas O. Olwal[c]
[a]Department of Information Technology, Faculty of Information and Communication Technology, Tshwane University of Technology, Pretoria, South Africa
[b]Department of Computer Science and Informatics, Faculty of Natural and Agricultural Sciences, University of the Free State, Bloemfontein, South Africa
[c]Department of Electrical Engineering/F'SATI, Faculty of Engineering and the Built Environment, Tshwane University of Technology, Pretoria, South Africa

1. Introduction

Over the past few years, we have seen an increase in the number of users with the desire and need to be connected and reachable anywhere at any time. As a result, we have witnessed a growing figure of more cutting-edge mobile devices/mobile nodes (MNs), services on demand, services on offer, evaluation of wireless standards, and evaluation of wireless network access technologies in order to fulfill users' desire and needs [1–26].

Heterogeneous wireless networks (HWNs) are made up of different wireless network access technologies namely fifth-generation (5G) network, fourth-generation (4G) network/long-term evolution (LTE), long-term evolution-advanced (LTE-advanced), and IEEE 802.11 wireless local area network (WLAN), among others [1–26].

The existence of HWNs enables MNs users to access network services and applications globally at anytime and anywhere, between these different wireless network access technologies. The availability of handover/handoff services such as intersystem handover enables MNs/user equipment (UE) to switch between different radio access technologies (RATs) with different operational parameters and mobility management entity (MME)/serving gateway (S-GW) in HWNs [1–26].

Several handover algorithms, namely analytic hierarchy process (AHP) and adaptive neuro-fuzzy inference system (ANFIS), among others that use multiple input parameters for handover decision in HWNs have been proposed previously. These formerly recommended weighting-based handover algorithms select the target network/target base station (TBS) using multiple input parameters namely: received signal strength (RSS), service cost, delay, security, mobile speed, and battery status, among others [5–10]. However, these previously proposed handover algorithms have failed to predict the future values of the measured RSS needed for rapid handover processes.

Comprehensive Guide to Heterogeneous Networks
https://doi.org/10.1016/B978-0-323-90527-5.00001-0

As a result, these previously proposed handover algorithms experience lengthy handover delay as well as packet loss. This is because these handover algorithms started the network discovery process only when the RSS of the serving base station (SBS) was poor/very weak. In addition, existing handover algorithms are not adaptable to the changes of the network conditions and user's preferences based on a traffic type, i.e., real-time application and nonreal-time application. Hence, these handover algorithms are unable to select a target network capable of satisfying both the user's preferences and the application's quality of services (QoS) requirements. These previously proposed handover algorithms experience erroneous network selection, packet loss, and ping-pong effect due to the high-ranking abnormality as a result [5–10].

A careful pairing of the multiple-attribute decision-making (MADM) technique and network attributes' weighting algorithm is therefore necessary, in order to alleviate the identified problem. We proposed an intelligent intersystem handover (IIH) algorithm in order to ensure an optimum network selection, while balancing the trade-offs that exist between network conditions, user's preferences, and the application's QoS requirements. We designed the IIH algorithm by integrating gray prediction theory (GPT), MADM, fuzzy analytic hierarchy process (FAHP), and multiobjective optimization ratio analysis (MOORA) in order to lessen handover delay, packet loss, and ping-pong effect during the handover process in HWNs.

When designing the proposed IIH algorithm, we used the following multiple input parameters: predicted received signal strength (PRSS), signal-to-interference and noise ratio (SINR), jitter, network load, bandwidth, delay, service cost, mobile speed, and security for handover decision in HWNs [1–26]. During the handover decision, other existing schemes also consider the UE battery status. However, we deliberately overlooked this parameter as the user can control this parameter by connecting a UE to a battery charger while on the move.

We used GPT to predict the future values of the measured RSS of all available base stations (BSs)/networks. The proposed IIH algorithm utilized the obtained PRSS values to determine the proper time for the forthcoming handover and to preserve the continuity, and the quality of the current session. Hence, the proposed IIH algorithm minimalized the network discovery process and handover delay during the handover process in HWNs. Moreover, we proposed FAHP to calculate the weight of each network based on the input network parameters mentioned in this section. This is because FAHP can handle the imprecise information or uncertain data that come with the use of multiple parameters during weighting matrices. This resulted in minimalized inaccurate network selection and the consistency of network selection improved.

In addition, we proposed MOORA in order to calculate the QoS factor of each network so as to select the best suitable network with the highest QoS factor and maintain QoS requirements of ongoing services. MOORA allows the identification of beneficial parameters/attributes/criteria and nonbeneficial parameters during QoS factor estimation. This is because the users prefer a higher value for beneficial parameters and a lower value for nonbeneficial parameters. As a result, the ranking abnormality is lessened, hence decreased inaccurate network selection, packet loss, and ping-pong effect.

The assimilation of GPT, MADM, FAHP, and MOORA provided intelligent handover decision-making by selecting a target network capable of satisfying both the user's preferences and the application's QoS requirements. The proposed IIH algorithm can adapt to the changes of network conditions and users' preferences based on a traffic type. This resulted in reduced ranking abnormality and as a result, the proposed IIH algorithm further diminished erroneous network selection, packet loss, and ping-pong effect.

The remainder of this chapter is organized as follows: in Section 2, we provide the related works. In Section 3, we give the design of the proposed IIH algorithm. In Section 4, we present the simulation results. In Section 5, we conclude the chapter and provide the future work.

2. Related work

Recent related works have proposed numerous handover algorithms, which consider diverse aspects of network performance optimization such as the minimization of handover signaling, ping-pong effect, handover delay diminishing, and data/packet loss [1–26].

In this section, we discuss existing handover algorithms and various gaps identified in these algorithms.

2.1 Fuzzy logic-based algorithms

Zhang et al. [11] designed a handover algorithm by integrating gray relational analysis (GRA), FAHP, and entropy weight for handover decision in HWNs. They used mobile speed, service cost, RSS, bandwidth, security, and delay as input parameters for the handover decision. In their algorithm, they used FAHP and entropy theory to perform objective weight calculation and further used GRA to rank the available networks based on the obtained network weight. Their scheme reduced the call blocking probability while balancing the network load efficiently [11].

Zineb et al. [14] proposed an ANFIS algorithm that uses any given input data set to construct a fuzzy inference system and adjust the membership function parameters using a neuro-adaptive learning algorithm. They used mobile speed, bandwidth, security, network load, delay, and RSS as input parameters. They implemented their ANFIS algorithm using MATLAB/SIMULINK and the simulation results showed that their algorithm reduced inaccurate network selection.

Zineb et al. [16] proposed a quality of experience (QoE) fuzzy-based scheme to optimize QoS by selecting the best target network during the handover process in HWNs. Their algorithm used the following network parameters namely: RSS, bandwidth, mobile speed, delay, security, and battery level as input parameters for the handover decision. They used network simulator-2 (NS-2) to set up a HWNs environment using the following wireless network access technologies namely: UMTS, 4G, WLAN, and wireless wide area network (WWAN). The obtained simulation results showed that their QoE scheme maximized QoS by reducing high-ranking abnormality.

However, the algorithms presented by Zhang et al. [11], Zineb et al. [14,16] did not reduce the handover delay efficiently. This is because their algorithms failed to predict the future values of the measured RSS needed for rapid handover processes. As a result, their algorithms experienced a lengthy network discovery process, which led to a prolonged handover delay. This is because these handover algorithms started the network discovery process only when the RSS of the SBS was very weak. We proposed GPT to predict the future values of the measured RSS in order to minimalize handover delay by starting the network discovery process, while the RSS of the SBS was still good.

Azzali et al. [17] designed a vertical handover decision algorithm by integrating fuzzy logic and IEEE 802.21 Media independent handover (MIH) in order to improve the handover process in heterogeneous vehicular ad hoc networks (VANETs). Their algorithm used the following input parameters: throughput, handover rate, bandwidth, service cost, packet loss ratio, RSS, noise signal ratio, signal-to-interference ratio, carrier to interference ratio, bit error ratio, battery status, and mobile speed to select the best suitable network. They used NS-2 version 29 (NS-2.29) to evaluate the performance of their algorithm. Simulation results exhibited that their algorithm minimized handover delay, and packet loss, respectively.

Abdullah and Zukarnain [19] proposed a fuzzy logic-based vertical handover (FLBVH) algorithm for a trunking system. Their algorithm utilized RSS, mobile speed, delay, network load, security, and battery utilization as input parameters to design a fuzzy logic system and calculate the weight of each network. Their algorithm selected the network with the highest network weight as the target network. MATLAB simulation results showed that their algorithm ensured better QoS by reducing packet loss and handover delay.

However, the handover algorithms proposed by Azzali et al. [17], and Abdullah and Zukarnain [19] are not adaptable to the changes of the network conditions and users' preferences based on a traffic type. Hence, these algorithms are unable to select a target network capable of satisfying both the user's preferences and the application's QoS requirements. As a result, these previously proposed handover algorithms experience erroneous network selection and ping-pong effect due to the high-ranking abnormality.

We proposed FAHP to calculate the weight of each network based on the input parameters stated in Section 1. Furthermore, we used MOORA to calculate the QoS factor of each network by processing the weighted decision and, thereafter, select the best suitable target network with the highest QoS factor value. The proposed IIH algorithm could adapt to the changes of the network conditions and user's preferences based on a traffic type. The proposed IIH algorithm achieved this by using FAHP and MOORA for weighting matrices and QoS factor calculation, respectively. As a result, the proposed IIH algorithm reduced high-ranking abnormality, hence, minimized ping-pong effect and packet loss during the handover process in HWNs.

2.2 Dwell timer-based schemes

Omoniwa et al. [25] proposed the use of the amoebic-based geometric model to reduce unnecessary handovers in HWNs. In their algorithm, they used estimated dwell time along with the threshold values in order to ensure an optimal handover decision. They used Monte Carlo simulations to assess the behavior of their model in HWNs. Simulation results showed that their model was more robust and capable of reducing unnecessary handovers and handover failure probability.

However, the algorithm proposed by Omoniwa et al. [25] did not decrease the handover delay efficiently. This is because their algorithm failed to predict the future values of the measured RSS needed for rapid handover processes. As a result, their algorithm experienced a prolonged network discovery process, which led to a lengthy handover delay and packet loss. We used GPT to predict the future values of the measured RSS in order to start with the network discovery process, while the RSS of the SBS was good. As a result, the proposed IIH algorithm diminished the handover delay and packet loss during the handover process in HWNs.

2.3 Fuzzy logic-based algorithms

Mahira and Subhedar [24] proposed a multilayer feedforward artificial neural network algorithm for handover decision in HWNs. Their artificial neural network algorithm used data rate, service cost, bandwidth, RSS, and velocity of the mobile device as input parameters when calculating the network weight. Computer simulation results showed that their handover algorithm minimized the number of unnecessary handovers effectively.

The algorithm proposed by Mahira and Subhedar [24], however, did not reduce the handover delay efficiently. This is because their algorithm has failed to predict the future values of the measured RSS needed for rapid handover processes. As a result, their scheme experienced an extensive network discovery process, which led to an extended handover delay. This is mainly because their handover scheme started the network discovery process only when the RSS of the SBS was very weak. We introduced the GPT in the proposed IIH algorithm to predict the future values of the measured RSS in order to lessen handover delay and packet loss. The proposed IIH algorithm achieved this by starting the network discovery process, while the RSS of the SBS was good.

2.4 MULTIMOORA-based algorithms

Obayiuwana and Falowo [23] proposed a multicriteria handover decision-making scheme named multiplicative form with multiobjective optimization ratio analysis (MULTIMOORA), to select the best target network in HWNs. They integrated ratio system, reference point system, and multiplicative form in order to calculate the network weight and rank the available network. Their algorithm used mobile speed, bandwidth, jitter, network load, service cost, and delay as input parameters for a handover decision. They validated the performance of MULTIMOORA by comparing it with TOPSIS across the speed range of 1–100 km/h. The simulation results showed that MULTIMOORA performed better than TOPSIS by minimizing inaccurate network selection, packet loss, and ping-pong effect during the handover process in HWNs.

However, the algorithm presented by Obayiuwana and Falowo [23] did not reduce the handover delay efficiently. This is because their algorithm has failed to predict the future values of the measured RSS needed for rapid handover processes. As a result, their scheme experienced elongated handover delay and packet loss. This is because their handover algorithm started the network discovery process only when the RSS of the SBS was very weak. In the proposed IIH algorithm, we used GPT to predict the future values of the measured RSS in order to minimalize handover delay and packet loss. The proposed IIH algorithm achieved this by starting the network discovery process, while the RSS of the SBS was good.

2.5 Context-aware-based algorithms

Goudar et al. [22] proposed multicriteria-based handover algorithm for HWNs. In their work, they used RSS, data rate, battery status, packet loss, delay, jitter, and mobile speed as input parameters for a handover decision. They used a context-aware AHP mechanism to obtain network weight and rank the available networks according to the obtained network weights. They implemented and validated their proposed handover algorithm using MATLAB. The obtained simulation findings demonstrated that their handover algorithm outperformed classic MADM methods namely: technique for order preferences by similarity to the ideal solution (TOPSIS), simple additive weighting (SAW), and GRA with respect to the number of handover failures. However, the handover algorithm proposed by Goudar et al. [22] is not adaptable to the changes of the network conditions and user's preferences based on a traffic type. Their algorithm is unable to select a target network capable of satisfying both the user's preferences and the application's QoS requirements. Their handover algorithm experiences erroneous network selection, packet loss, and ping-pong effect due to the high-ranking abnormality, as a result. The proposed IIH algorithm could adapt to the changes of the network conditions and user's preferences based on a traffic type. The proposed IIH algorithm achieved this by using FAHP and MOORA for weighting matrices and QoS factor calculation. As a result, the proposed IIH algorithm reduced high-ranking abnormality hence minimized ping-pong effect and packet loss during the handover process in HWNs.

2.6 TOPSIS and AHP-based algorithms

Maaloul et al. [21] proposed a handover model that used TOPSIS to select the target network. For the handover decision, their algorithm used bandwidth, RSS, mobile speed, service cost, and delay as input parameters. Their model used TOPSIS to calculate network weight and rank the available networks based on obtained weight. The simulation results showed that their algorithm selected the best suitable network and improved network throughput.

Guo and Li [20] proposed the use of AHP to calculate the network weight in order to select the target network based on the obtained weight. Their algorithm used the following parameters namely: service cost, security, power consumption, mobile speed, RSS, bandwidth, and signal delay for a handover decision. MATLAB simulation results showed that AHP selected the best target network and improved user experience in HWNs.

However, both TOPSIS and AHP suffer from high-ranking abnormality, which occurs when the deletion of the lowest graded network affects the grading of the top-graded ones and therefore obscures decision-making. This led to packet loss and ping-pong effect during the handover process in HWNs. We proposed FAHP and MOORA to calculate the weight and QoS factor, respectively. The proposed IIH algorithm selected the network with the highest QoS as the target network. As a result, the proposed IIH algorithm minimized the ping-pong effect and packet loss during the handover process in HWNs.

2.7 Media independent handover-based algorithms

Kwon et al. [18] proposed the integration of the access network discovery and selection function (ANDSF) algorithm with the MIH algorithm. The aim was to minimize inaccurate network selection and packet loss during the handover process in HWNs. NS-2 simulation results showed that an integration of these two mechanisms minimized inaccurate network selection and packet loss during the handover process in HWNs.

However, their solution failed to minimize handover delay efficiently. This is because their solution has failed to predict the future values of the measured RSS needed for rapid handover processes. In addition, they did not show how they calculated the QoS factor and if they used all QoS-related parameters, i.e., bandwidth, delay, and jitter during weighting matrices and QoS factor calculation. The proposed IIH algorithm used GPT to predict the future value of the measured RSS needed for rapid handover processes. Hence, lessened the network discovery process and minimized handover delay as a result. Furthermore, the proposed IIH algorithm considered the following

QoS-related parameters namely: bandwidth, delay, and jitter during network weighting metrics and QoS factor calculation, respectively. This resulted in decreased inaccurate network selection, packet loss, and ping-pong effect.

2.8 Comparison of related handover schemes

The literature review showed that other researchers have designed and implemented handover algorithms in order to lessen handover delay, packet loss, and ping-pong effect during the handover process in HWNs. However, more research is still required in this area because the existing handover algorithms result in unfavorable situations such as unnecessary delays, long communication disruptions, loss of signal, service degradation, poor handover decision-making, inappropriate handover triggering, and mobility management.

As a result, we proposed an IIH algorithm in order to ensure an optimum network selection while balancing the trade-offs that exist between network conditions, user's preferences, and the application's QoS requirements. We designed an IIH algorithm by integrating GPT, MADM, FAHP, and MOORA in order to lessen handover delay, packet loss, and ping-pong effect. For the handover decision, the proposed IIH algorithm used multiple input parameters mentioned in Section 1 [5–10]. To the best of our knowledge, this chapter is the first attempt to incorporate GPT, MADM, FAHP, and MOORA in order to perform handover decision/network selection in the HWNs environment. We present a summary of the identified advantages and disadvantages of the existing handover algorithms in Table 1.

3. Design of Intelligent Intersystem Handover Algorithm

We assimilated existing solutions namely GPT, MADM, FAHP, and MOORA when designing the proposed IIH algorithm to minimalize handover delay, packet loss, and ping-pong effect during the handover process in HWNs. In addition, the proposed IIH algorithm was designed to select a target network capable of satisfying both the user's preferences and the application's QoS requirements.

3.1 Gray prediction theory

To improve the network discovery speed and accuracy while reducing handover delay, we used GPT to predict the future values of the measured RSS and then used the obtained PRSS values to initiate the handover process. GPT uses a predefined mathematical model to make precise predictions and the model can deal with the systems that have unfixed parameters and it is more accurate with its prediction. In addition, this model predicts the future values based only on a set of the most recent data depending on the window size of the predictor. As a result, the model provides prediction results that

Table 1 Advantages and disadvantages of existing handover algorithms [9].

Category of handover decision scheme handover decision scheme	Advantages of the scheme	Disadvantages of the scheme
ANN-based schemes	• Reduced handover failure • Better network selection • Lower handover processing delay	• High latency • Slow training and learning • Supplementary resource consumption
Fuzzy logic-based schemes	• Reduced handover processing delay • Reduced packet loss • User satisfaction for QoS	• Increased complexity
MIH-based schemes	• Reduced packet loss • Optimal network selection • Embedded security • Optimized throughput	• Supplementary signaling • Higher resource consumption
Mobility prediction-based schemes	• Reduced ping-pong for low mobile speed • Suitable in uncertain network environments	• Instability for variable speeds • Longer handover delay • Higher packet loss
AHP-based schemes	• Reduced packet loss • Better throughput	• Resource consuming • Might compromise on QoS if a low-cost network is available
Dwell timer-based schemes	• Reduced handover failure	• Increased packet loss • Increased signaling • Unsuitable for real-time applications
RSS threshold-based schemes	• Reduced false hand-over initiation • Reduced handover failure	• Increased handover failures • Wastage of network resources

are more accurate. We used Eq. (1) to predict the future values of the measured RSS of all available networks.

$$PRSS^0(n+1) = \left(RSS^0(1) - \frac{b}{c}\right)e^{-cn}\left(1 - e^{c}\right) \tag{1}$$

where c represents the development gray number and b represents the internal control gray number. The intersystem handover was initiated if the RSS level of the SBS drops below a threshold, and the PRSS level of the TBS is greater than that of the SBS and hysteresis margin as illustrated by.

$$\begin{gathered} \text{if } PRSS_{TBS} > RSS_{SBS} + HYS\&RSS_{SBS} < TH \\ \text{or} \\ \text{if } PRSS_{TBS} > RSS_{SBS} + HYS\&PRSS_{TBS} > TH \end{gathered} \tag{2}$$

After initiating the intersystem handover, we used multiple parameters to design a multiattribute decision matrix as discussed in the following section. Later, in Section 3.4, we show the integration of all the equations used when designing the proposed IIH algorithm.

3.2 Multiattribute decision-making

We introduced multiattribute decision-making (MADM) into the proposed IIH algorithm in order to provide context awareness during handover decision-making in HWNs. This means the network selection process is more intelligent considering varying needs of the application's QoS requirements and users' preference in order to recommend the best suitable network in fluctuating network conditions. In order to design a multiattribute decision matrix, the following network attributes namely: PRSS, SINR, jitter, network load, bandwidth, delay, service cost, mobile speed, and security, were used, respectively, as exemplified by.

$$A = \left(a_{ij}\right)_{n \times m} = \begin{bmatrix} a_{11} & a_{12} & \cdots & a_{1n} \\ a_{21} & a_{22} & \cdots & a_{2n} \\ \vdots & \vdots & \ddots & \vdots \\ a_{m1} & a_{m2} & \cdots & a_{mn} \end{bmatrix} \tag{3}$$

where n represents the number of networks and m is the number of the attribute value of each network, whereas, a_{ij} is the j_{th} attribute value of the i_{th} network. In the following section, we calculate the weight of each network using FAHP and the QoS factor using MOORA, respectively.

3.3 Fuzzy analytic hierarchy process and multiobjective optimization ratio

3.3.1 Analysis

Due to the dynamic nature of the HWNs environment, accurate measurement of the network parameters used to calculate the weight of each TBS is a difficult task. As a result, we used FAHP and MOORA for weighting matrices and QoS factor calculation. We used FAHP because the technique can handle the imprecise information or uncertain data that come with the use of multiple parameters during weighting matrices. In order to ensure that the correct values of the multiple parameters are used, FAHP uses TFNs to clean up the uncertain data that come with the use of multiple parameters for weighting matrices. This ensured that the correct values are used for weighting matrices and QoS factor calculation, respectively. In addition, TFNs are used to represent the membership function of the multiple parameters used for decision marking. TFNs are valued in the real unit interval and this permits the gradual calculation of the membership of multiple parameters in a set. As a result, we used FAHP to minimalize inaccurate network selection and improve the consistency of network selection. Therefore, after designing multi-attribute decision matrix using Eq. (3), we used TFNs to construct a fuzzy comparison matrix as $x = (L, M, U)$, where L and U are the lower and upper limits of each attribute, while the parameter M is the most possible median value as illustrated by.

$$\mu(x) = \left\{ \begin{array}{cc} \frac{1}{M-L}x - \frac{L}{M-L}, & x \in [L, M], \\ \frac{1}{M-U}x - \frac{U}{M-U}, & xe[M, U], \\ 0 & \text{other} \end{array} \right\} \tag{4}$$

Continuously, we constructed a FAHP comparison matrix that contains TFNs representing pair-wise comparisons as exhibited by.

$$\tilde{A}\left(a_{ij}\right)_{n \times m} = \begin{bmatrix} (1,1,1) & (L_{12}, M_{12}, U_{12}) & \cdots & (L_{1n}, M_{1n}, U_{1n}) \\ (L_{21}, M_{21}, U_{21}) & (1,1,1) & \cdots & (L_{2n}, M_{2n}, U_{2n}) \\ \vdots & \vdots & \ddots & \vdots \\ (L_{m1}, M_{m1}, U_{m1}) & (L_{m2}, M_{m2}, U_{m2}) & \cdots & (1,1,1) \end{bmatrix} \tag{5}$$

After creating the FAHP comparison matrix using Eq. (5), we calculated the weight W of each network i as given by.

$$W_i = \frac{\sum_{j=1}^{m} a_{ij}}{\sum_{i=1}^{n}\sum_{j=1}^{m} a_{ij}} \tag{6}$$

Here, the fuzzy triangle number a_{ij} represents all the elements of the fuzzy pair-wise matrix. Successively, we used MOORA to normalize the obtained weight by Eq. (6) in order to calculate the QoS factor of each network as exemplified by Eq. (7). We used MOORA to calculate the QoS factor of each network with the intention of selecting the best suitable network, with the highest QoS factor. Hence, the proposed IIH algorithm maintained the QoS requirements of ongoing services in the UE. MOORA allows the identification of beneficial parameters and nonbeneficial parameters during QoS factor estimation. This is because network users prefer a higher value for beneficial parameters and a lower value for nonbeneficial parameters. As a result, the proposed IIH algorithm could adapt to the changes of network conditions and users' preferences based on a traffic type.

$$Q_{ij} = \sum_{j=m+1}^{n} \left(\frac{R_j - R_{ij}^{\min}}{R_j^{\max}} \right) + \sum_{j=m-1}^{n} \left(\frac{R_j^{\max} - R_{ij}}{R_j^{\max} - R_j^{\min}} \right) \tag{7}$$

where $i=1, 2, \ldots, n$ and $j=1, 2, \ldots, m$ represent the number of networks and attributes, respectively. $R_j^{\max}$ and $R_j^{\min}$ are the maximum and minimum values of an attribute for a particular service. R_{ij} is the actual value of the attribute in network i, whereas $m+1$ and $m-1$ represent the number of attributes to be maximized and the number of attributes to be minimized, respectively. In order to normalize the beneficial criteria and nonbeneficial criteria for nonreal-time applications (i.e., HTTP and e-mail), we expanded Eq. (7) to obtain Eq. (8). In Eq. (8), D, J, NL, and SC are downward criteria, while PRSS, SINR, BW, MS, and S are upward criteria.

$$\begin{aligned} Q_{ij} = \sum_{j=m+1}^{n} & \left(\begin{array}{l} \dfrac{PRSS_{ij} - PRSS_j^{\min}}{PRSS_j^{\max} - PRSS_j^{\min}} + \dfrac{SINR_{ij} - SINR_j^{\min}}{SINR_j^{\max} - SINR_j^{\min}} + \\ \dfrac{BW_{ij} - BW_j^{\min}}{BW_j^{\max} - BW_j^{\min}} + \dfrac{MS_{ij} - MS_j^{\min}}{MS_j^{\max} - MS_j^{\min}} + \dfrac{S_{ij} - S_j^{\min}}{S_j^{\max} - S_j^{\min}} \end{array} \right) \\ + \sum_{j=m-1}^{n} & \left(\begin{array}{l} \dfrac{D_j^{\max} - D_{ij}}{D_j^{\max} - D_j^{\min}} + \dfrac{J_j^{\max} - J_{ij}}{J_j^{\max} - J_j^{\min}} + \\ \dfrac{NL_j^{\max} - NL_{ij}}{NL_j^{\max} - NL_j^{\min}} + \dfrac{SC_j^{\max} - SC_{ij}}{SC_j^{\max} - SC_j^{\min}} \end{array} \right) \end{aligned} \tag{8}$$

In order to normalize the beneficial criteria and nonbeneficial criteria for real-time applications (i.e., video streaming), we expanded Eq. (7) to obtain Eq. (9). In Eq. (9), D, J, and NL are downward criteria, while PRSS, SINR, BW, MS, and S are upward criteria. Service cost is not part of the handover decision for real-time applications as we assume users are willing to pay for better QoS when accessing this traffic type.

$$\begin{aligned} Q_{ij} = \sum_{j=m+1}^{n} & \left(\begin{array}{l} \dfrac{PRSS_{ij} - PRSS_j^{\min}}{PRSS_j^{\max} - PRSS_j^{\min}} + \dfrac{SINR_{ij} - SINR_j^{\min}}{SINR_j^{\max} - SINR_j^{\min}} + \\ \dfrac{BW_{ij} - BW_j^{\min}}{BW_j^{\max} - BW_j^{\min}} + \dfrac{MS_{ij} - MS_j^{\min}}{MS_j^{\max} - MS_j^{\min}} + \dfrac{S_{ij} - S_j^{\min}}{S_j^{\max} - S_j^{\min}} \end{array} \right) \\ + \sum_{j=m-1}^{n} & \left(\dfrac{D_j^{\max} - D_{ij}}{D_j^{\max} - D_j^{\min}} + \dfrac{J_j^{\max} - J_{ij}}{J_j^{\max} - J_j^{\min}} + \dfrac{NL_j^{\max} - NL_{ij}}{NL_j^{\max} - NL_j^{\min}} \right) \end{aligned} \tag{9}$$

After normalizing the beneficial criteria and nonbeneficial criteria given by Eqs. (8) and (9), we calculated the QoS factor given by Eq. (10) in order to rank the available networks. During the QoS factor calculation by Eq. (10), the value of R_j^{max} is replaced by the value obtained by Eq. (8) for nonreal-time applications. While for real-time applications, the value of R_j^{max} is replaced by the value obtained by.

$$QoS_i = \frac{Q_{ij} \times W_i}{\sqrt{\sum_{i=1}^{n} Q_{ij}^2}} \quad i = 1, 2, \cdots n \tag{10}$$

We used the obtained QoS factor value to rank the available networks in the descending order. Thereafter, the UE selects the network with the highest QoS factor as the target network. The use of FAHP and MOORA for weighting matrices and QoS factor calculation ensured that the proposed IIH algorithm adapts to the changes of network conditions and users' preferences based on a traffic type. Next, in Section 3.4, we present the developed IIH algorithm.

3.4 Intelligent intersystem handover algorithm

We integrated GPT, MADM, FAHP, and MOORA when designing the proposed IIH algorithm with the aim of selecting a target network capable of satisfying both the users' preferences and the application's QoS requirements as given by Algorithm 1.

Algorithm 1: Intelligent intersystem handover algorithm

Initialization:

1. $PRSS, SINR, BW, MS, S, D, J, NL, SC=0$

Process:

2. UE scans for available BSs

3. Apply Eq. (1)

4. While $PRSS \neq 0$

5. If $PRSS > RSS + HYS$ & $RSS < TH$

6. Apply Eq. (3)

7. Apply Eqs. (4)-(6)

8. If nonreal-time applications

9. Apply Eq. (8)

10. Else Apply Eq. (9)

11. End If

12. Apply Eq. (10)

13. Else UE remains in its SBS

14. End If

Output:

15. return QoS_i

16. End while

After the TBS was selected, the TBS reserves the available resources, then sends the handover acknowledgement message to the SBS so that it can perform the handover. After receiving the acknowledgement message, the SBS transmits a handover command to the UE. In the proposed IIH algorithm, before the UE detaches itself from the SBS, it synchronizes with the TBS first and be configured with new IP address by the DHCP server. This allowed the proposed IIH algorithm to maintain the QoS requirements of ongoing services on the UE. The new IP address is assigned on the second interface of the UE. Thereafter, the MME/S-GW exchanges messages with TBS. Upon the reception of the release message sent by TBS, the SBS releases the control of related resources. Subsequently, TBS transmits the downlink packet data to UE in order to complete the intersystem handover procedure as exemplified in Fig. 1.

In the next section, we present the simulation results recorded during the several simulations conducted to evaluate the performance of the proposed IIH algorithm.

4. Simulation results

Network simulators allow many researchers to evaluate protocols and applications under varying network configurations. Simulators allow researchers to model the network in order to get the approximate behavior of the real communication networks, as it is expensive and time consuming to set up a real network. We implemented the proposed IIH algorithm using the IEEE 802.11 ac model developed in NS-2 version 35 (NS-2.35). We ran NS-2 on Linux Ubuntu 14.04 operating system, 512 RAM, and 10 GB of storage. In addition, we installed an LTE patch in order to set up LTE nodes and the LTE base stations. We further installed Ns2viop patch to allow nodes to send and receive voice/video packets. We used the tool command language (Tcl) script to simulate the network topology. In addition, since NS-2 uses C++ to implement network algorithms, it was easier to implement GPT, MADM, FAHP, and MOORA methods in NS-2 as a result. We configured a network topology of 500 × 500 m with fixed located two LTE base stations (labeled LTE eNB1 and LTE eNB2), one LTE MME/S-GW, two WLAN base stations (labeled WLAN BS1 and WLAN BS2), and 10 randomly located nodes (labeled node 4–14) (see Fig. 2).

We configured the simulation to start transmitting packets at 10 s and stop transmitting at 100 s. In the simulations, we used an FTP traffic load with a file size of 500 bytes for email services. We further used the FTP traffic load with a file size of 1000 bytes for HTTP services. In addition, the size of the online video-conferencing frame was 84,480 bytes in the simulations. We configured email and HTTP applications as nonreal-time traffic, while online video conference as real-time traffic. We recorded the performance results gained in out.tr script, then we used R programming language to display the simulation results. We used R programming because the source code for the R software environment is written primarily in C++, and is freely available under the GNU General Public License. In addition, precompiled binary versions are provided for

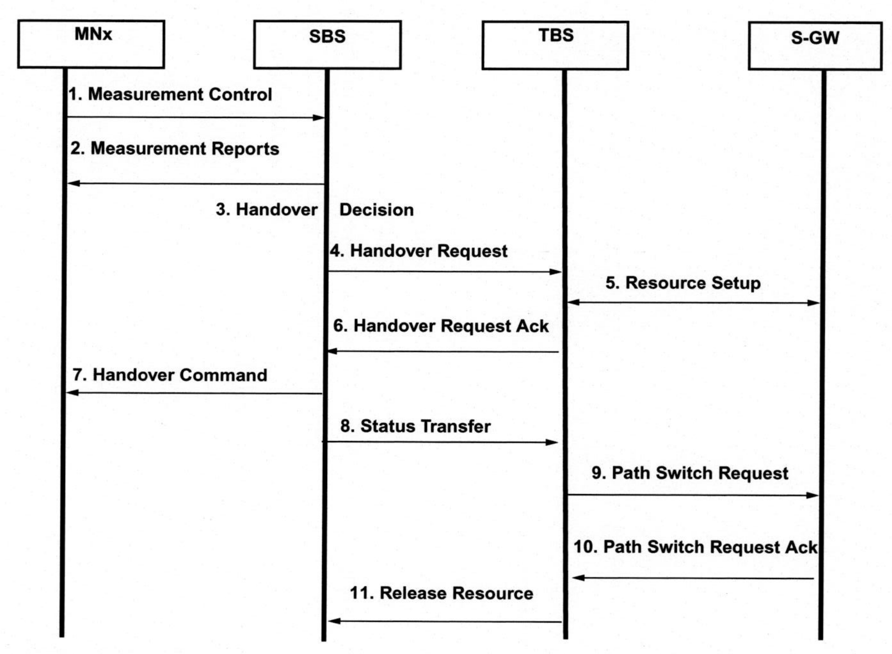

Fig. 1 Intelligent intersystem handover procedure.

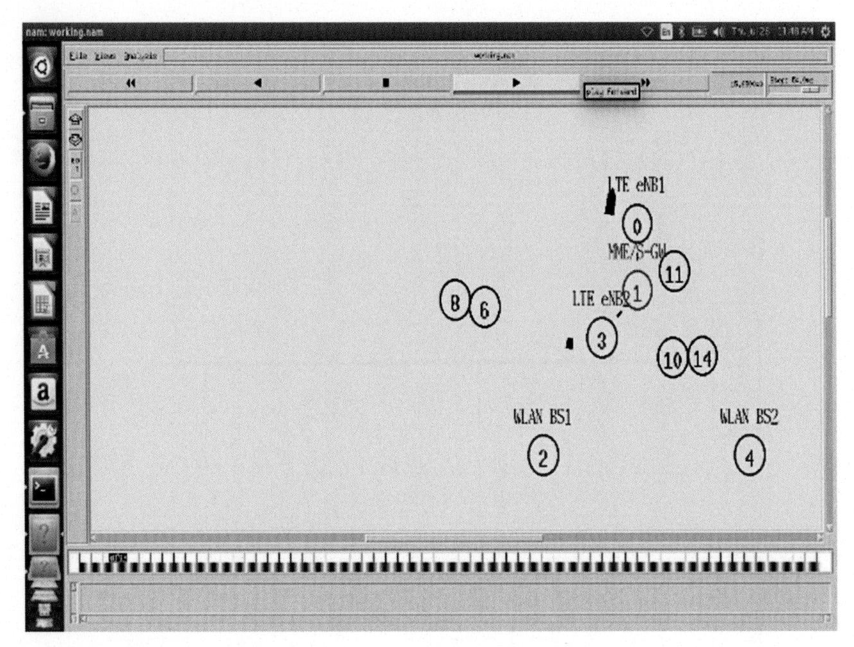

Fig. 2 HWNs simulation scenario.

various operating systems. Although R software has a command-line interface, there are several graphical user interfaces, such as RStudio, which is an integrated development environment that makes it easy for a researcher to analyze the results.

Moreover, we configured the MNs to use multiple interfaces in order to allow the MNs to communicate with two BSs during the handover process. Thereafter, we validated the performance of the three algorithms using four scenarios. In scenario 1, the MNs were moving at varying speeds between 1 and 120 km/h. In scenario 2, the MNs were moving at varying slow speeds between 1 and 30 km/h. In scenario 3, the MNs were moving at varying high speeds between 100 and 120 km/h. In scenario 4, the MNs were moving at a constant speed of 60 km/h. The reason we used the different speeds is that these speeds are normally recommended in South African roads, i.e., when driving near a school is 30 km/h, when driving in a township/urban area is 60 km/h, and lastly, when driving on a freeway is 120 km/h. We performed the simulations numerous times in order to achieve very convincing results.

We used the ad hoc on-demand distance vector (AODV) routing protocol this is because the AODV routing protocol enables dynamic, self-starting routing between MNs that are communicating. AODV creates routes only when they are needed, which makes it easier to be implemented in HWNs. In addition, AODV supports both unicast and multicast routing packet transmissions. AODV builds routes between MNs when source nodes need to establish and maintain communication with other MNs on a wireless network [27]. In addition, we used 802.11 ac because it is a dual-band wireless technology and supports simultaneous connections on both the 2.4 and 5 GHz Wi-Fi bands. Furthermore, 802.11ac offers backward compatibility to 802.11b/g/n and bandwidth rated up to 1300 Mbps on the 5 GHz band plus up to 450 Mbps on 2.4 GHz [28].

We used the IEEE 802.21 handover standard in order to allow MIH. This standard enables seamless handover between wired and wireless networks of the same type as well as handover between different wired and wireless networks referred to as MIH or intersystem handover. In addition, the IEEE 802.21 standard specifies link-layer intelligence and other related network information to upper layers in order to optimize handovers between networks of different types, such as WiMAX, Wi-Fi, and 4G [28]. This made it easier to implement the proposed IIH algorithm in NS-2.

For the radio propagation model, we chose to use a two-ray-ground reflection model, since it is more suitable for our computer simulation as it makes use of both the ground reflection and direct path for communication. In addition, the two-ray-ground reflection model was used because it does not require lengthy simulation time as compared to free-space model. We configured our MNs to use an omnidirectional antenna for both transmission and reception. For the processing and forwarding of packets, we used AODV protocol in the link layer (LL). We further used LossMonitor, a packet sink that checks for packet losses. We summarize the simulation parameters and values used in the simulations in Table 2.

4.1 Findings and discussions

In this section, we discuss the performance of the proposed IIH algorithm based on the results we obtained from the conducted simulation tests with respect to the four performance metrics that we considered: handover delay, packet loss, ping-pong effect, and throughput. This is because QoS and QoE rely on these four performance metrics during the handover in HWNs. We used trace files to record the results obtained during simulations in order to calculate average handover delay, probability of ping-pong effect, average packet loss, and average throughput.

We compared the IIH algorithm with the FLBVH algorithm and the ANFIS algorithm. We chose the FLBVH algorithm mainly because the algorithm uses FAHP only for network selection and we used FAHP in designing the proposed IIH algorithm. We chose the ANFIS algorithm because this algorithm uses any given input data set to construct a fuzzy inference system. It also adjusts the membership function of the parameters using a neuro-adaptive learning algorithm.

Besides, both the FLBVH algorithm and the ANFIS algorithm use multiple parameters to select the TBS and have been previously implemented in HWNs. Abdullah and Zukarnain [19] and Zineb et al. [14] claimed that the FLBVH algorithm and the ANFIS algorithm are the best when it comes to network selection. We, therefore, wanted to see if our proposed IIH algorithm could outperform them. It is worth noting that the time used to predict the future values of the measured RSS is not included/accounted for in the handover delay measurement.

Table 2 Simulation parameters.

Parameters	Values
Routing protocol	AODV
MAC protocol	IEEE 802.11 ac
Handover standard	IEEE 802.21
Channel type	Wireless Channel
Network interface type	WirelessPhy
Propagation model	TwoRayGround
Antenna model	OmniAntenna
Queue type	Queue/DropTail/PriQueue
Link layer type	LL
Number of nodes	10
Max packets in ifq	100
Mobility mode	Random waypoint
Bandwidth	20 Gpbs for each BS
Simulation area	500 × 500 m
Simulation time	100 s
Agent Trace	ON
Router Trace	ON
Mac Trace	OFF
Movement Trace	ON

4.1.1 Scenario 1: Performance evaluation for real-time traffic (1–120 km/h)

Average handover delay

Handover delay is the time elapsed from the transmission of a packet from the old access network to the new access network after a successful handover in HWNs. We present the real-time traffic average handover delay of the three compared algorithms when the MNs were moving at varying speeds between 1 and 120 km/h as shown in Fig. 3. The proposed IIH algorithm, the FLBVH algorithm, and the ANFIS algorithm have shown an average handover of 1.9, 2.8, and 3.5 s, respectively.

Average probability of ping-pong effect

The ping-pong effect occurs when the SBS handover to TBS, which quickly handovers the UE back to the SBS. This happens if TBS does not have sufficient bandwidth to support the application or service running on the UE. During performance evaluation, the proposed IIH algorithm, the FLBVH algorithm, and the ANFIS algorithm experienced 4.6%, 6.75%, and 8.1% probability of ping-pong effect, respectively, when the MNs were moving at varying speeds between 1 and 120 km/h as illustrated in Fig. 4.

Average packet loss

Packet loss occurs when one or more packets fail to reach its intended destination, during packet transmission. It is, therefore, important to minimize packet loss during handover

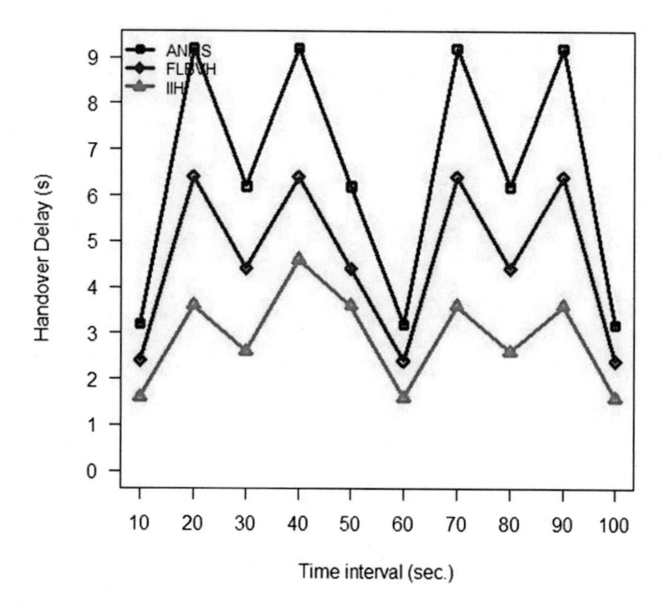

Fig. 3 Real-time traffic average handover delay at varying speeds.

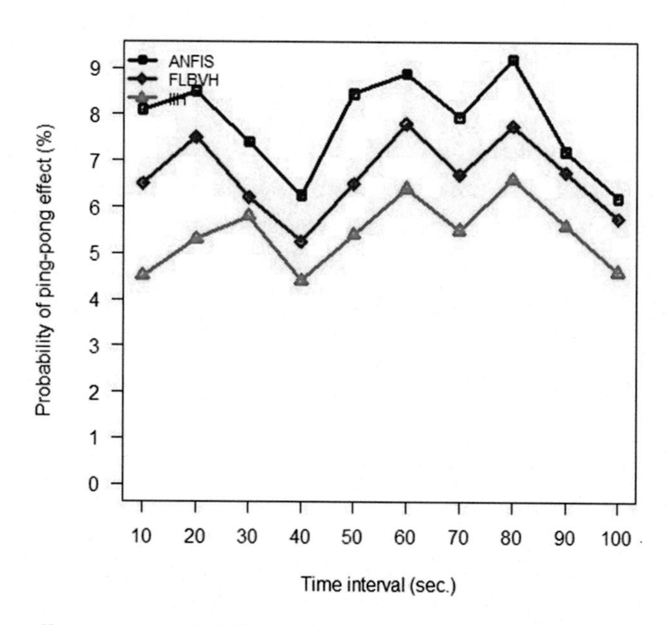

Fig. 4 Real-time traffic average probability of ping-pong effect at varying speeds.

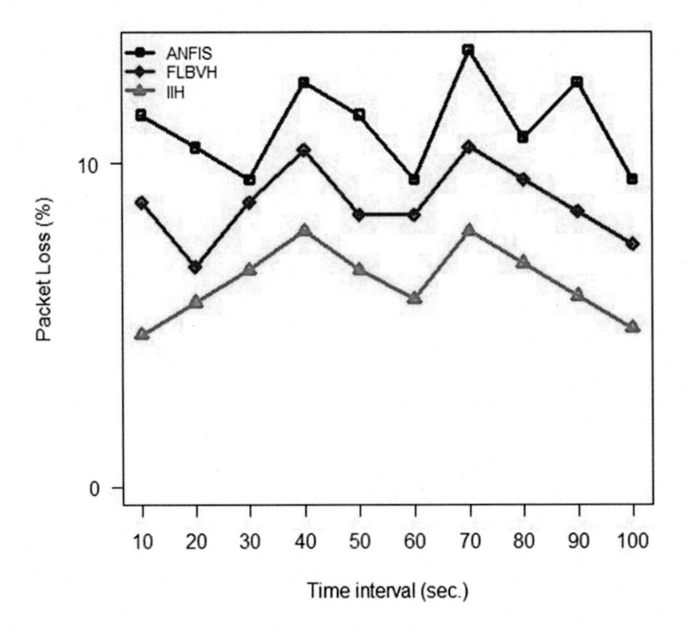

Fig. 5 Real-time traffic average packet loss at varying speeds.

in HWNs. During performance evaluation, the proposed IIH algorithm, the FLBVH algorithm, and the ANFIS algorithm experienced an average packet loss of 4.9%, 9.5%, and 10.5%, respectively, when the MNs were moving at varying speeds between 1 and 120 km/h as exemplified in Fig. 5.

Average network throughput

Network throughput refers to the rate at which the data are transferred from a UE to its destination. We compared the average network throughput for the three algorithms when the MNs were moving at varying speeds between 1 and120 km/h as shown in Fig. 6. On average, the proposed IIH algorithm, the FLBVH algorithm, and the ANFIS algorithm have shown 95.1%, 90.5%, and 89.5% throughput performance, respectively, at a time interval of 100 s as exhibited in Fig. 6.

4.1.2 Scenario 2: Performance evaluation for real-time traffic (1–30 km/h)

Average handover delay

We compared the three algorithms and we present the average handover delay when the MNs were moving at varying slow speeds between 1 and 30 km/h as shown in Fig. 7. The IIH algorithm, the FLBVH algorithm, and the ANFIS algorithm have shown an average handover delay of 1.4, 2.1, and 2.8 s, respectively.

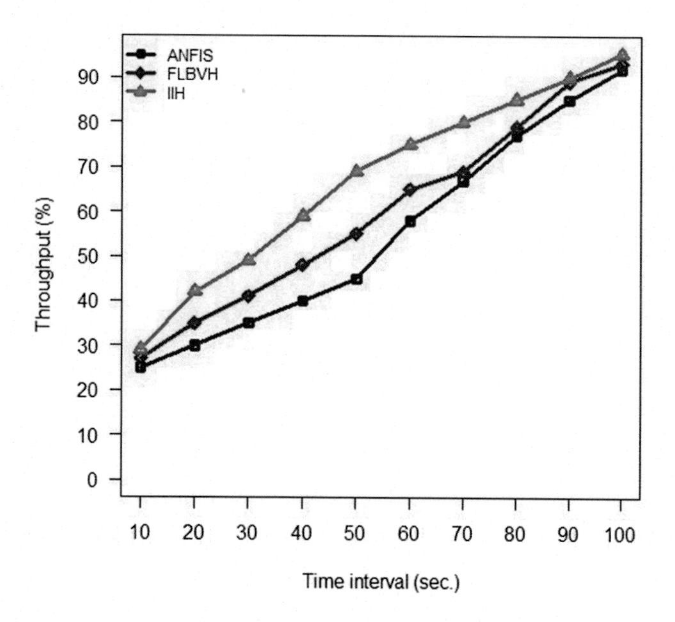

Fig. 6 Real-time traffic average network throughput at varying speeds.

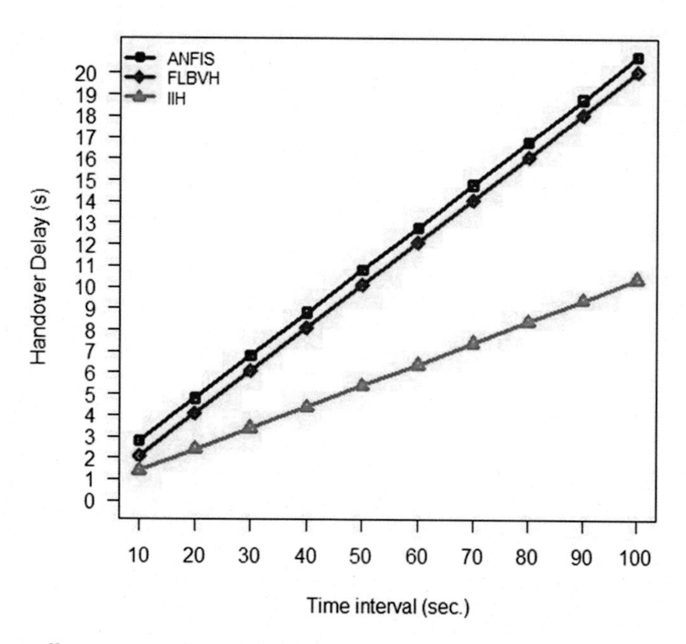

Fig. 7 Real-time traffic average network throughput at varying speeds.

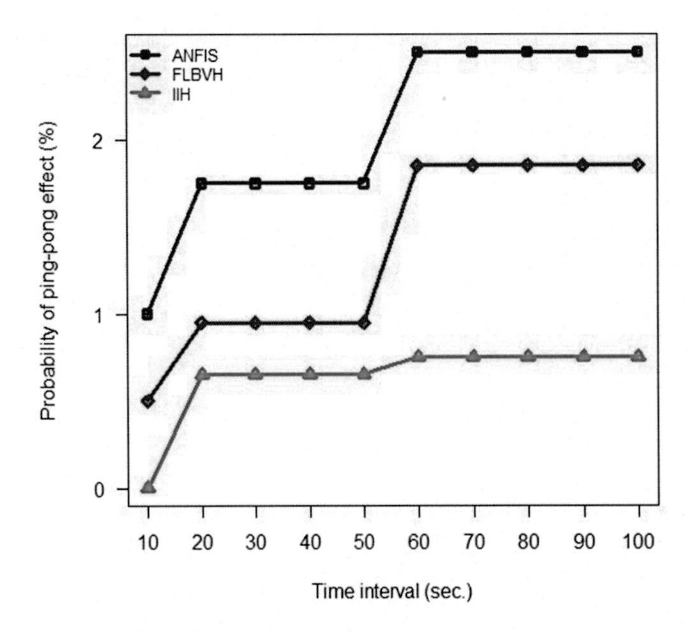

Fig. 8 Real-time traffic average probability of ping-pong effect at varying slow speeds.

Average probability of ping-pong effect

We evaluated the three algorithms and we present the average probability of the ping-pong effect when the MNs were moving at varying slow speeds between 1 and 30 km/h as shown in Fig. 8. On average, the IIH algorithm, the FLBVH algorithm, and the ANFIS algorithm have shown an average probability of ping-pong effect of 0.75%, 1.85%, and 2.5%, respectively.

Average packet loss

During performance evaluation, the proposed IIH algorithm, the FLBVH algorithm, and the ANFIS algorithm experienced 2.5%, 4.5%, and 6.5% average packet loss, respectively, when the MNs were moving at varying slow speeds between 1 and 30 km/h as illustrated in Fig. 9.

Average network throughput

During performance evaluation, on average the proposed IIH algorithm, the FLBVH algorithm, and the ANFIS algorithm have shown 97.5%, 95.5%, and 93.5% throughput performance, respectively, at a time interval of 100 s when the MNs were moving at varying slow speeds between 1 and 30 km/h as exhibited in Fig. 10.

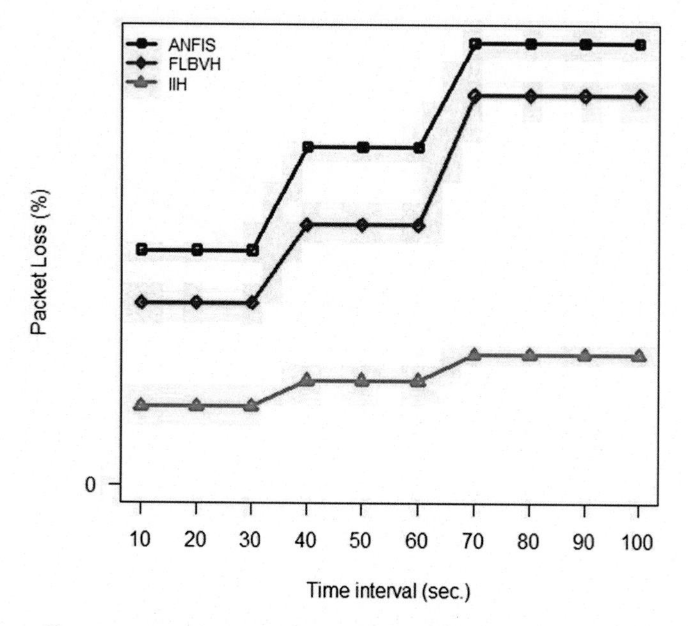

Fig. 9 Real-time traffic average packet loss at varying slow speeds.

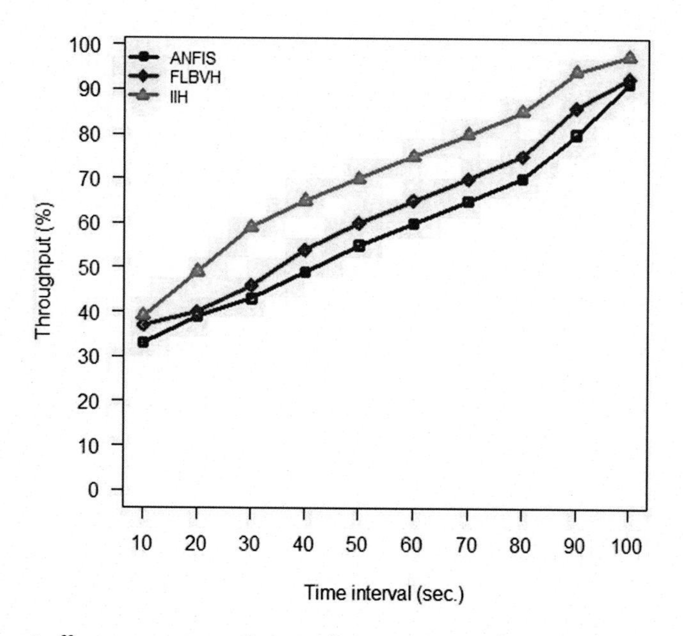

Fig. 10 Real-time traffic average network throughput at varying slow speeds.

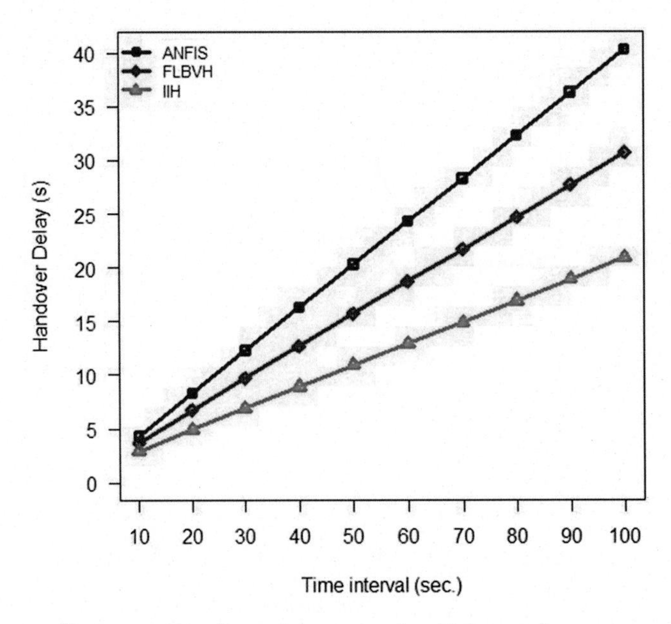

Fig. 11 Real-time traffic average handover delay at varying high speeds.

4.1.3 Scenario 3: Performance evaluation for real-time traffic (100–120 km/h)

Average handover delay

We evaluated the handover delay of the three compared algorithms, and we present the results as shown in Fig. 11 when the MNs were moving at varying high speeds between 100 and 120 km/h. The proposed IIH algorithm, the FLBVH algorithm, and the ANFIS algorithm have shown an average handover delay of 2.9, 3.7, and 4.3 s, respectively.

Average probability of ping-pong effect

We evaluated the three algorithms and we present the average probability of the ping-pong effect when the MNs were moving at varying high speeds between 100 and 120 km/h as shown in Fig. 12. The proposed IIH algorithm, the FLBVH algorithm, and the ANFIS algorithm have shown an average probability of ping-pong effect of 6.85%, 8.35%, and 9.6%, respectively.

Average packet loss

During performance evaluation, the proposed IIH algorithm, the FLBVH algorithm, and the ANFIS algorithm experienced an average packet loss of 6%, 11.5%, and 12.5% respectively, when the MNs were moving at varying high speeds between 100 and 120 km/h as illustrated in Fig. 13.

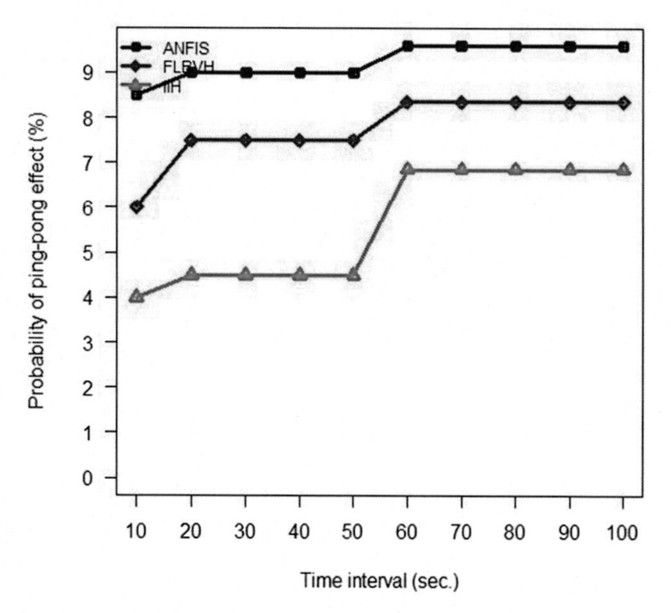

Fig. 12 Real-time traffic average probability of ping-pong effect at varying high speeds.

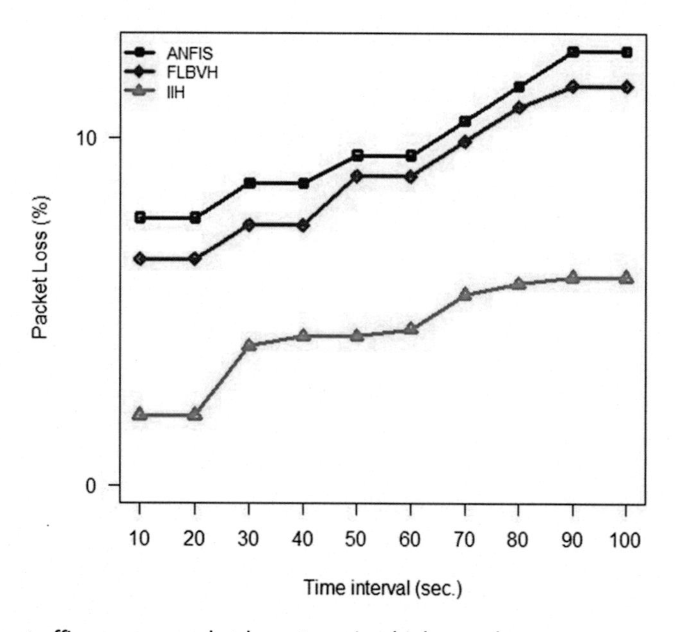

Fig. 13 Real-time traffic average packet loss at varying high speeds.

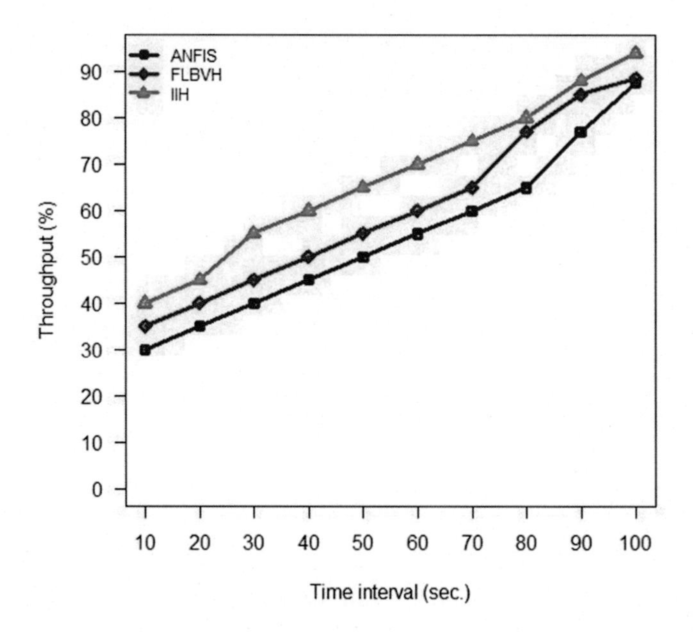

Fig. 14 Real-time traffic average network throughput at varying high speeds.

Average network throughput

We compared the average network throughput for the three algorithms as shown in Fig. 14 when the MNs were moving at varying high speeds between 100 and 120km/h. On average, the proposed IIH algorithm, the FLBVH algorithm, and the ANFIS algorithm have shown 94%, 88.5%, and 87.5% throughput performance, respectively, at a time interval of 100s as exemplified in Fig. 14.

4.1.4 Scenario 4: Performance evaluation for real-time traffic (constant speed—60km/h)

Average handover delay

We evaluated the handover delay of the three compared algorithms, and we present the results as shown in Fig. 15 when the MNs were moving at a constant speed of 60km/h. The proposed IIH algorithm, the FLBVH algorithm, and the ANFIS algorithm have shown an average handover delay of 1.6, 2.3, and 3.2s, respectively.

Average probability of ping-pong effect

During performance evaluation, the proposed IIH algorithm, the FLBVH algorithm, and the ANFIS algorithm experienced 1.25%, 2.4%, and 2.9% probability of ping-pong effect, respectively, when the MNs were moving at a constant speed of 60km/h as shown in Fig. 16.

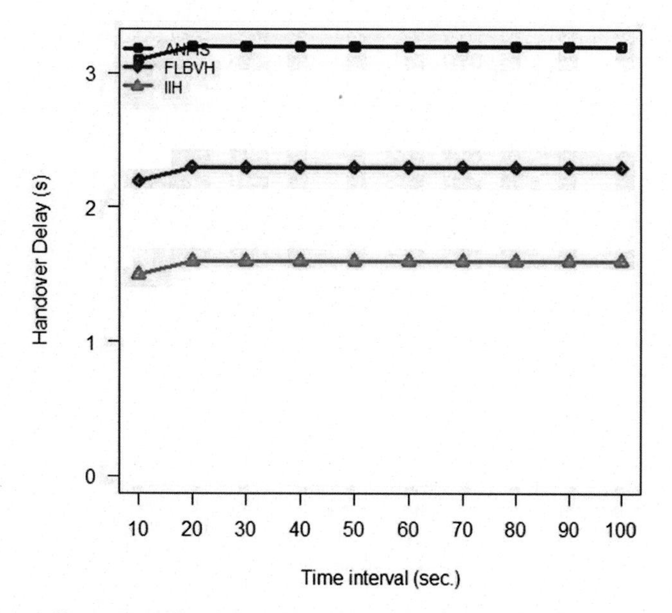

Fig. 15 Real-time traffic average handover delay at a constant speed of 60 km/h.

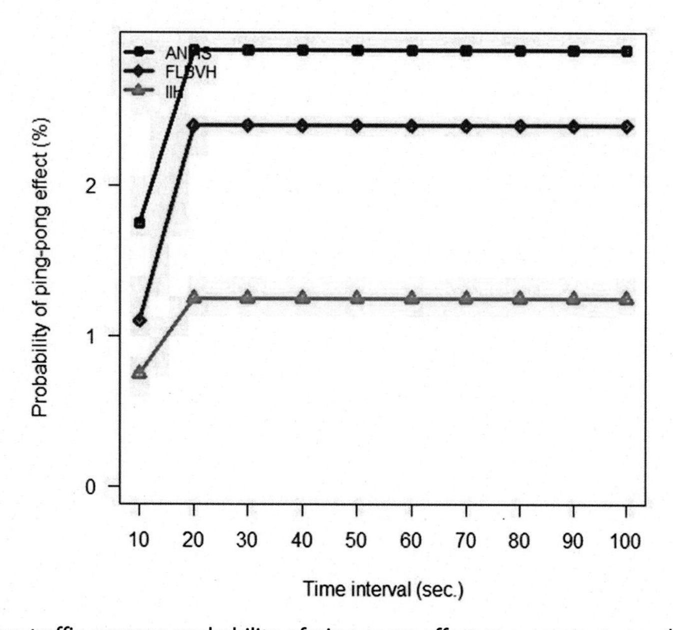

Fig. 16 Real-time traffic average probability of ping-pong effect at a constant speed of 60 km/h.

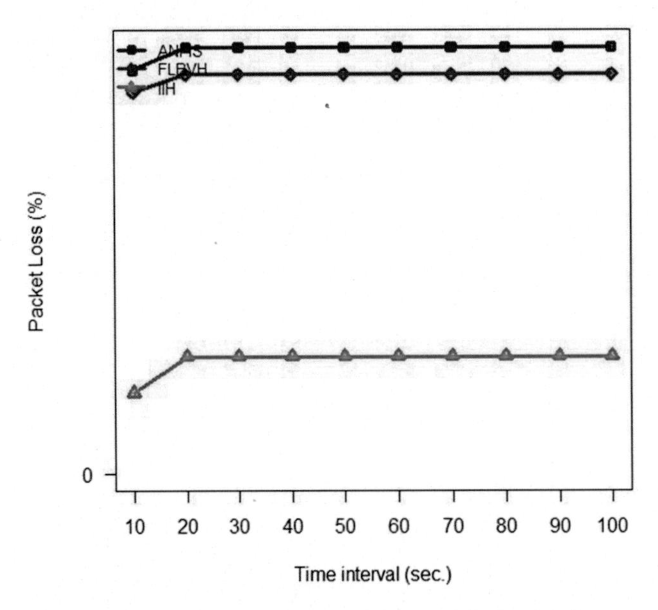

Fig. 17 Real-time traffic average packet loss at a constant speed of 60 km/h.

Average packet loss

The proposed IIH algorithm, the FLBVH algorithm, and the ANFIS algorithm experienced an average packet loss of 3.6%, 8.9%, and 9.5%, respectively, when the MNs were moving at a constant speed of 60 km/h as illustrated in Fig. 17.

Average network throughput

We compared the average network throughput for the three algorithms as shown in Fig. 18 when the MNs were moving at a constant speed of 60 km/h. On average, the proposed IIH algorithm, the FLBVH algorithm, and the ANFIS algorithm have shown 96.4%, 91.1%, and 90.5% throughput performance, respectively, at a time interval of 100 s as exemplified in Fig. 18.

4.1.5 Scenario 1: Performance evaluation for nonreal-time traffic (1–120 km/h)

Average handover delay

We present the nonreal-time traffic average handover delay of the three compared algorithms when the MNs were moving at varying speeds between 1 and 120 km/h as shown in Fig. 19. The proposed IIH algorithm, the FLBVH algorithm, and the ANFIS algorithm have shown an average handover delay of 1.6, 2.4, and 3.2 s, respectively.

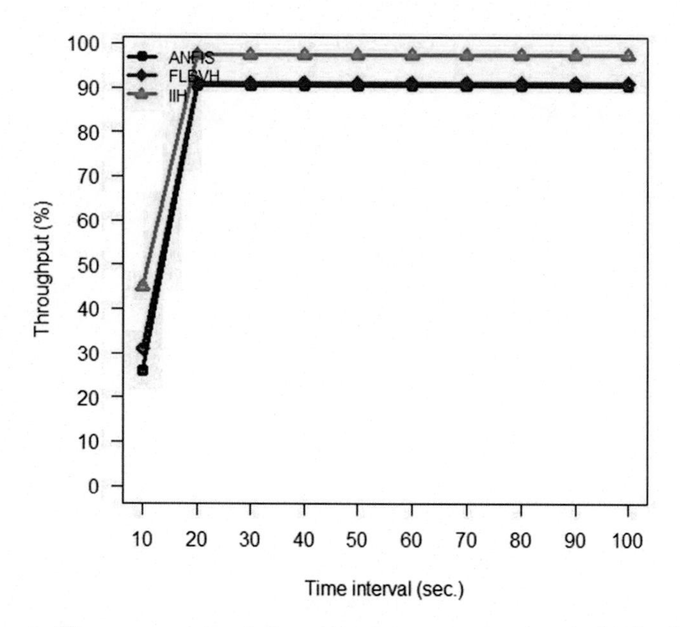

Fig. 18 Real-time traffic average network throughput at a constant speed of 60 km/h.

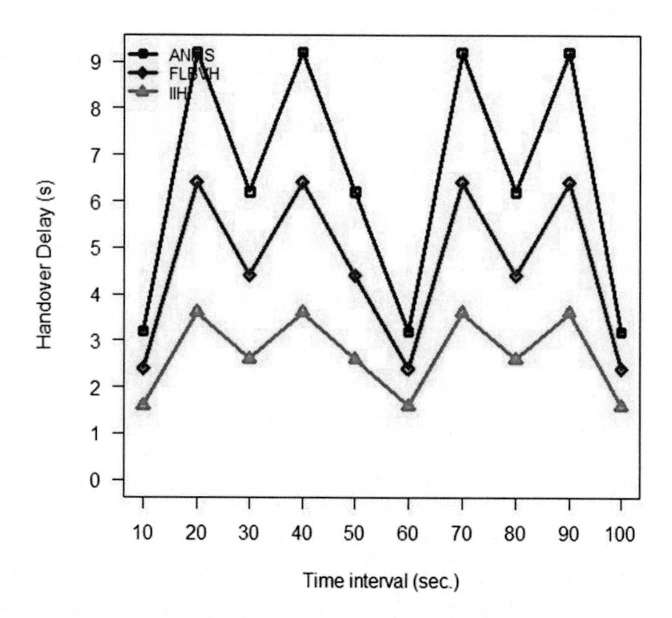

Fig. 19 Nonreal-time traffic average handover delay at varying speeds.

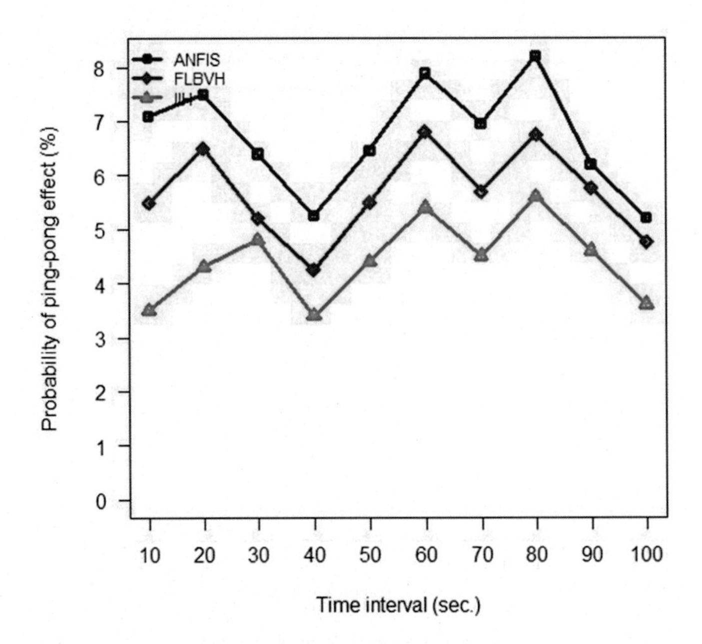

Fig. 20 Nonreal-time traffic average probability of ping-pong effect at varying speeds.

Average probability of ping-pong effect

During performance evaluation, the proposed IIH algorithm, the FLBVH algorithm, and the ANFIS algorithm experienced 3.5%, 4.55%, and 5.3% probability of ping-pong effect, respectively, when the MNs were moving at varying speeds between 1 and 120 km/h as exhibited in Fig. 20.

Average packet loss

The proposed IIH algorithm, the FLBVH algorithm, and the ANFIS algorithm experienced an average packet loss of 3.5%, 7.58%, and 8.25%, respectively, when the MNs were moving at varying speeds between 1 and 120 km/h as illustrated in Fig. 21.

Average network throughput

We compared the average network throughput for the three algorithms when the MNs were moving at varying speeds between 1 and 120 km/h as shown in Fig. 22. On average, the proposed IIH algorithm, the FLBVH algorithm, and the ANFIS algorithm have shown 96.5%, 92.42%, and 91.75% throughput performance, respectively, at a time interval of 100 s as exemplified in Fig. 22.

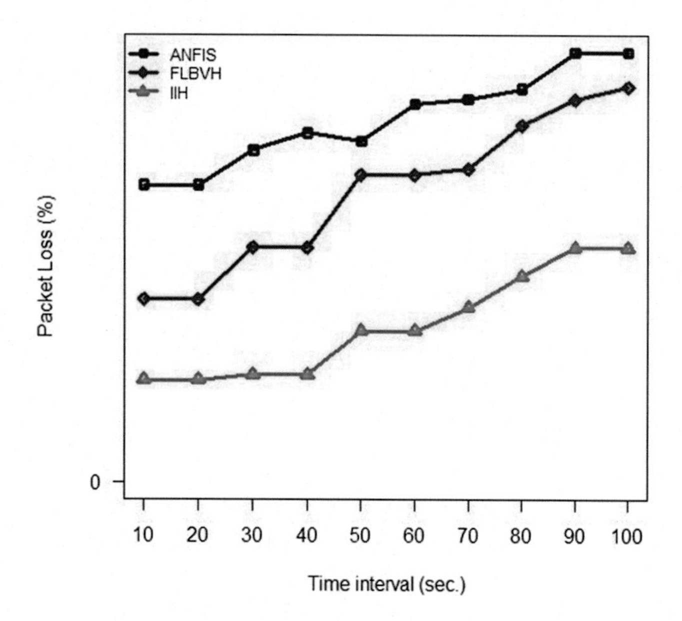

Fig. 21 Nonreal-time traffic average packet loss at varying speeds.

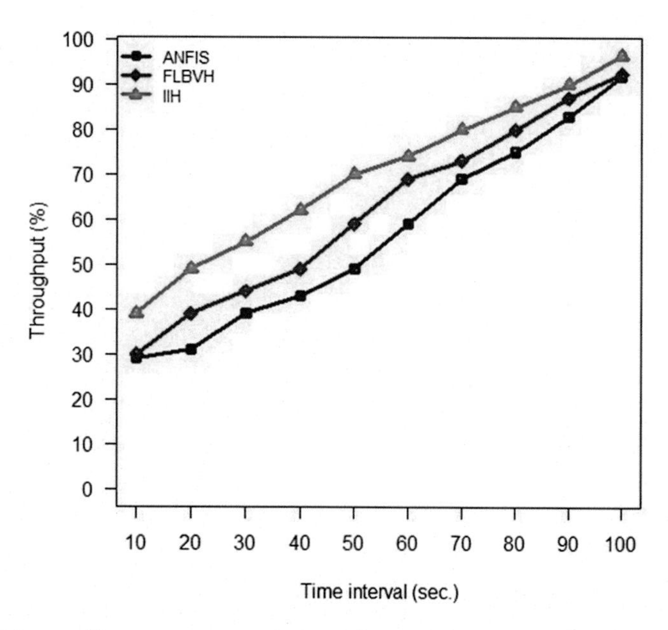

Fig. 22 Nonreal-time traffic average network throughput at varying speeds.

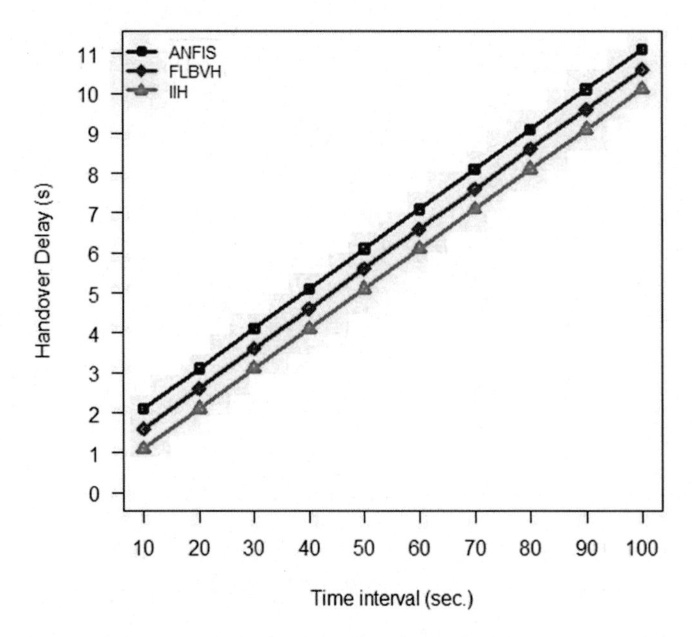

Fig. 23 Nonreal-time traffic average handover delay at varying slow speeds.

4.1.6 Scenario 2: Performance evaluation for nonreal-time traffic (1–30 km/h)

Average handover delay

We compared the three algorithms and we present the average handover delay when the MNs were moving at varying slow speeds between 1 and 30 km/h, as shown in Fig. 23. On average, the IIH algorithm, the FLBVH algorithm, and the ANFIS algorithm have shown an average handover delay of 1.1, 1.6, and 2.1 s, respectively.

Average probability of ping-pong effect

We evaluated the three algorithms and we present the average probability of the ping-pong effect when the MNs were moving at varying slow speeds between 1 and 30 km/h, as shown in Fig. 24. On average, the IIH algorithm, the FLBVH algorithm, and the ANFIS algorithm have shown an average probability of ping-pong effect of 0.65%, 1.05%, and 1.55%, respectively.

Average packet loss

During performance evaluation, the proposed IIH algorithm, the FLBVH algorithm, and the ANFIS algorithm experienced 1.95%, 5.55%, and 6.45% average packet loss, respectively, when the MNs were moving at varying slow speeds between 1 and 30 km/h, as shown in Fig. 25.

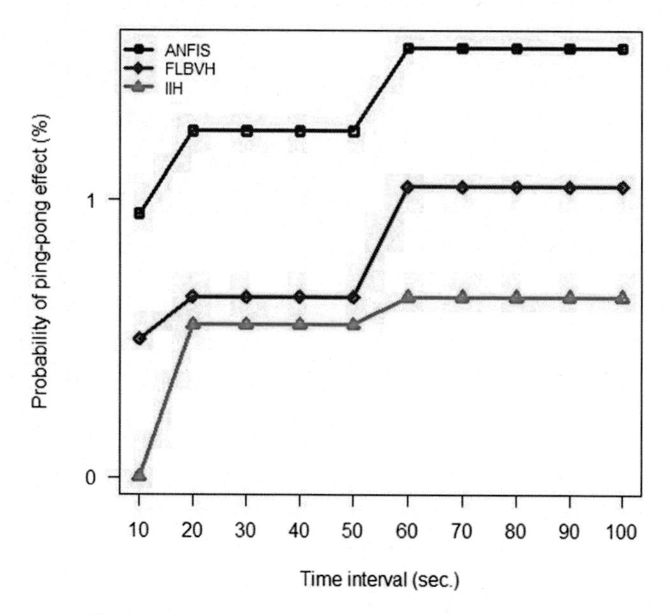

Fig. 24 Nonreal-time traffic average probability of ping-pong effect at varying slow speed.

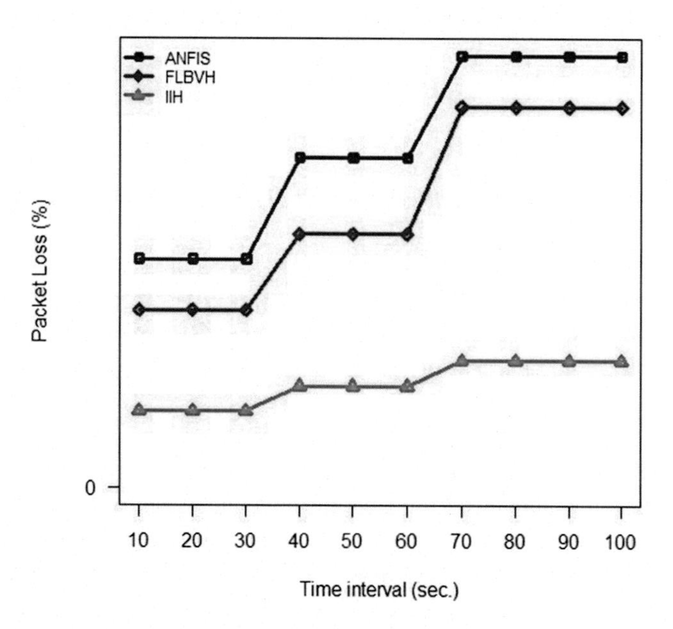

Fig. 25 Nonreal-time traffic average packet loss at varying slow speed.

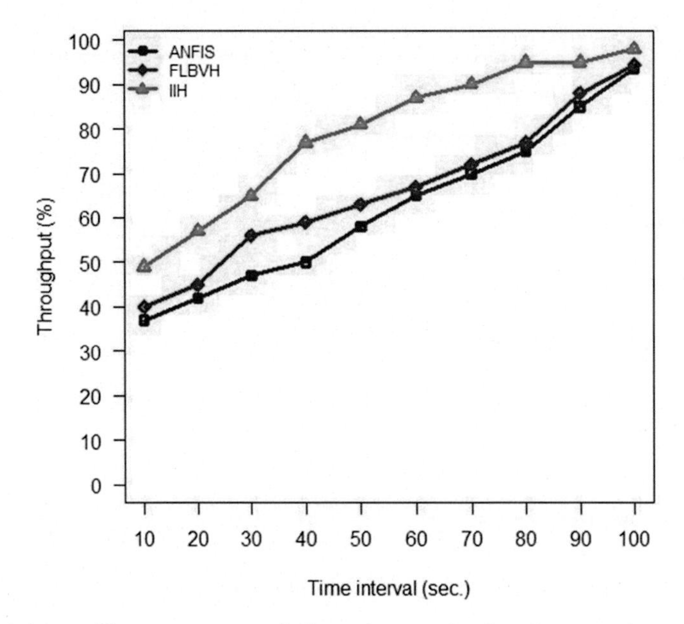

Fig. 26 Nonreal-time traffic average network throughput at varying slow speeds.

Average network throughput

During performance evaluation, on average, the proposed IIH algorithm, the FLBVH algorithm, and the ANFIS algorithm have shown 98.05%, 94.45%, and 93.55% throughput performance, respectively. The time interval was 100 s and the MNs were moving at varying slow speeds between 1 and 30 km/h, as illustrated in Fig. 26.

4.1.7 Scenario 3: Performance evaluation for nonreal-time traffic (100–120 km/h)

Average handover delay

We evaluated the handover delay of the three compared algorithms, and we present the results as shown in Fig. 27, when the MNs were moving at varying high speeds between 100 and 120 km/h. The proposed IIH algorithm, the FLBVH algorithm, and the ANFIS algorithm have shown an average handover delay of 2.1, 2.9, and 3.8 s, respectively.

Average probability of ping-pong effect

We evaluated the three algorithms, and we present the average probability of the ping-pong effect when the MNs were moving at varying high speeds between 100 and 120 km/h, as shown in Fig. 28. The proposed IIH algorithm, the FLBVH algorithm, and the ANFIS algorithm have shown an average probability of ping-pong effect of 4.1%, 5.6%, and 6.4%, respectively.

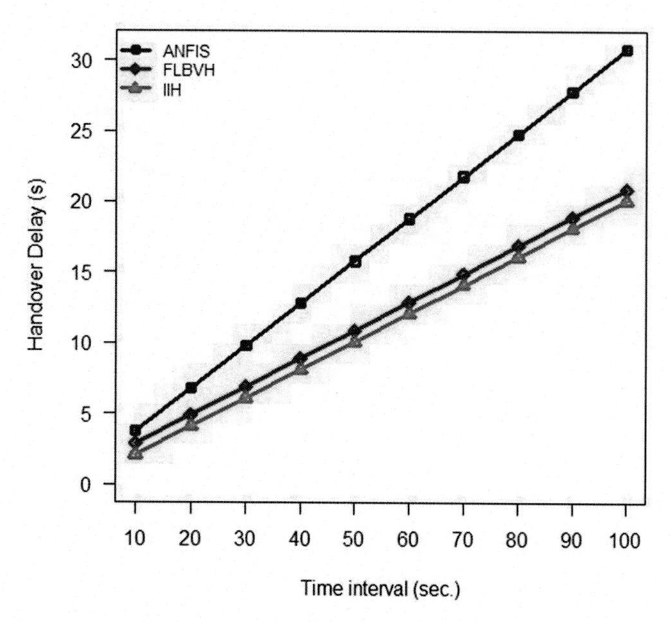

Fig. 27 Nonreal-time traffic average handover delay at varying high speeds.

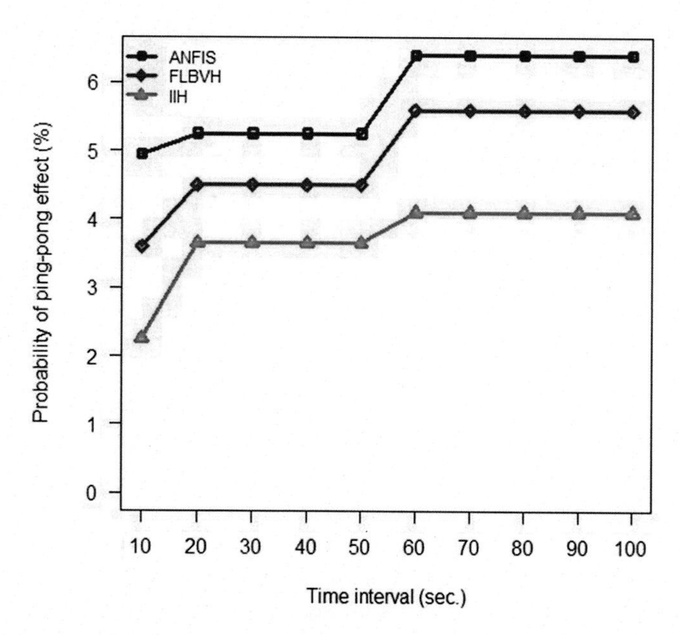

Fig. 28 Nonreal-time traffic average probability of ping-pong effect at varying high speeds.

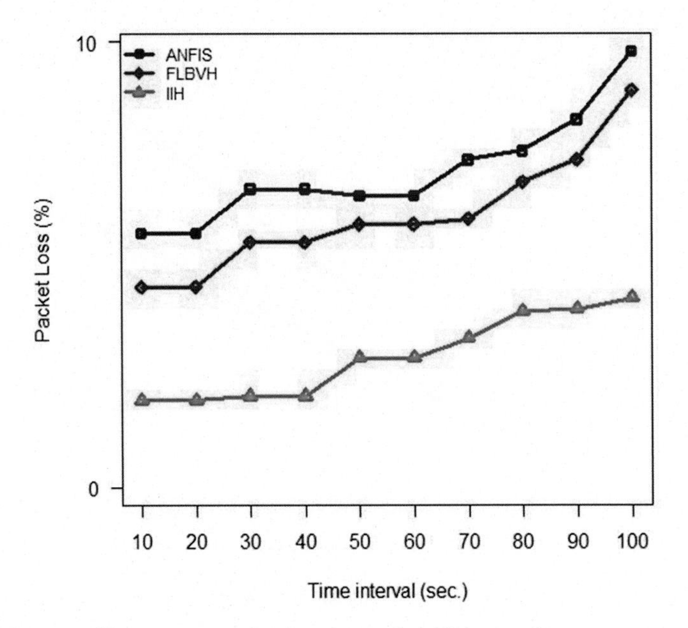

Fig. 29 Nonreal-time traffic average packet loss at varying high speeds.

Average packet loss

During performance evaluation, the proposed IIH algorithm, the FLBVH algorithm, and the ANFIS algorithm experienced an average packet loss of 4.25%, 8.9%, and 9.78%, respectively. The MNs were moving at varying high speeds between 100 and 120 km/h, as exemplified in Fig. 29.

Average network throughput

We compared the average network throughput for the three algorithms as shown in Fig. 30 when the MNs were moving at varying high speeds between 100 and 120 km/h. On average, the proposed IIH algorithm, the FLBVH algorithm, and the ANFIS algorithm have shown 95.75%, 91.1%, and 90.22% throughput performance, respectively, at a time interval of 100 s as exhibited in Fig. 30.

4.1.8 Scenario 4: Performance evaluation for nonreal-time traffic (constant speed—60 km/h)

Average handover delay

We evaluated the handover delay of the three compared algorithms, and we present the results as shown in Fig. 31, when the MNs were moving at a constant speed of 60 km/h. The proposed IIH algorithm, the FLBVH algorithm, and the ANFIS algorithm have shown an average handover delay of 1.3, 1.8, and 2.1 s, respectively.

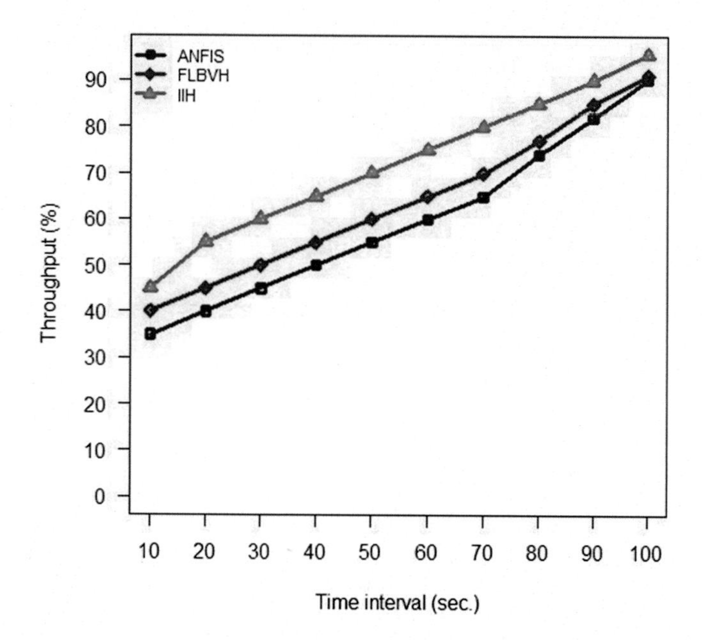

Fig. 30 Nonreal-time traffic average packet loss at varying high speeds.

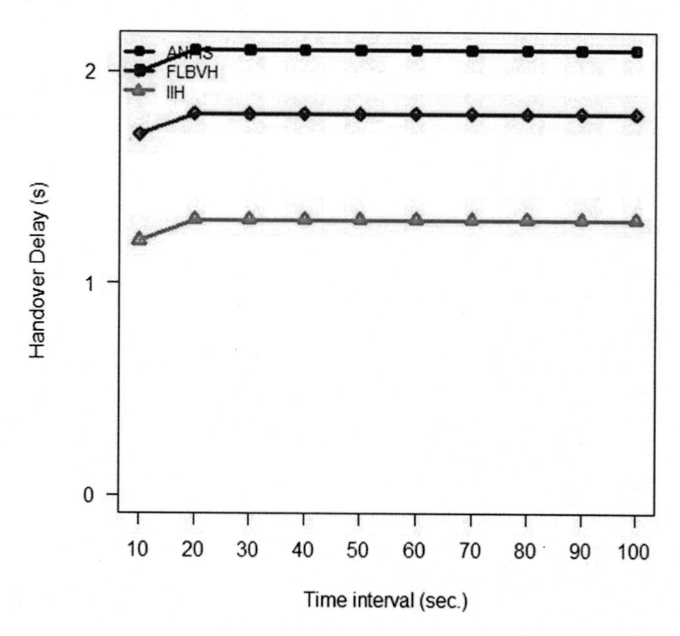

Fig. 31 Nonreal-time traffic average handover delay at a constant speed of 60 km/h.

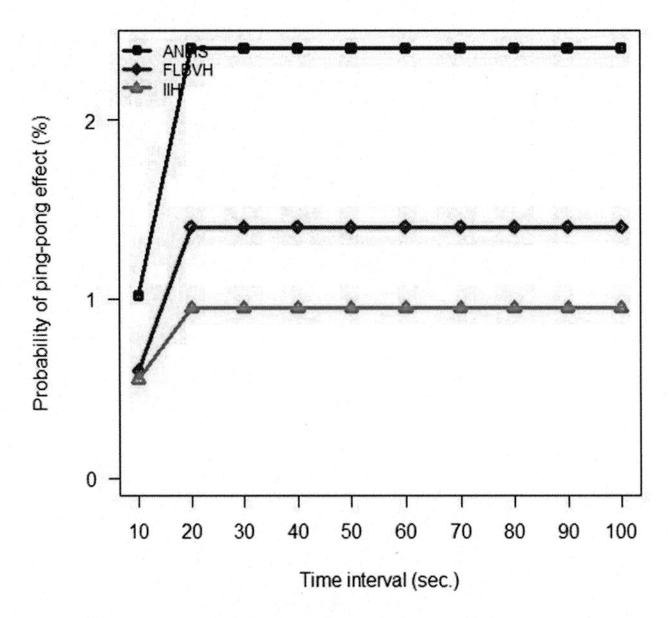

Fig. 32 Nonreal-time traffic average probability of ping-pong effect at a constant speed of 60 km/h.

Average probability of ping-pong effect

During performance evaluation, the proposed IIH algorithm, the FLBVH algorithm, and the ANFIS algorithm experienced 0.95%, 1.4%, and 2.4% probability of ping-pong effect, respectively, when the MNs were moving at a constant speed of 60 km/h as illustrated in Fig. 32.

Average packet loss

The proposed IIH algorithm, the FLBVH algorithm, and the ANFIS algorithm experienced an average packet loss of 2.35%, 6.75%, and 8.45%, respectively, when the MNs were moving at a constant speed of 60 km/h as exemplified in Fig. 33.

Average network throughput

We compared the average network throughput for the three algorithms as shown in Fig. 34, when the MNs were moving at a constant speed of 60 km/h. On average, the proposed IIH algorithm, the FLBVH algorithm, and the ANFIS algorithm have shown 97.65%, 93.25%, and 91.55% throughput performance, respectively, at a time interval of 100 s as exhibited in Fig. 34.

The simulation results showed that the proposed IIH algorithm experienced minimal handover delay for both real-time and nonreal-time traffic as compared to the FLBVH algorithm and the ANFIS algorithm as displayed in Figs. 3, 7, 11, 15, 19, 23, 27, and 31, respectively. This is because, during the design of the proposed IIH algorithm, we used

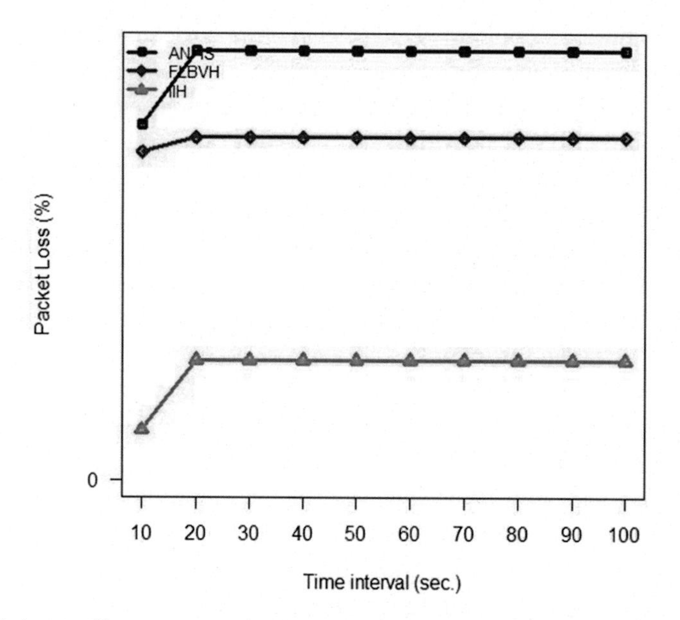

Fig. 33 Nonreal-time traffic average packet loss at a constant speed of 60 km/h.

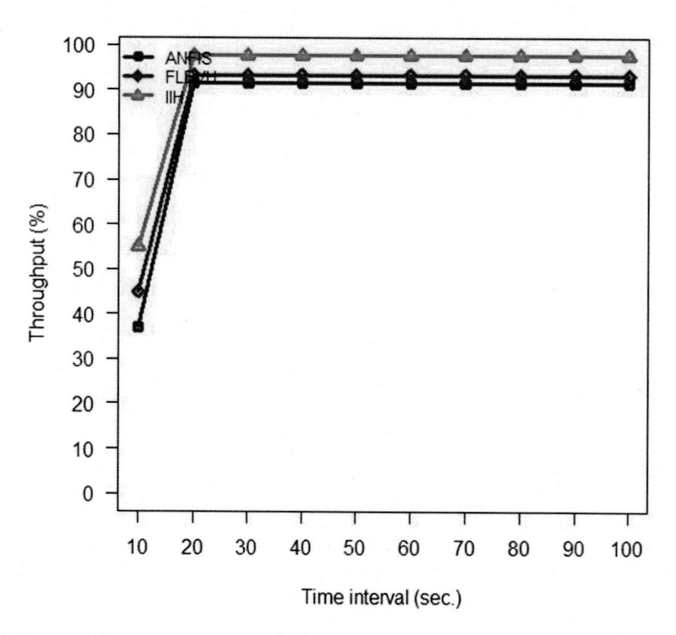

Fig. 34 Nonreal-time traffic average network throughput at a constant speed of 60 km/h.

GPT to predict future values of measured RSS in order to determine the correct time for forthcoming handover. This reduced handover delay as a result. The proposed IIH algorithm achieved this by starting with the network discovery process, while the RSS of the SBS was good. The FLBVH algorithm and the ANFIS algorithm failed to predict the future values of the measured RSS and thus, these algorithms experienced a lengthy network discovery process, leading to elongated handover delay. This is because these handover algorithms started the network discovery process only when the RSS of the SBS was very weak. During performance evaluation, real-time traffic experienced longer handover delay as compared to nonreal-time traffic; reason being, real-time traffic is delay sensitive.

The proposed IIH algorithm achieved a lower probability of ping-pong effect for both real-time and nonreal-time traffic as exhibited in Figs. 4, 8, 12, 16, 20, 24, 28, and 32. The proposed IIH algorithm achieved this by considering the independence of multiple parameters used for handover decision during weight and QoS factor calculation, using FAHP and MOORA, respectively. As a result, the proposed IIH algorithm was able to adapt to the change of network conditions and user's preference. The IIH algorithm, therefore, managed to handle the imprecise information that comes with the use of multiple parameters during weighting matrices.

The FLBVH algorithm experienced minimal probability of ping-pong effect when compared with the ANFIS algorithm. This is because the FLBVH algorithm uses FAHP during weighting matrices. This allowed the FLBVH algorithm to handle the imprecise information that comes with the use of multiple parameters during the weighting of matrices. The ANFIS algorithm failed to handle the imprecise information that comes with the use of multiple parameters during weighting matrices. As a result, the ANFIS algorithm experienced a higher probability of ping-pong effect as compared to the proposed IIH algorithm and the FLBVH algorithm, as illustrated in Figs. 4, 8, 12, 16, 20, 24, 28, and 32.

In addition, both the FLBVH algorithm and the ANFIS algorithm failed to identify beneficial criteria and nonbeneficial criteria during QoS factor estimation. As a result, these algorithms could not check if the TBS has adequate bandwidth to support the application or service running on the UE. This led to inaccurate network selection and amplified the ping-pong effect for both the FLBVH algorithm and the ANFIS algorithm.

The proposed IIH algorithm experienced lower packet loss for both real-time and nonreal-time traffic as compared to the FLBVH algorithm and the ANFIS algorithm, as exemplified in Figs. 5, 9, 13, 17, 21, 25, 29, and 33. This is because we used FAHP for weighting matrices as mentioned in Section 1. We used FAHP because it can handle the imprecise information that comes with the use of multiple parameters during weighting matrices. This leads to minimalized inaccurate network selection and as a result, the consistency of network selection was improved.

In addition, MOORA was used to calculate the QoS factor of each network in order to select the best suitable network with the highest QoS factor, in order to maintain the

QoS requirements of ongoing services. MOORA allows the identification of beneficial criteria and nonbeneficial criteria during QoS factor estimation. This is because the users prefer a higher value for beneficial criteria and a lower value for nonbeneficial criteria. As a result, the proposed IIH algorithm experienced lessened ranking abnormality, which decreased inaccurate network selection and packet loss. Both the FLBVH algorithm and the ANFIS algorithm experienced higher average packet loss as compared to the proposed IIH algorithm. This is because both the FLBVH algorithm and the ANFIS algorithm failed to identify beneficial criteria and nonbeneficial criteria during QoS factor estimation. This led to erroneous network selection and increased packet loss due to high-ranking abnormality.

The proposed IIH algorithm produced a higher average network throughput for both real-time and nonreal-time traffic as compared to the FLBVH algorithm and the ANFIS algorithm, as exhibited in Figs. 6, 10, 14, 18, 22, 26, 30, and 34. This is because the proposed IIH algorithm shortened handover delay by using GPT to predict the future values of the measured RSS. Thereafter, the PRSS values were used to determine the correct time for forthcoming handover, hence, it reduced the handover discovery process and handover delay. In addition, the integration of FAHP and MOORA for weighting matrices and QoS factor calculation minimalized ranking abnormality, hence, reduced erroneous network selection, packet loss, and ping-pong effect. The minimization of the erroneous network selection, packet loss, and ping-pong effect are attributed to the capability of the proposed IIH algorithm to adjust better to the changes in network conditions and user preferences than the FLBVH algorithm and the ANFIS algorithm.

The presented simulation results exemplified that the proposed IIH algorithm performed well as compared to the FLBVH algorithm and the ANFIS algorithm in terms of handover delay, packet loss, and the probability of ping-pong parameters. The results showed that the proposed IIH algorithm did tremendously well in terms of throughput; thus, the work will find applications in industries, society, and academia.

5. Conclusion and future work

We designed and implemented an IIH algorithm in order to reduce prolonged handover delay, packet loss, and ping-pong effect during the handover process in HWNs. We designed the proposed IIH algorithm by integrating GPT, MADM, FAHP, and MOORA in order to provide better QoS and QoE in HWNs. MADM was introduced into the proposed IIH algorithm in order to provide context awareness during handover decision-making. This means the network selection process is more intelligent considering varying needs of the application's QoS requirements and user preference in order to recommend the best suitable network in fluctuating network conditions. The computer simulation results showed that the proposed IIH algorithm was slightly superior to the FLBVH algorithm and the ANFIS algorithm in terms of handover delay, packet loss, and the probability of ping-pong parameters and did exceptionally well in terms of throughput. Further improvements may be achieved by using real propagation data, and by testing the proposed IIH algorithm performance utilizing real wireless network conditions.

Acknowledgments

The authors thank the Tshwane University of Technology for financial support. The authors declare that there is no conflict of interest regarding the publication of this chapter.

Appendix: Table of notations

Notations	Meaning
a_{ij}	All the elements of the fuzzy pair-wise matrix
b	Control gray number
c	Development gray number
m	Number of the attribute value of each network
MN_x	Mobile node
n	Number of networks
$R_j^{\max}$	The maximum value of an attribute for a particular service
$R_j^{\min}$	The minimum value of an attribute for a particular service
R_{ij}	The actual value of the attribute in the network

References

[1] S. Panev, P. Latkoski, Performance analysis of handover delay and buffer capacity in mobile OpenFlow-based networks, Int. J. Commun. Syst. 33 (15) (2020) 1–28.

[2] D. Xinhang, Z. Zhilong, L. Danpu, Low-delay secure handover for space-air-ground integrated networks, in: 2020 IEEE 31st Annual International Symposium on Personal, Indoor and Mobile Radio Communications Personal, Indoor and Mobile Radio Communications, 2020, pp. 1–6.

[3] H.E.I. Jubara, An efficient handover procedure in vehicular communication, in: 2020 2nd International Conference on Computer and Information Sciences (ICCIS), 2020, pp. 1–5.

[4] N. Islam, S. Kandeepan, K.G. Chavez, J. Scott, A MDP-based energy efficient and delay aware handover algorithm, in: 2019 13th International Conference on Signal Processing and Communication Systems (ICSPCS), 2019, pp. 1–5.

[5] T.E. Mathonsi, O.P. Kogeda, T.O. Olwal, Enhanced intersystem handover algorithm for heterogeneous wireless networks, J. Adv. Comput. Intell. Intell. Inform. 23 (6) (2019) 1063–1072.

[6] T.E. Mathonsi, O.P. Kogeda, T.O. Olwal, Intersystem handover decision model for heterogeneous wireless networks, in: The Proceedings of IEEE Open Innovations Conference 2018 (IEEE OIC 2018), 2018, pp. 1–7.

[7] T.E. Mathonsi, T.M. Tshilongamulenzhe, B.E. Buthelezi, An efficient resource allocation algorithm for heterogeneous wireless networks, in: 2019 Open Innovations (OI), 2019, pp. 15–19.

[8] T.E. Mathonsi, O.P. Kogeda, Handoff delay reduction model for heterogeneous wireless networks, in: The Proceedings of IST-Africa 2016 Conference, 2016, pp. 1–7.

[9] T.E. Mathonsi, O.P. Kogeda, T.O. Olwal, A survey of intersystem handover algorithms in heterogeneous wireless networks, Asian J. Inform. Technol. 16 (6) (2017) 422–439.

[10] T.E. Mathonsi, O.P. Kogeda, T.O. Olwal, Intersystem handover delay minimization model for heterogeneous wireless networks, in: The Proceedings of the South African Institute of Computer Scientists and Information Technologists 2017 (SAICSIT 2017), 2017, p. 377.
[11] D. Zhang, Y. Zhang, L.V. Na, Y. He, An access selection algorithm based on GRA integrated with FAHP and entropy weight in hybrid wireless environment, in: Processing of Seventh International Conference on Application of Information and Communication Technologies, 2013, pp. 1–5.
[12] L. Zhang, S. Gao, A novel weights generating approach for multiple attribute decision making under interval-valued intuitionistic fuzzy environment, in: IEEE, 2017 29th Chinese Control and Decision Conference (CCDC), 2017, pp. 4406–4409.
[13] L. Zhang, L. Ge, X. Su, J. Zeng, Fuzzy Logic Based Vertical Handover Algorithm for Trunking System, IEEE, 2017, pp. 1–5.
[14] A.B. Zineb, A. Mohamed, T. Sami, Fuzzy MADM based vertical handover algorithm for enhancing network performances, in: IEEE, 2015 23rd International Conference on Software, Telecommunications and Computer Networks (SoftCOM), 2015, pp. 1–7.
[15] A.B. Zineb, M. Ayadi, S. Tabbane, Cognitive Radio Networks Management Using an ANFIS Approach with QoS/QoE Mapping Scheme, IEEE, 2015, pp. 1–6.
[16] A.B. Zineb, M. Ayadi, S. Tabbane, QoE-fuzzy VHO approach for heterogeneous wireless networks (HWNs), in: IEEE 30th International Conference on Advanced Information Networking and Applications, 2016, pp. 949–956.
[17] F. Azzali, O. Ghazalo, M.H. Omar, Fuzzy logic-based intelligent scheme for enhancing QoS of vertical handover decision in vehicular ad-hoc networks, IOP Conf. Ser.: Mater. Sci. Eng. (2017) 12–81.
[18] J. Zhang, Y. Lou, Water level prediction based on improved grey RBF neural network model, in: 2016 IEEE Advanced Information Management, Communicates, Electronic and Automation Control Conference (IMCEC), 2016, pp. 775–779.
[19] Y.M. Kwon, J.S. Kim, J. Gu, M.Y. Chung, ANDSF-based congestion control procedure in heterogeneous networks, in: IEEE, ICOIN 2013, 2013, pp. 547–550.
[20] R.M. Abdullah, Z.A. Zukarnain, Enhanced handover decision algorithm in heterogeneous wireless network, Sensors (Basel) 17 (7) (2017) 1–14.
[21] D. Guo, X. Li, An adaptive vertical handover algorithm based on the analytic hierarchy process for heterogeneous networks, in: IEEE, 12th International Conference on Fuzzy Systems and Knowledge Discovery (FSKD), 2015, pp. 2059–2064.
[22] S. Maaloul, M. Afif, S. Tabbane, Handover decision in heterogeneous networks, in: The Proceedings of the 2016 IEEE 30th International Conference on Advanced Information Networking and Applications (AINA), 2016, pp. 588–595.
[23] S.I. Goudar, A. Habbal, S. Hassan, Context-aware multi-criteria framework for RAT selection in 5G networks, Adv. Sci. Lett. 23 (6) (2017) 5163–5167.
[24] E. Obayiuwana, O. Falowo, A new network selection algorithm for group calls over heterogeneous wireless networks with dynamic multi-criteria, in: 2016 13th IEEE Annual Consumer Communications and Networking Conference (CCNC), 2016, pp. 491–494.
[25] A.G. Mahira, S.S. Subhedar, Handover decision in wireless heterogeneous networks based on feedforward artificial neural network, in: H. Behera, D. Mohapatra (Eds.), Computational Intelligence in Data Mining, Advances in Intelligent Systems and Computing, 556, 2017, pp. 663–669.
[26] B. Omoniwa, R. Hussain, J. Ahmed, A. Iqbal, A. Murkaz, Q. Ul-hasan, S.A. Malik, A novel model for minimizing unnecessary handover in heterogeneous networks, Turk. J. Electr. Eng. Comput. Sci. 26 (4) (2018) 1771–1782.
[27] S. Singh, A. Mishra, U. Singh, Detecting and avoiding of collaborative black hole attack on MANET using trusted AODV routing algorithm, in: 2016 Symposium on Colossal Data Analysis and Networking (CDAN), 2016, pp. 1–6.
[28] K. Kosek-szott, G. Cuka, Consequences of performing DL MU-MIMO transmissions with TXOP sharing for QoS provisioning in IEEE 802.11ac networks, IEEE Commun. Lett. 22 (3) (2018) 606–609.

Index

Note: Page numbers followed by *f* indicate figures, *t* indicate tables, and *b* indicate boxes.

E

F

G

H

I

J

K

L

M

N

O

P

Q

R

S

T

Printed in the United States
by Baker & Taylor Publisher Services